AF740097

# INDUSTRIAL WASTE

Edited by **Kuan-Yeow Show** and **Xinxin Guo**

**Industrial Waste**

Edited by Kuan-Yeow Show and Xinxin Guo

**Published by InTech**

Janeza Trdine 9, 51000 Rijeka, Croatia

**Copyright © 2012 InTech**

All chapters are distributed under the Creative Commons Attribution 3.0 license, which allows users to download, copy and build upon published articles even for commercial purposes, as long as the author and publisher are properly credited, which ensures maximum dissemination and a wider impact of our publications. After this work has been published by InTech, authors have the right to republish it, in whole or part, in any publication of which they are the author, and to make other personal use of the work. Any republication, referencing or personal use of the work must explicitly identify the original source.

As for readers, this license allows users to download, copy and build upon published chapters even for commercial purposes, as long as the author and publisher are properly credited, which ensures maximum dissemination and a wider impact of our publications.

**Notice**

Statements and opinions expressed in the chapters are these of the individual contributors and not necessarily those of the editors or publisher. No responsibility is accepted for the accuracy of information contained in the published chapters. The publisher assumes no responsibility for any damage or injury to persons or property arising out of the use of any materials, instructions, methods or ideas contained in the book.

**Publishing Process Manager** Vana Persen

**Technical Editor** Teodora Smiljanic

**Cover** InTech Design Team

Additional hard copies can be obtained from orders@intechopen.com

Industrial Waste, Edited by Kuan-Yeow Show and Xinxin Guo
p. cm.
ISBN 978-953-51-0253-3

# We are IntechOpen, the world's leading publisher of Open Access books

# Built by scientists, for scientists

4,100+
Open access books available

116,000+
International authors and editors

120M+
Downloads

151
Countries delivered to

Our authors are among the
Top 1%
most cited scientists

12.2%
Contributors from top 500 universities

BOOK CITATION INDEX
CLARIVATE ANALYTICS
INDEXED

WEB OF SCIENCE™

Selection of our books indexed in the Book Citation Index in Web of Science™ Core Collection (BKCI)

Interested in publishing with us?
Contact book.department@intechopen.com

Numbers displayed above are based on latest data collected.
For more information visit www.intechopen.com

For EU product safety concerns:
IN TECH d.o.o., Prolaz Marije Krucifikse Kozulić 3, 51000 Rijeka, Croatia,
info@intechopen.com or visit our website at intechopen.com.

# Meet the editors

Dr Show Kuan Yeow is currently serving as Professor and Head, Department of Environmental Engineering, Faculty of Engineering and Green Technology, University Tunku Abdul Rahman (UTAR), Malaysia. He is also serving as Chair, SP Setia Professor of Environmental Engineering & Green Technology. He has been involved in numerous industrial R&D projects and serving as a consultant in Singapore, Taiwan, China, Vietnam and Malaysia. He has received several professional awards, including the prestigious National Technology Award in Singapore, an US Patent 6793822 and an International Patent WO 2003/070649 on a biogranulation technology for wastewater treatment. He has published over 140 technical papers in refereed international journals and conferences, 70 research reports and short courses, 2 journal/book editorships, 1 co-authored book and 18 book chapters. A citation record search by Scopus on Dr Show's journal publications indicated a total citation count of 872 and an h-index score of 16.

Dr Guo Xinxin was born in Jilin, China. After completing a bachelor degree in Environmental Chemistry at Jilin University, She went to the UK for further study and obtained a master degree in Environmental Sciences from University of Nottingham, UK, in 2004, and then she did a PhD in University of East Anglia to study the aerosol chemistry of phenols in the atmosphere. In 2010, she joined the Faculty of Engineering and Green Technology, Universiti Tunku Abdul Rahman, Malaysia. She currently holds the position of Assistant Professor at the Department of Environmental Engineering. Her research interests include indoor and outdoor air pollution, chemistry of atmospheric aerosols, chemistry of organic acids and metal-organic complexes in the atmosphere, but now she concentrates on atmospheric surfactants, their characteristics, concentration, composition and sources around the tin mine lakes ecosystems in Malaysia.

# Contents

# Preface

Ever since the Industrial Revolution, industrial activities have been accompanied by a problem: industrial waste. The commensurate increase in industrialization, urbanization and population growth are leading to production of enormous quantities of industrial wastes that may cause degradation in environment and health hazards. On the other hand, the desire for a healthy environment increases, which leads to the need for better ways of waste minimization, pollution prevention and better use of resources in achieving the required industrial and environmental standards.

This book is intended to fulfil the need for state-of-the-art development on the industrial wastes from different types of industries. Most of the chapters are based upon the ongoing research, how the different types of wastes are most efficiently treated and minimized, technologies of wastes control and abatement, and how they are released to the environment and their associated impact. A few chapters provide updated review summarizing the status and prospects of industrial waste problems from different perspectives. The book is comprehensive and not limited to a partial discussion of industrial waste, so the readers are acquainted with the latest information and development in the area, where different aspects are considered. The user can find both introductory material and more specific material based on interests and problems. For additional questions or comments, the users are encouraged to contact the authors.

The book was a result of effort by many experts from different industrial disciplines in the world. We would like to acknowledge the authors for their contributions to the book and we would also appreciate the assistance from Ms Vana Persen throughout the whole publishing process. We hope that this book will be helpful for graduate students, environmental professionals, engineers, researchers, or policy makers.

**Dr. Kuan-Yeow Show**
Faculty of Engineering and Green Technology, Universiti Tunku Abdul Rahman,
Malaysia

**Dr. Xinxin Guo**
Department of Environmental Engineering, Faculty of Engineering and Green
Technology, Universiti Tunku Abdul Rahman (Perak Campus), Jalan University,
Bandar Barat, Kampar, Perak,
Malaysia

1

# Industrial Discharge and Their Effect to the Environment

Y.C. Ho[1], K.Y. Show[1], X.X. Guo[1], I. Norli[2], F.M. Alkarkhi Abbas[2] and N. Morad[2]
*[1]Universiti Tunku Abdul Rahman*
*[2]Universiti Sains Malaysia*
*Malaysia*

## 1. Introduction

Industrialization has become an important factor to the development of a country's economy, through the establishment of plants and factories. However, the waste or by-products discharged from them are severely disastrous to the environment consists various kind of contaminant which contaminate the surface water, ground water and soil. There are a number of reasons the waste are not safely treated. One of the reasons is mainly due to the lacking of highly efficient and economic treatment technology. The focus of this chapter is to give a detail illustration at the effect of industrial discharge and on the environment and human health. Some corrective actions shall also be illustrated in the later part of this chapter, to overcome the contamination of industrial discharge. The content of this chapter will be as follows:-

In section 2, there is an illustration of the source and types of industrial contaminants in many parts of the world. It is essential to understand the characteristic of industrial discharge in order to have an idea for ways to reduce or remove the contaminants for a sustainable tomorrow.

It is required to understand the impacts of industrial waste to the environment (freshwater, seawater, land) in order to design highly efficient treatment and developing effective remedial methods. Section 3-5 explains how such contamination is going to affect living organism, as well as, abiotic compartment such as sediment through case studies.

Section 6 depicts the possible health problem which caused by the contaminant from industrial discharge. Some of the contaminants merely cause mild discomfort to human being, while others might be detrimental to human health. The latter is usually caused by carcinogenic contaminants.

Section 7 discusses the treatments to the contaminant. Corrective actions that can be taken to solve the current environmental issue caused by industrial discharge are discussed as well. Currently, many researchers all over the world use remediation which is a method using biological agent in treating the contaminant in surface water, groundwater and land. Hence, various bioremediation and phytoremediation case studies are also outlined in this section.

Suggestion of some future work can be done on the industrial discharge is included in section 8 as part of the summary to this book chapter.

## 2. Industrial discharge

The contaminant from the discharge is directly related to the nature of the industry. For example, in textile industry, the discharge is usually high chemical oxygen demand (COD), biochemical oxygen demand (BOD) and colour point; tannery industry is on the other hand, produces discharges which have high concentration of metal such as cadmium, and etc. Table 1 presents a summary of the types of contaminants discharged from the industries at different parts of the world by random case studies.

| Location | Contaminant | Type of industry | Reference |
|---|---|---|---|
| Malaysia, to Juru River | **Metalloid**<br>Arsenic (As)<br><br>**Metal**<br>Chromium (Cr), Cadmium (Cd), Zinc (Zn), Copper (Cu), Lead (Pb), Mercury (Hg)<br><br>**Organic/ inorganic matter and parameter**<br>Phosphate ($PO_4^{3-}$)<br>Ammonia ($NH_3$)<br>Nitrate ($NO_3^-$)<br>Sulphates ($SO_4^{2-}$)<br>Chloride ($Cl^-$)<br>Aluminium (Al) | Chemical products<br>Papers and printings<br>Batteries<br>Electroplating<br>Textile and leathers<br>Fertilizers, pesticides, insecticides<br>Plastic-based products<br>Rubber-based products<br>Wood based products<br>Electric and electronic industries<br>Cosmetics<br>Fungicides<br>Fluorescent lights<br>Dental amalgams<br>Art supplies<br>Mining and siltation<br>Cement and cement products<br>Iron, steel and tin workshops<br>Welding fumes<br>Medical equipment<br>Smelting plants<br>Metal fabrications<br>Oil refineries<br>Quarries<br>Beverages and food | DOE, 1998<br>Salman A. Al-Shami et al., 2010 |
| Bangladesh, to lagoon | **Metalloid**<br>As<br><br>**Metal**<br>Zn, Cu, Strontium (Sr), Pb, Nickel (Ni), Cr, Lithium (Li), Vandadium (V), Silver (Ag), Cobalt (Co), Selenium (Se)<br><br>**Organic/inorganic matter and parameter**<br>Biochemical<br>oxygen demand (BOD), chemical | 300 industries included textile, dyeing to plastics, metal fabrications, semiconductor goods, lather tanning etc. | Ahmed et al., 2011 |

| Location | Contaminant | Type of industry | Reference |
|---|---|---|---|
| | oxygen demand (COD), electrical conductivity, pH, total alkalinity, total hardness, total organic carbon (TOC), Turbidity ($Cl^{-}$), total suspended solids (TSS) and total dissolved solids (TDS). | | |
| Japan, to Nishitakase River | 2-[2-(acetylamino)-4-[bis(2-methoxyethyl)amino -5-methoxyphenyl]-5-amino-7-bromo-4-chloro-2-H-benzotriazole (PBTA-1) | Textile industry | Shiozawa et al., 1999 |
| Germany, to three rivers of North Rhine-Westphalia | **Organic/ inorganic matter and parameter**<br>(i) Chemical process site 1<br>Dichloroaniline<br>Tetramethylbutanedinitrile<br>Tributylphosphate<br>Triethylphosphate<br>Diisopropylnaphthalenes<br>Benzoic acid<br>2,2,4-Trimethyl-1,3 pentanedioldiisobutyrat (TXIB)<br><br>(ii) Meat production site and chemical site<br>N, N-dibenzylamine<br>1-methyl-2-indolinone<br>N,N-Dibenzylamine<br>Triethyl phosphate (TEP)<br>Trimethyl- and 4-tert-butylbenzoic<br>2-(Chloromethyl)-1,3-dioxolan<br>1-Methyl-2-indolinon<br>Trimethylbenzoic acid<br>Tris(chloro-propyl) phosphat (TCPP)<br><br>(iii) Oil production sites and chemical complex<br>Tributylamine Dimethylpyridine<br>Dimethylpyrazine<br>Indole<br>Methylindole<br>1-Ethylpyrrolidone Thioanisole<br>Methylphenyl sulfone TCPP, Isomer 1<br>TCPP, Isomer 2<br>C1 Benzoic acid<br>C2 Benzoic acid<br>2,4,6-Trimethylbenzoic acid | Petrochemical site, paper production, meat production | Botalova & Schwarzbaue et al., 2011 |

| Location | Contaminant | Type of industry | Reference |
|---|---|---|---|
| Kingdom of Saudi Arabia, to Red Sea | **Metal**<br>Cd, Cr, Cu, Iron (Fe), Ni, Pb, Zn, Aluminium (Al), Barium (Ba), Molybdenum (Mo), Sr<br><br>**Organic/ inorganic matter and parameter**<br>Benzene, styrene, toluene, indene, Naphthalene, 1, 4-dioxane, Ethyl Benzene, Xylene, O&G | Two petrochemicals, three refineries | Ahmad et al., 2008 |
| India, to agriculture field | **Metal**<br>Cd, Cu, Fe, Ni, Pb, Zn<br><br>**Organic/ inorganic matter and parameter**<br>BOD, COD, TDS, dissolved solids (DS)<br>Chloride, sulphate, phosphate | Paper Industry | Devi et al., 2011 |
| India, to unlined lagoon | **Organic/ inorganic matter and parameter**<br>Sodium, chloride, calcium, COD, BOD | Cystine production industry | Srivivasa Gowd & Kotaiah et al., 2000 |
| Croatia, to Sava River | **Metal**<br>Fe, Zn, Cu, Ni, Pb, Cr | Pharmaceutical and food industries | Radić et al., 2009 |
| India, to Uppanar river | **Metal**<br>Fe, Mn, Pb, Zn, Cu, Ni, Cr, Cd , Co<br><br>**Organic/inorganic matter and parameter**<br>DO, COD | Chemicals, beverage manufacturing, tanneries, oil, soap, paint production, paper, and metal processing plants | Jonathan et al., 2008 |
| India, to Bandi River | **Metal**<br>Cu, Fe, Zn and Mn<br><br>**Organic/inorganic matter and parameter**<br>TDS, TSS, COD, BOD, chlorides, sulfates, carbonates and sodium, calcium and magnesium. | Dyeing and printing industries | Nepal Singh et al., 2000 |

Table 1. Contaminants discharge from different types of industry and location found from the studies conducted

Through the studies, we can deduce that most of the industrial discharge carries toxic substances. Due to the presence of high amount of toxic, carcinogen, and teratogen of metals, researchers are highly concerned with its effect on the environment and health of mankind. Rigorous investigations are currently being carried out to study the consequences of the contamination on the surface water, groundwater, and surface land due to industrial discharge. The result of these case studies will then be presented as a solid evident for the effects of metals ions, organic and inorganic matters to environment. The interactions and impacts which caused by these chemical contaminants towards the environment will be further explained.

## 3. The effects of industrial discharge to the freshwater

The industrial discharge carries various types of contaminants to the river, lake and groundwater. The quality of freshwater is very important as it is highly consumed by human for drinking, bathing, irrigation and etc. The presence of contaminants from industrial contaminant within the water may reduce the yield of crops and the growth of plant and it will harmful to the aquatic living organism too.

### 3.1 River

River is a system which includes the main course and its tributaries. It is responsible in carrying the load of dissolved and particulate phases from both natural and anthropogenic sources along with other contents. This substance moves downstream and will be experiencing chemical and biological changes. Thus, the water chemistry of a river is affected by the lithology of the reservoir, atmospheric, and anthropogenic inputs. Furthermore, the transport of natural and anthropogenic sources to the oceans and their state during land–sea interaction can be determined by the water quality from rivers and estuaries. Estuaries could be categorized as a geochemical reactor and its heterogeneous reaction could bring the understanding on the fate of metals, organic and inorganic matters along the river to the ocean. Through the studies conducted by Jonathan et al. (2008) noted that there is relationship between the water-particle interactions and solution chemistry, such as flocculation, organic and inorganic complexation, adsorption, and sediment resuspension.

#### 3.1.1 Metals

The contamination of metals is a major environmental problem and especially in the aquatic environment. Some metals are potentially toxic or carcinogenic even at very low concentration and are thus, hazardous to human if they enter the food chain. Metals are usually dissolved into the aquatic system through natural or anthropogenic sources. Metal ions are distributed thoroughly during their transport in different compartments of the aquatic ecosystems, in biotic or abiotic compartment such as fishes, water, sediment, plant. Metals remain in contaminated sediments may accumulate in microorganisms which in return entering into the food chain and eventually affect human well being (Shakeri & Moore, 2010).

In 2010, Shakeri & Moore conducted a study to evaluate the distribution and average concentrations of Cu, Zn, Ni, Mo, Pb, V, As and Co in Chenar Rahdar river sediment. The result concluded that there is a strong association of elements such as Zn, Co, Ni, Sc, Cu, Al, and Fe in the sediment at the study site. The authors indicated that Al and Fe hydroxides and clay content play a significant role in the distribution and sorption of metals in sediments. This study noted that metal inputs have brought negative impact to the freshly deposited sediments and the accumulation of the metal on the sediment surface.

Metal in sediment is affected by mineralogical and chemical composition of suspended material, anthropogenic influences by deposition, sorption, and enrichment in living organism or aquatic plant (Jain et al., 2005). Naturally, suspended and bed sediment are an important compartment to buffer metal concentration in an aquatic system especially by adsorption or precipitation (Jain & Ali, 2000; Jain, 2001; Jain & Sharma, 2002; Jain et al., 2004). However, the metal discharges from industry may change the role of sediment as it may not be able to act as a sink and buffer to higher concentration of metal. Metals contributed by man-made sources are possible to associate with organic matter in the thin fraction of the sediments, or adsorbed on metal hydrous oxides, or precipitated as hydroxide, sulfides and carbonates (Singh et al., 2005; Shakeri & Moore, 2010).

In India, the discharge from fertilizer industry has not undergone any treatments, is one of the major sources of pollution to water reservoirs such as lakes, ponds, rivers and ocean. The discharge contains certain toxic components such as metals, nitrates and ammonia which might be responsible for causing metabolic impairment in the aquatic organisms. At times, the toxic components could even cause fatality in aquatic living organism (Bobmanuel et al., 2006; Yadav et al., 2007; Ekweozor et al., 2010). A study has been conducted by Yadav et al. (2007) on freshwater fish, *Channa striatus* which are exposed to fertilizer industry discharge. It is found that the toxicity of the fertilizer discharge on the fish tissues could be due to metals and ammonia. Heavy metals such as Zn, Cr, Cu and Pb in the fertilizer industry discharge can bind with certain proteins in fish and disrupting membrane integrity, cellular metabolism and ion-transports that will bring harm to the maintenance of homeostasis. The result showed that the average protein concentration in various tissues of the control fish is, in descending order: gills>liver>brain>muscle>kidney>heart. However, the protein level reduced in all fish tissues at a higher sublethal concentration of industrial discharge at 7% higher than control fishes, the values of tissue reduced in descending order: liver, brain, muscle, gills, kidney, and heart at 76.23%, 55.95%, 52.16%, 50.06%, 49.28%, and 42.86%, respectively.

When metals associate with other chemicals compound in the fertilizer discharge may cause distortion in the cell organelles and inhibit the activity of various enzymes (Valarmathi & Azariah, 2003; Yadav et al., 2007), which may greatly disturb the physiological state of the exposed living organism. The heavy metals present in the fertilizer industry discharge are usually in dissolved state which could easily be uptaken by fish and enter human food chain. There have been studies showed that metals will cause damage to the human kidney and liver even at low concentration. The early studies suggested that higher concentration in metals can be carcinogenic and teratogenic (O'Brien et al., 2003; Yadav et al., 2007). Generally, carbohydrate metabolism is a major source of energy production and the activity of Lactate dehydrogenase (LDH). It has been a target for the action of various xenobiotics. The activity of LDH in different part of body tissues of *C. striatus* after exposing to the fertilizer industry discharge has been examined. In this study, the result showed that the exposure of *C. striatus* to fertilizer industry discharge resulted in a drastic reduction in the enzyme activity.

Rai & Tripathi (2009a) added that most metals in aquatic environment associated with particulate matter, then settled and accumulated in the bed sediments. The accumulation of contaminant in the bed sediments and the remobilization of contaminant are the most important mechanisms of contaminant in an aquatic ecosystem regulation. Furthermore, under certain circumstances such as deficit in dissolved oxygen or decreased in pH, the bed sediments can be another source of secondary water pollution when the heavy metals from bed sediments are released.

Another study conducted in a kaolin refinery industry produces hazardous by-product such as Al, Fe, and Zn. In kaolin processing, sulphuric acid is used to improve the whitening (Jordao et al., 2002) is discharged to the river waters. This will influence the well being of aquatic organisms that adapted well at close to neutral pH. Also, in order not to affect the colour and whiteness of paper, impurities such as iron oxides is needed to be removed. This can be made through the reduction of Fe(III) to Fe(II) with metallic Zn. Therefore, Zn, Fe, Al are usually present in the discharge. The study examined the pH, conductivity and hardness values and various metals such as Al, Ca, Cd, Cr, Cu, Fe, Mg, Ni, Pb, and Zn. Excessive concentration of these parameters in discharges that flows into rivers may also cause adverse effects to human health (Jordao et al., 2002). The discharges from the industry without proper treatment will decreased the pH of the river water. This is due to the usage of sulphuric acid in the kaolin processing. The pH values will bring effects in flora and fauna nearby, change the taste of water and lead to heavy corrosion in pipe lines. High conductivity naturally indicates the presence of ionic substances dissolved in the river water. However, the result showed that 90% of the study site exceeded the data reported for non-contaminated rivers due to excessive metal ions within the water. At the site nearer to kaolin industry the conductivity is 852 times higher than the non-polluted study site. The industrial discharge also changed the hardness in river water. However, the result showed that the study site is not exceeded the maximum limit (500 mg $CaCO_3$ $L^{-1}$) of hardness for drinking water as recommended by the Brazilian government (Jordao et al., 2002).

### 3.1.2 Organic/inorganic matter

The study conducted by Hiller et al. (2011) to investigate the concentrations, distributions, and hazards of polychlorinated biphenyls (PCBs) and polycyclic aromatic hydrocarbons (PAHs). PCBs are used mainly as coolant and electronic industries (capacitors, transformers), paints, sealants for wood, cutting and lubricating fluids, plasticizers, and as dielectric fluids. Therefore, at the former site of PCB manufacturing area in Slovakia, high concentrations of PCBs are detected in soils, sediments, humans, and wildlife (Kocan et al., 2001; Petrik et al., 2001; Hiller et al., 2011).

Due to their low aqueous solubilities, the PCBs and PAHs lay on the surface of soils and waters. PCBs and PAHs adsorb strongly to the organic fraction of soils (Girvin & Scott, 1997; Hiller et al., 2011). Soils contaminated with PCBs and PAHs are transported directly or indirectly by rivers to the water reservoir and are subsequently converted into the bed sediments. Therefore, soils could be considered as the primary sinks for these organic contaminants. PCBs and PAHs are persistent in the environment, resistant to degradation process, and accumulate in food chain. This will eventually bring health hazard to living organisms, including mutagenicity and carcinogenicity (Hiller et al., 2011).

Yadav et al. (2007) studied on fertilizer industrial discharge showed that some components in the discharge may interact with each other and produce toxic to aquatic organisms. For instance, the interaction between dissolved oxygen and ammonia changed the respiratory physiology in fresh water fish. In addition, results showed that the toxicity of the effluent in fish depends on concentration and duration of exposure. Several study showed that the excess concentration of ammonia, whether is ionized or unionized, is one of the major contaminants in fertilizer discharge and it is toxic to aquatic living organism. It could cause impairment to the cerebral energy in fish, such as *O. niloticus* and a hybrid catfish (Yadav et al., 2007; Ekweozor et al., 2010).

Surprisingly the toxicity level of fertilizer industry discharge may influence by the environmental factors such as conductivity, temperature, pH, cardon dioxide ($CO_2$), oxygen and elements. Through studies, these factors will influence the behavior and certain biochemical indices of the fish, such as *C. striatus*, by acting either in synergistic, antagonistic or simple additive manner (Yadav et al., 2005; Bobmanuel et al., 2006; Yadav et al., 2007). A high conductivity value indicates high concentration of dissolved ion within the industrial discharge. However, the conductivity value which recorded in this study was slightly below the required limit. Moreover, in this study the increased in fish mortality may due to the increased in water temperature, the increased uptake of industrial discharge components and low dissolved oxygen in the water (Yadav et al., 2007).

Carmago et al. (1992) found that the rivers nearer industrial discharge point have adverse impact to the environment as well as to macrobenthic communities. Toxic contaminants from surface runoff, sewage discharges and industrial discharge have caused negative impacts towards the freshwater macrobenthic communities. The presence of substance chemical such as ammonia, chlorine, cyanide, metals, PCBs, pesticides and phenols would caused a decline pattern on the number of species and changes in the species composition. Furthermore, when industrial discharge and river regulation interact, benthic macroinvertebrates will be highly exposed to the toxic contaminants. The living organism which will be deeply affected are shredders, which feed on coarse sedimentary detritus, and collector-gatherers, which feed on fine sedimentary detritus, were the macroinvertebrate functional feeding groups are most adversely affected. Furthermore, during the industrial process, high amount of hydrofluoric acid (HF) are used to separate different sandy materials which are subsequently used for manufacturing glass at industrial plants. Therefore, high concentration of fluoride ion and suspended inorganic matter discharged by the industrial into the study site. Carmago et al. (1992) noted that short-term flow fluctuations, low concentration of dissolved oxygen and also the siltation of suspended inorganic matter caused by industrial discharge contribute greatly to the changes in sediment and directly affect the structure of macroinvertebrate community. The high siltation of suspended inorganic matter caused significant reductions in taxa richness and abundance of zoobenthic communities as it changes the natural structure of the substratum. Other than that, the fluoride pollution which generated by the industrial discharge was also contributed to adverse effects on the macrobenthic community.

### 3.2 Groundwater

Groundwater is regarded as the largest reservoir of drinkable water for mankind. To many countries, groundwater is one of the major sources of water supply for domestic, industrial and agricultural sectors. In India, groundwater supplies more than 50% for irrigation, and 80% for drinking water (Singh et al., 2009). It is estimated that approximately one-third of the world's population are using groundwater for drinking purposes. Pollution of ground water due to industrial effluents is a major issue (Vasanthavigar et al., 2011). Poor groundwater quality brings negative impact to human health and plant growth. In developing countries like India, it is estimated that around 80% of all diseases are directly related to poor drinking water quality and unhygienic conditions (Olajire & Imeokparia, 2001; Vasanthavigar et al., 2011). Human activities like industrialization are responsible to the groundwater quality and the groundwater contamination and spread of contaminant are amongst the major factor lead to human hazards.

### 3.2.1 Metals

Khe´rici et al. (2009) reported that there is high concentration of Cr(VI) during low precipitation and upper aquifer in the groundwater at the town of Annaba, which is also an important location for heavy industries

During high precipitation, infiltration of sulphates and chromates occurred, and subsequently when low precipitation, the aquifer without recharge becomes a confined environment favourable to a reduction of sulphates to sulphide ($H_2S$ and $HS^-$) by a complex biochemical process (a phenomenon called sulphatoreducing), due to the bacterial activity. Subsequently, this reduction results from the sulphur rejected (Khe´rici et al., 2009) by a sulphato-reducing bacteria (Desulfovibrio desulfricans), which can transform of Cr(VI) to Cr(III) which is a stable substance. Cr(III) is stable at pH 7 with the presence of oxygen. While, chromate in the thermodynamic form is stable under these conditions, however, it is toxic to many organisms at very low concentrations. Chromate is further subjected to microbiological reduction.

In upper aquifer, the chromate concentrations increase during high precipitation. Long precipitation resulted in higher concentration of Cr(VI) in the upper aquifer. In contrast, during low precipitations, chromate concentrations in the upper aquifer are lesser. Similarly, during high precipitation, the infiltration of chromates, sulphate and dissolved oxygen is higher in the upper aquifer (Khe´rici et al., 2009).

In the study performed by Purushotham et al. (2011) showed that about 28% of total hardness in the study site exceeded the desirable limits at 600 mg $L^{-1}$. Naturally, the water hardness is due to the presence of alkaline earths such as calcium (Ca) and magnesium (Mg). Excess of magnesium affects the soil quality, which results in poor crops yield. The high concentration of magnesium and calcium can cause decrease in water quality and may cause encrustation in the water supply structure. High concentrations of sodium can deteriorate soil quality and damage the sensitive vegetation due to its phytotoxcity. Water containing high concentrations of carbonate and bicarbonate ions tends to precipitate Ca and Mg as their carbonates. As a consequence the relative proportion of sodium increases and settled in the soil, and decreased soil permeability.

### 3.2.2 Organic matter/inorganic matter

The salts present in the groundwater influence the soil structure, permeability and aeration, which indirectly affect the plant growth. Study conducted by Purushotham et al. (2011)

showed high sodium conductivity (>1500 µs $cm^{-1}$) around 17.5% of the groundwater samples and this probably due to high salinity in groundwater. Sodium concentration in irrigation water replaces calcium by the process of Base Exchange, therefore reduces soil permeability. Furthermore, excess salinity in groundwater used for irrigation decreased plants osmotic activity and interfere water absorption and nutrients from the soil.

Nearly 5% of groundwater from the study site exceeds the desirable limit (1000 mg $L^{-1}$) of chloride. The natural source of chloride is due to the weathering of phosphate mineral apatite present in granites. However, apart from natural sources, industrial discharge is one of the sources that contribute chloride in groundwater. Excessive chloride concentration leads to salinity, which deteriorate the soil (Purushotham et al., 2011).

## 4. The effects of industrial discharge to the seawater

Contaminants from the industry discharge flows through river. Some are accumulate, interact and settle with the living organism, plant and sediment and finally reach the coastal and ocean. Plants and living organism in the ocean are important food sources for human intake. Contaminants may then enter human food chain and accumulate in fishes, molluscs (octopus, shellfish, and cockle), crustaceans (shrimp, crab, and lobster), seaweed, sea cucumber and etc. Therefore, it is essential to understand the effect to aquatic environment.

### 4.1 Metals

Metals which have altered biogeochemically along the flows from river to estuaries and coastal area are transported to the ocean and the original composition of seawater and sediments is altered (Jonathan et al., 2008).

Abbas et al. (2008) studied on the blood cockles (*Anadara granosa)* found in two rivers in Penang state, Malaysia, namely, Juru River and Jejawi River nearby Prai industrial zone. The result showed that the average content of arsenic and metals found in cockles are arranged in the order: As >Hg >Cd > Zn >Cu >Cr > Pb and As >Hg > Zn >Cd >Cu >Cr > Pb for Juru and Jejawi River, respectively. The mean concentration of As was higher than the permissible limit (1 mg $kg^{-1}$ wet weight) established by the Malaysian Food Act 1983 and Food Regulations 1985 Fourteen Schedule. The result for this study is important as there is an extensive culture of cockles is being carried out in the coastal areas. The excessive contamination of this bivalve sea food is not safe for human intake.

In the study conducted by Yap et al. (2002) at coastal areas adjacent to industrial areas, the authors noted that sediment plays an important role for the contaminant transport and metal repository. Impact to human health may possible to identify via sediment analyses as it is also a long term integrator of geochemical processes. There are four types of sediments that could results in different adsorption of metals in sediment, namely, 'easily, freely leachable and exchangeable' (EFLE), 'acid-reducible', 'oxidisable-organic' and 'resistant' types. Characteristic of sediment is important to determine in order to understand the chemical reaction of sediment and metal or organic matter. The study indicated that the affinity is lower for metals in the 'acid-reducible' fraction of the sediment for both offshore and intertidal sediment. The reducing conditions are mainly caused by decomposition of organic matter by microorganism activity (Yap et al., 2002). The 'acid-reducible' fraction involves metals associated with manganese and iron oxides and hydroxides and carbonates.

Apart from that, the 'oxidisable-organic' sediment proved to organically bind with metal and increase the adsorption of metals in the sediment. For instance, the adsorption of Cu and Pb in 'oxidisable organic' fraction is high and it is due to the organic matter and its physico-chemical properties of the sediments, respectively. The study also concluded that the 'bioavailabilities' of the metals on EFLE fraction and 'resistant' fraction are poor. On the other hand, the 'nonresistant' (non-lithogenous) fractions, tend to adsorb Cu and Pb into the 'oxidisable organic' fraction. Higher adsorption capability may due to higher affinities exist between metals and humic substances, which are the fractions of organic matters. They are chemically very active in complexing elements such as heavy metals. Furthermore, the metals may be associated with living organisms, and detritus or coating on mineral particles.

The organic phase of sediment could also play a dominant role in the transport of metals in natural water systems. The organic substances in sediments may break up to free the soluble metals in waters under oxidized conditions. The fate of metals in aquatic systems is significantly influenced by dissolved organic matter as dissolved organic matter is capable to alter the distribution between the oxidized and reduced forms of metals (Yap et al., 2002).

At Pakistan, Saifullah et al. (2002) studied on the mangrove which is the habitat for some marine organism and shrimp fishery. However, the mangroves are deteriorating rapidly due to several anthropogenic stresses and industrialization is one of the important reasons. In the Karachi area (the largest industrial city of Pakistan), there are more than 6000 different industrial units such as chemical industries, metal industries, oil refineries, petrochemicals, tanneries, pharmaceuticals, textiles etc. Due to the marine pollution cause by industrialization, the mangroves health is at risk. Metal contamination is the most significant source of pollution to the marine environment. Furthermore, high industrialization activity is partly contributed to around 2000 tons of untreated BOD that discharge to the seaside daily. This directly affects the marine organism and the shrimp fishery.

### 4.2 Organic matter/ Inorganic matter

Jonathan et al. (2008) conducted a study in Uppanar River. In their study, the loadings of industrial effluents were found to change the dissolve oxygen (DO) in water. Such changes have affected the environmental quality in the river and its coastal zone. In the coastal zone, the dissolved metals behave in a reversible manner with physicochemical parameters, indicating that fresh water input has changed the coastal water (Kuppusamy & Giridhar, 2006; Jonathan et al., 2008). The river water showed that low content of DO situated very close to the river bank. This is due to contamination of industrial discharge with high COD. The low DO is due to the discharge of industrial effluents and the toxicity of the combined effect from chemicals and heavy metals (Jonathan et al., 2008).

The study performed by Din & Ahamad (1995) on cockles' growth at coastal receiving industrial discharge from the nearby Prai Industrial Estate. Cockles are facing mortalities nearer to the discharge point. The surrounding industrial activity in year 1995 includes textile and chemical manufacturing, and electroplating work. The study conducted at different located given 2, 4, 6 and 8 weeks of exposure to the pollution by those industrial discharges. The result showed that the effects of on the cockles' growth seem significantly as early at week two. Several parameters were taking into account, namely, temperature, TSS, BOD, oil and grease. The parameters were found relatively high at the discharge point. However, parameters such as temperature, salinity, pH, DO and BOD directly affect the well being of all livings organism. They affect living organisms through the intake and activities of toxic materials.

## 5. The effects of industrial discharge to the land

Generally, the land is contaminated due to the uncontrolled and unplanned disposal of industrial waste onto the soil surface. The contaminant will infiltrate to the groundwater. The contamination on the soil surface may interrupt human daily activity and bring adverse effect to the growth of plant as well as human health.

### 5.1 Metals

Last few decades, the drainage basin of the Gangetic plain has been used for the disposal of domestic and industrial wastes which bring negative impact to the water quality, sediments and soil surrounding (Ansari et al., 1999). Unplanned disposal of industrial waste directly on the land is one of the factors that caused the pollution to surface water and groundwater. A study conducted by Singh et al. (2009) showed high concentration of Cr(VI) in ground water from various industrial cities in India. The Cr in the groundwater resulted from on-land burial of Cr sludge (5-10% $Cr_2O_3$ by wt) by various industries engaged in the manufacture of Basic Chrome Sulphate (BCS) ($Cr(OH)_2SO_4$), an important input for local tanneries. The sludge resulting from the initial stage of BCS manufacture is a product of incineration of chromite ore ($FeCr_2O_4$) and has remnant of Cr(VI) which is water soluble. The leachate of Cr (VI) in the sludge ultimately reached the groundwater.

Kisku et al. (2000) noted that for soil contaminated by industrial discharge, the plant could accumulate the toxic metal such as iron (Fe), zinc (Zn), copper (Cu), and manganese (Mn) are essential trace elements to plant life while lead (Pb), chromium (Cr), nickel (Ni), and cadmium (Cd). They are toxic even at a very low concentration. They indicated that the concentration of metal in the soils and plants of the polluted field were significantly higher than non-polluted field. Soil-plant bioaccumulation relationships are varies by element and plant. The highest accumulation of Fe, Mn, Zn, Cu, Pb, Ni, Cr, and Cd were found in *Spinacea oleracea, Raphanus sativus, Amaranths viridis* and *Lycopersicon esculentum, Coriandrum sativum, Solanum melongena, Spinacea oleracea, Lycopersicon esculentum,* and *Coriandrum sativum,* respectively. Furthermore, results reflected that *Spinacea oleracea* having more affinity to Fe and Ni, *Lycopersicon esculentum* showed higher affinity to Zn and Cr and *Coriandrum sativum* showed high affinity to Cu and Cd. The toxicity symptom is different by plant species, however, the most common and nonspecific symptoms are chlorosis, intervenial chlorosis, necrosis, stunted growth, shorter root length, and narrow leaves. For example, fertilization and fruiting phenomena of brinjal plant was significantly affected, tomatoes grew smaller and lesser, cabbage showed abnormal growth. Usually, plants do not always show visible morphology symptoms, they may have hidden injury due to contaminant or a change in metabolic pathways. Eventhough Fe, Mn, Zn concentration in plant tissues are well below the critical concentration, while Cu, Pb, Ni, Cr and Cd are within the prescribed plant critical concentration range, these toxic metals are obviously affecting the plant life and reduce the yield capacity about 10%. Generally, metals could cause a decrease in total chlorophyll content and therefore changing the metabolic pathways of the plant. However, weeds in exception where it grows luxuriously with heavy metal contamination. It can be suggest an ideal agent for metal remediation.

Govil et al. (2008) conducted a study at an industrial zone which has 300 various types of industry nearby such as manufacturing chemicals, pharmaceutical, batteries, metal alloys, metal plating and plastic products, dyeing, edible oil production, battery manufacturing, metal plating, chemicals, etc. There are three sources of contamination exist within the industrial area, dumpsites of solid waste, untreated industrial discharge, and emission from smokestacks. And most of these industries directly release their discharge into nearby ditches and streams and the solid waste is randomly dumped on open land, and along roads and lakes. The soil contamination is suggested to be the main causes from the random dumping of solid waste from the industry and it could be spread by rainwater and wind. The result showed that the land around the industrial area is heavily contaminated by As and then Pb, Zn. Furthermore, there is an analysis conducted at the pre- and postmonsoon over two hydrological cycles in 2002 and 2003 indicated that As, Cd and Pb contaminants are more mobile. Subsequently, these heavy metals can cause groundwater pollution through the infiltration by soil.

## 6. The impact to human health

The impact to human health is the utmost important criteria to look into apart from the effect to surface water and groundwater on the living organism and sediments. Metals, as described in the above case studies showed the potential for health risk. However, the organic matter also will bring adverse health impact to human. The health hazard to human is further described in the following.

### 6.1 Metals

#### 6.1.1 Aluminium

High concentrations of Al can cause hazard to brain function such as memory damage and convulsions. In addition, there are studies suggested that Al is linked to the Alzheimer disease (Jordao et al., 2002).

#### 6.1.2 Cadmium

Cd is harmful to both human health and aquatic ecosystems. Cd is carcinogenic, embryotoxic, teratogenic, and mutagenic and may cause hyperglycemia, reduced immunopotency, and anemia, as it interferences with iron metabolism (Rehman & Sohail Anjum, 2010). Furthermore, Cd in the body has been shown to result in kidney and liver damages and deformation of bone structures (Abbas et al., 2008).

#### 6.1.3 Chromium

Cr(III) is essential nutrient for animal and essential to ensure human and animal lipids' effective metabolism but Cr(VI) is carcinogenic. Cr(VI) is the most toxic form of chromium and having equivalent toxicity to cyanides. It can cause skin ulcer, convulsions, kidney and liver damage. Moreover, it can generate all types of genetic effects in the intact cells and in the mammals in vivo (Khe´rici-Bousnoubra et al., 2009). It has also been reported that intensive exposure to Cr compounds may lead to lung cancer in man (Jordao et al., 2002).

#### 6.1.4 Iron

Iron is an essential element in several biochemical and enzymatic processes. It involved the transport of oxygen to cells. However, at high concentration, it can increase the free radicals production, which is responsible for degenerative diseases and ageing (Jordao et al., 2002).

#### 6.1.5 Lead

Lead could accumulate in kidney, liver, bone, and brain. Chronic intoxication can lead to encephalopathy mainly in children (Jordao et al., 2002).

#### 6.1.6 Mercury

Mercury can cause brain damage, heart, and kidney and lung disease in human. At very low concentration, Hg can permanently damage to the human central nervous system (Rai & Tripathi, 2009a). Inorganic and mercury through biological processes, can converted into MeHg. MeHg is organic, toxic, and persistent (Wang et al., 2004; Rai & Tripathi, 2007). Furthermore, MeHg can cross the placental barriers and lead to foetal brain damage (Rai & Tripathi, 2009a).

#### 6.1.7 Nickel

Nickel is an essential element to both plant and human, but high exposure to this metal can lead to cancer in organs of the breathing system, cardiovascular and kidney diseases (Jordao et al., 2002).

#### 6.1.8 Zinc

Zinc is an essential element to human and plant (Jordao et al., 2002). Recent studies indicated that Zn is also involved in bone formation. However, elevated intake of Zn can cause muscular pain and intestinal haemorrhage (Honda et al., 1997; Jordao et al., 2002).

### 6.2 Organic/inorganic matters

#### 6.2.1 Fluoride

High concentration of fluoride can cause dental and skeletal fluorosis such as mottling of teeth, deformation of ligaments and bending of spinal cord (Janardhana Raju et al., 2009).

#### 6.2.2 Nitrate

High concentrations of nitrate cause methemoglobinemia in infants and could cause cancer. In the blood, nitrate convert hemoglobin to methemoglobin, where it does not carry oxygen to the body cells, which may lead to death from asphyxiation (Purushotham et al., 2011).

#### 6.2.3 Potassium

High potassium concentration may cause nervous and digestive disorders (Purushotham et al., 2011), kidney heart disease, coronary artery disease, hypertension, diabetes, adrenal insufficiency, pre-existing hyperkalaemia. Infants may also experience renal reserve and immature kidney function (WHO, 2009).

#### 6.2.4 Sulphate

Excessive sulphate concentration may lead to laxative effect (Purushotham et al., 2011) and it affects the alimentary canal (WHO, 2004).

## 7. Corrective action

There are several ways to solve the environmental problem caused by industrial discharge. Some of the methods are bioremediation, biosorption, phytoremediation, application of green chemistry and green monitoring. Many studies have been conducted on bioremediation using bacteria, fungi and yeast. Bioremediation is the use of microbial in remediating the contaminant while phytoremediation uses plant. The examples of bioremediation are land farming, composting, bioreactors, bioventing, biofilters, bioaugmentation, biostimulation, intrinsic bioremediation, pump and treat of groundwater (Boopathy, 2000). Brown algae and yeast are examples of the application in biosorption. The examples for phytoremediation could be the use of plant in surface and submerged aquatic plant. Green chemistry and green monitoring are alternative option to prevent and to monitor the contamination of industrial discharge. Types of corrective actions are illustrated in the following with selective case studies.

### 7.1 Bioremediation

Many research have been performed on bioremediation for other types of contaminant such as pesticide like DDT (Purnomo et al., 2011), herbicide like Pendimethalin (Venkata Mohan et al., 2007) petroleum and diesel oil contamination (MacNaughton et al., 1999; Watanabe, 2001), a few to name.

Conventional methods to cleanup pollutants usually involve physical treatment such as sedimentation and filtration, and chemical treatments such as flocculation, neutralization, and electro-dialysis. Very often, the treatment efficiency does not meet the regulation limits. Hence, further treatments are to be applied as well. Of all the technologies that have been investigated, bioremediation has emerged to be the most desirable approach for cleaning up contaminant from industrial discharge (Shedbalkar & Jadhav, 2011). Bioremediation can be defined as a technology that utilizes the metabolic potential of microorganisms to clean up contaminated environments (Wanatabe, 2001). Indeed, numerous studies have been carried out to search for the appropriate and useful bioremediation agent, such as bacteria, yeast and filamentous fungi.

#### 7.1.1 Bacteria

A study conducted by Taheri et al. (2008) on organic sulphur compounds in petroleum and other fossil fuels. Disulfide oil (DSO) may be channeled to water ecosystem and surface land mainly by refinery activities from its reservoir tanks. For refineries site near in seaside, DSO can pollute sea water through penetration of sandy soil. This study worked on seven microbial strains which is isolated from DSO contaminated soil. However, only two strains, namely, *Paenibacillus* and *Rhodococcus* have high potential to remove DSO from contaminated soil. The result showed that *Rhodococcus* and *Paenibacillus* were found to be suitable for bioremediation of DSO contaminated soil. It removed DSO with 19.3% and 24.3% for *Rhodococcus* and *Paenibacillus*, respectively, at after 48 hr. The author suggested the best condition for bioremediation which is to provide ample amount of water 20 $\mu L\ g^{-1}$ soil day at and d-glucosa at 4 $g\ L^{-1}$ (added water) as nutrient. Adding more than 4 $g\ L^{-1}$ glucose added water showed an adverse effect.

Faisal & Hasnain (2004) reported that there are two chromium-resistant bacterial strains, CrT-1 and CrT-13. CrT-1 successfully reduced 82%, 28% and 16% of Cr(VI) after 24 hr at 100, 500, and 1000 $\mu g\ mL^{-1}$ respectively. While CrT-13 reduced 41%, 14% and 9%, at 100, 500, and 1000 $\mu g\ mL^{-1}$ respectively, after 24 hr. Cr(VI) at 150 and 300 $\mu g\ mL^{-1}$ reduced by 87% and 71%, respectively, with CrT-1 and by 68% and 47% with CrT-13. Uptake and reduction of toxic Cr(VI) into Cr(III) by the bacterial strains *O. intermedium* CrT-1 and *Brevibacterium* CrT-13 showed a very high-level resistance to chromate. In addition, the chromate resistance level of these strains is very high in minimal medium relative to strains reported by other researcher (Megharaj et al., 2003).

### 7.1.2 Yeast

Machado et al. (2010) demonstrated a study on bioremediation using sub-product of fermentative industries, such as wine or brewing industry. The biomass can be obtained in large quantity and cheap. This study has shown that brewing flocculant yeast cells set a new approach for heavy metals bioremediation. According to Wang and Chen (2006), yeast cells of *Saccharomyces cerevisiae* is among the different types of biomass suggested for metals bioremediation. Hence, *S. cerevisiae* were used in the bioremediation, in a batch mode, of a real electroplating wastewater containing metals like Cu, Ni, and Cr. In their approach, no pretreatment of Cr(VI) to Cr(III) was required. Using heat-inactivated cells of *S. cerevisiae*, the result showed that 98% of Cr(VI) is removed after the first batch at an initial pH 2.3.

### 7.1.3 Protozoa

A study has been conducted by Haq et al. (1998) on protozoa and algae tannery discharge. It showed that the presence of protozoa in these industrial discharge suggest their strategies to tolerate, resist or detoxicate organic substances and metals. Therefore, protozoa are suitable to be utilized as an agent in bioremediation. Furthermore, protozoa may cooperate with other microbial during remediation. Since, protozoan can tolerate and resist heavy metals and toxic compounds, it can therefore provide an insight into detoxication mechanisms in human cells as protozoa are true eukaryotes and the results can be easily and safely extrapolated to higher eukaryotic organisms, even to human being. The results showed that *Stylonychia mytilus*, and *Tetrahymena pyriformis* is found after inoculation. The number of cell almost doubled for *Stylonychia mytilus* while *Tetrahymena pyriformis* did not show any duplication and disappear in the medium containing chromium after 15 days. *Stylonychia mytilus* showed the ability to tolerate Cr(VI) at a concentration of 1 $\mu g\ mL^{-1}$. The disappearance of *Tetrahymena pyriformis* with the presence of Cr may be due to cannibalism among the spesies (Haq et al., 1998).

### 7.1.4 Fungus

Sani et al. (1998) studied on the degradation of dye wastewater using white-rod fungus, *P. chrysosporium*. The researchers added that biodegradation becomes popular as it is economical. The decolourization of industrial discharge from dyeing industry could be due to either adsorption on fungal biomass or to biodegradation (Sani et al., 1998). In addition, previous studies have shown the degradation of recalcitrant xenobiotic compounds, including azo dyes using *P. chrysosporium*. The result has shown that more than 80 % colour reduced in most of the dyes (Red HE-8B, Malachite Green and Navy Blue HE-2R) by *P. chrysosporium* in shake culture. Malachite Green was more readily decolourized than Crystal Violet and Magenta, eventhough they have similar dye structures.

The result showed that decolorization ability increased with inoculum concentration with 5, 10, 15 and 20 % inoculum, maximum decolorizations recorded as 24, 68, 95 and 98 %, respectively, in shake culture. 20, 56, 55 and 57 % colour were removed, respectively for static culture. It proved that the *P. chrysospotium* is an agent to decolourize dye wastewater (Sani et al., 1998).

Morales-Barrera & Cristiani-Urbina (2008) conducted a study on leather factory discharge. The remediation in using fungus, the strain identified as *Trichoderma inhamatum*. The experimental results showed that this fungus is capable to transform Cr(VI) to Cr(III) where Cr(III) having lower toxicity level. The fungus could reduce Cr(VI) up to 2.43 mM. The author suggested this strain of fungal to be potential apply in bioremediation of Cr(VI) containing wastewater. The other genera is *Aspergillus* (Gouda, 2000; Acevedo-Aguilar et al., 2006), *Penicillium* (Acevedo-Aguilar et al., 2006), and *Phanerochaete* (Pal, 1997). The authors suggested that this fungus could be a better agent for bioremediation as at the highest concentration of Cr(VI) reduced by *T. inhamatum* was much higher than concentrations commonly found to be reduced by bacteria, fungi, and yeast.

As bioremediation uses living systems, especially microorganisms, in the degradation of wastes without disruption of the environment, thus, fungal systems appear to be most appropriate for the treatment of colored effluents (Ezeronye & Okerentugba, 1999; Shedbalkar & Jadhav, 2011). Another study conducted on textile industry discharge by Shedbalkar & Jadhav (2011) to reduce the colour, COD and BOD using *Penicillium ochrochloron*. This study is mainly targeting Malachite green. Malachite green and in its reduced form, leucomalachite green, remain in food chain as it may persist in edible fish tissues for a long period of time (Mitrowska & Posyniak, 2004; Shedbalkar & Jadhav, 2011). Also, Zahn & Braunbeck (1995) reported that in vitro studies proved that malachite green target the nuclei and mitochondria are major cellular in fishes and mammals (Shedbalkar & Jadhav, 2011). Malachite green was detoxified into p-benzyl- N,N-dimethylaniline and N,N-dimethyl-aniline hydrochloride by Penicillium ochrochloron. It successfully reduces the colour to 93% in czapek dox broth after 14 h with an optimum condition of pH 7 at 30°C. In addition, high BOD, COD, TDS and TSS values pose major environmental problems. These values were found lower in all of the treated samples than the control.

### 7.2 Biosorption

Esteves et al. (2000) used a waste biomass *Sargassum* sp. (brown macroalga) from an alginate extraction industry from Rio de Janeiro, Brazil to cleanup metals in wastewater via biosorption. Biosorption is sorption of metals onto biological materials. It could become an alternative way to remove toxic and recover precious metals from industrial discharge. In this work, the ability of *Sargassum* sp. to remove cadmium and zinc from a real effluent was investigated as well as its suitability to be reused in several consecutive metal adsorption-desorption cycles were investigated. The results showed that it could biosorbed 100% of $Cd^{2+}$ and 99.4% of $Zn^{2+}$ from a 3 and 98 mg $L^{-1}$ industrial discharge at pH 4.5, respectively, using HCl, $CaCl_2$, and NaOH as adsorbent-desorbent. On the other hand, when $CaCl_2$ was used as desorbent, the metal removal efficiency after the first cycle can achieve recovery at about 40%. The potential for metal recovery suggests that *Sargassum* sp.-based adsorption can be used as an alternative treatment method.

The biosorption may chemically occur due to the function of sulphated polysaccharides, which are absent in land plants, it enabled marine aquatic vegetation like algae to selectively adsorb trace metal ions (Chu et al., 1997; Esteves et al., 2000).

For brown algae, the carboxyl groups of alginate and sulfonate groups of polysaccharides were identified as the two important groups for binding metal (Kratochvil et al., 1997; Esteves et al., 2000). The other factors which are believed to contribute to the metal sorption capacities are electro-chemical properties of the metal ions (Esteves et al., 2000; Valdman & Leite, 2000) and the characteristic of a real wastewater.

*Sargassum* sp. has demonstrated its potential to be reused in metal-adsorption or recovery and it is cost effective. It showed a strong indication that such a system could be carried out commercially for the wastewater treatment as this biosystem could be fully regenerated and showed good treatment ability (Esteves et al., 2000).

Another study conducted by Rehman & Sohail Anjum (2010) used a yeast, *Candida tropicalis* to uptake the Cd from industrial discharge. *C. tropicalis* could reduce $Cd^{2+}$ at 57%, 69%, and 80% from the medium after 48, 96, and 144 hr, respectively, and, reduction of $Cd^{2+}$ 56% and 73% from the wastewater after 6 and 12 days, respectively. Apart from Cd, the yeast showed tolerance towards $Zn^{2+}$ (3100 mg $L^{-1}$), $Ni^{2+}$ (3,000 mg $L^{-1}$), $Hg^{2+}$ (2400 mg $L^{-1}$), $Cu^{2+}$ (2300 mg $L^{-1}$), $Cr^{6+}$ (2000 mg $L^{-1}$), and $Pb^{2+}$ (1200 mg $L^{-1}$).

*C. tropicalis* was found to be resistant to Cd up to a concentration of 2,800 mg $L^{-1}$. Cd-resistant yeast was also found to be resistant to other metal ions, in descending order was $Zn^{2+}>Ni^{2+}> Hg^{2+}>Cu^{2+}>Cr^{6+}>Pb^{2+}$. Zafar et al. (2007) reported the metal tolerance and biosorption potential of filamentous fungi having tolerance of metal in descending order of Cu>Cr>Cd>Co>Ni. Many Candida *spp.* have the capacity to absorb or accumulate metals. For instance, *C. albicans* and *C. tropicalis* are highly resistant to water-soluble ions $Hg^{2+}$, $Pb^{2+}$, $Cd^{2+}$, arsenate ($AsO_4^{3-}$), and selenite ($SeO_3^{2-}$) (Rehman & Sohail Anjum, 2010).

### 7.3 Green chemistry

A good management is important for a sustainable environment. Gar´cıa-Serna et al. (2007) has mentioned on the concept of sustainability which link between industries with philosophies and disciplines i.e. The Natural Step, Biomimicry, Cradle to Cradle, Getting to Zero Waste, Resilience Engineering, Inherently Safer Design, Ecological Design, Green Chemistry and Self-Assembly. Green chemistry is one of the possible sustainable environmental managements. Green chemistry is defined as the design of chemical products and processes that could reduce or eliminate the application and generation of hazardous substances (Kirchhoff, 2005; Tiwari et al., 2005, 2007, 2008; Seung-Mok & Tiwari, 2009; Rai, 2011).

There are three important areas in green chemistry:

i. Application of alternative synthetic pathways
ii. Application of alternative reaction conditions
iii. Design for safer chemicals compound that are less toxic or inherently safer with regards to accident potential (Gar´cıa-Serna et al., 2007; Rai, 2011).

For example, instead of using commercial-activated carbon, research based on inexpensive materials, such as chitosan, zeolites and other adsorbents, which have high adsorption capacity and economical have been conducted (Babel & Kurniawan, 2003). Yavuz et al. (2008) reported the adsorption of $Cu^{2+}$ and $Cr^{3+}$ onto pumice (Pmc) and polyacrylonitrile/Pmc composite (Rai, 2011). Similarly, Pehlivan & Arslan (2007) reported the adsorption of lignite (brown young coals) to remove copper, lead and nickel from aqueous solutions at the low concentrations. The implementation of green chemistry is essential in order to promote sustainability (Kirchhoff, 2005; Rai, 2011). Kirchhoff (2005) added that the collaboration between academia, industry and government are essential to expand and enhance the sustainability through the adoption of green chemistry (Rai, 2011).

### 7.4 Phytoremediation

A critical and complete review by Rai (2011) suggested another approaches to clean up the metals contaminant from industrial discharge. Green chemical processes could also reduce the impact to environment. However, they still pose threat in a long run. Therefore, the phytoremediation could be a better alternative where it could update the metals from contaminated areas a bio-cleanup technology.

Phytoremediation can be prepared from the naturally abundant plants which are very economical (Kratochvil & Volesky 1998; Rai, 2011). Wetland plants are important tools for heavy metals removal from aquatic ecosystems (Rai, 2011). Also, the wetland sediment provides metal reduction, chemically. According to Rai (2011), wetlands are proved to be effective in metal pollution abatement. *Typha, Phragmites, Eichhornia, Azolla, Lemna* and other aquatic macrophytes are important wetland plants for heavy metals contamination clean-up. The benefit of wetland plants is that it can absorb pollutants in their tissue and provide a surface and an environment for microorganisms to grow (Vymazal, 2002; Scholz, 2006; Vymazal et al., 2007; Rai, 2008; Rai, 2009b). In other words, wetland plants can stimulate aerobic decomposition of organic matter and the growth of nitrifying bacteria (Brix, 1997). However, in a wetland system, microorganism has a better capability in organic matters degradation compare to wetland plant. (Stottmeister et al., 2003; Rai, 2011). As the organic matter binds with metals directly, they provide a carbon and energy source for microbial metabolism. Microorganisms has the ability to grow in the presence of metals and metal uptake, hence, it can potentially bioremediate contaminant (Shakoori et al., 2004; Rehman & Sohail Anjum, 2010).

Water hyacinth (*Eichhornia crassipes*) is commonly used in constructed wetlands because of its rapid growth rate and its ability to uptake a lot of nutrients and contaminants (Rai 2008,

2009b). Due to its exotic invasive nature and multiplication, water hyacinth might also cause problems to constructed wetlands. Therefore, Rai (2011) recommended the use of artificial and high-rate algal ponds (HRAP) on industrial discharge before releasing the discharge to reservoir.

Hyperaccumulator, a term for plant species that could accumulate metals for 50- to 500-times concentration than normal plant, without any severe symptom of toxicity (Roosens et al., 2003; Kra¨mer, 2005; Dwivedi et al., 2008; Tiwari et al., 2008; Dwivedi et al., 2011). A hyperaccumulator has minimum threshold tissue concentration achieved for specific metal such as 0.001% for Hg whilst 0.1% for Cu, Pb, Ni, Al, or Se of the dry weight of plant (Pickering et al., 2003; Grata˜o et al., 2005; Dwivedi et al., 2011). However, slow growth and restricted distribution are the disadvantage of using hyperaccumulators. Dwivedi et al. (2011) conducted a study using *P. tuberosa* and *P. oleracea,* to accumulate metals such as Cu, Ni, Mo, Se, Hg, Pb, Al. The result showed that both of these two family member of Portulacaceae were able to hyperaccumulate Cu, Ni, Hg, and Pb. Selective hyperaccumulation of Se by *P. oleracea* and Al by *P. tuberosa* was were also reported in their study. Furthermore, these plants also showed higher accumulation of Mo than plants which could accumulate only in the range of 1–2 $\mu g\ g^{-1}$ dw (Hale et al., 2001; Dwivedi et al., 2011) and they could accumulate significant amount of Ni (Rooney et al., 2007; Dwivedi et al., 2011). These plants have shown to adapt and to grow naturally by uptaking, accumulating and detoxifying high load of metals and metalloids (Mishra et al., 2006; Dwivedi et al., 2008, 2010). In the study, Dwivedi et al. (2011) has also identified the parts of the plant which is able to accumulate better. The result showed that roots and shoots have the highest accumulation of Ni and Pb, respectively, whilst, Ni and Mo showed higher accumulation of metal in leaves than in stem. Adversely, other metals showed higher accumulation in stem than in leaves. They concluded that *P. oleracea* as a better accumulator species for various metals than *P. tuberose.*

Another study performed by Kisku et al. (2000) at Kalipur, Bangladesh, receiving industrial discharge of steel plant, thermal power plant, chemical plant, coke oven, pharmaceutical industry, and others. Bioaccumulation from soil to plant varies by element and plant. The result showed that the highest accumulation of Fe, Mn, Zn, Cu, Pb, Ni, Cr, and Cd were found in *Spinacea oleracea, Raphanus sativus, Amaranths viridis and Lycopersicon esculentum, Coriandrum sativum, Solanum melongena, Spinacea oleracea, Lycopersicon esculentum*, and *Coriandrum sativum*, respectively. Also, the results showed that *Spinacea oleracea, Lycopersicon esculentum and Coriandrum sativum* having more affinity to Fe and Ni, Zn and Cr and Cu and Cd, respectively. Furthermore, the growth of weeds at study site was similar to the growth of weeds at uncontaminated site, namely, Madhabpur. Weed grows better with positive growth for vegetative, root length, and number of rootless, number and colour of leaves and other but not crops and vegetables. This indicates that plant species has different capability in contaminant absorption, accumulation, and tolerance.

Rai & Tripathi (2009) studied on aquatic plant for effective phytoremediation at G.B. Pant Sagar located in Singrauli Industrial Region, India. Phytoremediation of *A. pinnata,* a water fern and *V. spiralis,* a submerged macrophyte, are studied. Result showed that the percentage removal of Hg was higher for *A. pinnata* (80–94%) than *V. spiralis* (70–84%). This indicated that free floating plant has higher capability than submerged macrophyte to remove Hg from wastewater. However, the authors suggested the possibilities of *V. spiralis* are higher to remove Hg remove from sediments. There is other research to study the removal of Hg from aquatic plants for phytoremediation purposes. Bennicelli et al. (2004) found that *Azolla caroliniana*, a small water fern could remove up to 93% of Hg contamination from water in 12 days. Kamal et al. (2004) also studied on aquatic plants parrot feather, creeping primrose and water mint and found to be able to remove 99.8% Hg from contaminated water after 21 days. However, a noticeable decrease showed in chlorophyll, protein, RNA, DNA and nutrients ($NO^{3-}$ and $PO_4^{3-}$) in *A. pinnata* more than *V. spiralis* which is due to the Hg toxicity. The decreased in chlorophyll may be due to the increased of Hg concentration and it is also observed to be having same result for several aquatic plant. The decline is caused by metal induced inhibition of chlorophyll biosynthesis. Furthermore, mercury inhibited biosynthesis of chlorophyll through targeting -SH groups of d aminolaevulinic acid dehydratase (ALAD) in seedlings of mung bean and bajra. At higher Hg concentrations, it can degrade pigments and result in great reduction (Rai & Tripathi, 2009).

There are a few studies suggested to use transgenic plants such as *Arabidopsis thaliana* Thal [Brassicaceae], and tobacco, *Nicotiana tabacum* L. [Solanaceae] to remove Hg from contaminated sites (Bennicelli et al., 2004; Rai & Tripathi, 2009). These plants have a modified bacterial mercuric reductase gene, merA, hence, it may convert Hg(II) to Hg(0) which is less in toxicity, or the bacterial organomercurial lyase gene, merB that convert methylmercury to sulfhydrylbound Hg(II) when taken up by plant roots (Rai & Tripathi, 2009).

Lesage et al. (2008) used *Myriophyllum spicatum* L. (Eurasian water milfoil, Haloragaceae) as an agent for bioremediation. *M. spicatum* L. is a rooted, perennial plant which reproduces primarily by vegetative fragmentation. Usually, deep root plant can survive at highly polluted condition and accumulated better while shallow roots are easily injured. The plant is able to tolerant wide water quality both in fresh and brackish water. In this study, the author studied the potential of *M. spicatum* L. in Co, Ni, Cu and Zn removal from industrial discharge. As comparison, the removal of Co and Ni by *M. spicatum* L was slower than of Cu and Zn. Co and Ni removal at 0.2 mg $L^{-1}$ need two weeks residence time, while Cu and Zn achieve 0.05 and 0.2 mg $L^{-1}$ removal only at 48 hr. The removal of about one third of Co and Ni removal involved precipitation with $CO_3^{2-}$ and $OH^-$ with the presence of the plant. As a submerged plant, *M. spicatum* L is having the ability to cleanup metals directly from the water. Hence, *M. spicatum* L is suggested as useful species in reducing metal concentrations from wastewaters. Furthermore, the toxic wastewater was not influence the growth of *M. spicatum* L. The use of aquatic plants for the removal of heavy metals from wastewater has been investigated by many authors. Other aquatic plants have been investigated in metal removal with good results are *Eichhornia crassipes* (Zhu et al., 1999), *Pistia stratiotes* and *Salvinia herzogii* (Maine et al., 2001, 2004).

However, there is a limitation in promting the use of *M. spicatum* L as full scale in a constructed wetland to cleanup industrial discharge. The reasons are explained by Kivaisi (2001) when countered the full-scale use of *E. crassipes* in developed countries, which also fit to the application of *M. spicatum* L:

i. The poor performance in temperate region like in northern-hemisphere winters
ii. Land acquisition
iii. The intensive harvesting schemes
iv. The construction and operation of systems for bioenergy recuperation.

### 7.5 Green monitoring

Phytotoxicity as part of green monitoring is a useful way for environmental risk assessment. For example, Seeds of *L. sativum* are very sensitive during seed germination and the early root growth period to change in environment (Ferrara et al., 2003; Montvydienė & Marčiulionienė, 2004; Montvydiene et al., 2008). Hence, it is suitable to use as screening tests and acute tests (seed germination and root growth). The other advantages using *L. sativum* is due to its simplicity, cheapness, and short duration. The turions of *S. polyrrhiza* at the first stage leafy stem development are very sensitive to change of surrounding, and the changes in meristematic tissue are reflected to the propagation of plant (Marčiulionienė et al., 2002; Montvydiene et al., 2008). *S. polyrrhiza* is suitable to apply as chronic tests (plant growth, various physiological parameters, and morphological alterations). *L. sativum* and *S. polyrrhiza* showed different sensitivity to tested samples of industrial discharge to water and sediments. However, it is hard to differentiate which is more sensitive as the duration of test, the time of plant contact with industrial discharge is different. Lastly, *Tradescantia* could determine genotoxicity of sample while other tests do not indicate the toxic effects. The investigations clearly indicated *L. sativum* and *Tradescantia* successfully used in the toxicity assessment of liquid or solid environmental component contaminated in different degree. *S. polyrrhiza* can be used in the examining various liquids (Montvydiene et al., 2008).

Shedbalkar et al. (2008) studied on phytotoxicity to determine the malachite green toxicity after treatment as part of green monitoring. The result showed that the treated wastewater was non-toxic to the plants of *Triticum aestivum* and *Ervum lens* Linn, and the amount of total chlorophyll was higher in plants with treated effluent when compared to control effluent. Phytotoxicity analysis showed high germination rate and significant growth in the shoots and roots for both plants grown after decolorization. Hence, phytotoxicity studies confirmed biodegradation of the dye by fungal culture, which resulted in detoxification. It is reported that the germination rate for *Triticum aestivum* and *Ervum lens* Linns was at 50% and 70%, respectively, using treated with effluent. Surprisingly, higher total chlorophyll contents of the leaves from *Triticum aestivum* and *Ervum lens* Linn plants were found in treated wastewater at pH 7. Furthermore, the result showed that there is low concentration of chlorophyll in leaves of plants grown in untreated wastewater as to compare with the control grown in water.

Radic´ et al. (2010) used Duckweed (*Lemna minor* L.) to monitor heavy metals and other aquatic pollutants over a 3-month period from the stream near the industrial estate of Savski Marof, Croatia. Duckweed is used due to its selective potential to accumulate certain chemicals. In previous study of Horvat et al. (2007) *L. minor* exposed to electroplating wastewater accumulated more Fe and Zn, and then followed by Pb and Ni. However, the authors also found that toxicity of metals in Lemna tissues damage in decreasing order: Zn>Ni>Fe>Cu>Cr>Pb (Radic´ et al., 2010). In this study, significant increase of duckweed growth which determined by Dry weight/ Fresh Weight (DW/FW) and also increased of inhibition of Relative Growth Rate (RGR) observed and suggested to be metals bioaccumulation. However, it has been shown that accumulation of heavy metals affect the plant water status which will results in osmotic stress and growth reduction (Perfus-Barbeoch et al., 2002; Radic´ et al., 2010). In the study, the increase of DW/FW after the treatment with industrial wastewater could be explained by the change of the plant's water status and accumulation of compatible solutes. In addition, according to Matysik et al. (2002) plants are shown to accumulate organic solutes such as carbohydrates and quaternary ammonium compounds.

In plants, pigment content (chlorophyll and carotenoids) and enzyme activities (like peroxidase) are commonly used as biomarker for toxicity tests (Mohan & Hosetti 1999; Radic´ et al., 2010). The result showed decreased in chlorophyll and carotenoids compared to the control sample plant. The carotenoids content affected lesser to chlorophylls. The decline in total chlorophyll and carotenoids contents and growth inhibition may be associated with metal toxicity (Radic´ et al., 2010). The growth reduction reported the study may be due to the decrease of chlorophyll content produced by heavy metals present in the industrial discharge. And Ku¨pper et al. (1996) proposed that photosynthesis and growth reduction caused by Zn, Ni, Cu and Pb. Lastly, from the results, the authors suggested that phyto- and genotoxicity tests with *L. minor* should be applied to biomonitoring of municipal, agricultural and industrial discharge due to its simplicity, sensitivity and economical (Radic´ et al., 2010).

## 8. Conclusion

This chapter gives an overview of the source and types of industrial discharge, the effects of industrial discharge to the environment and human health and lastly the corrective action that could be taken to minimize the negative impact of the discharge. Many ways have been proposed to protect the environment from contamination including enforcement of stringent rules and regulations. However, the discharge from some industries is still exceeding the permissible limits. Before designing good corrective actions, the knowledge of effects to environment and human health, and, the interaction between the contaminants and biotic and abiotic compartment are crucial to explore. In short, corrective actions for industries contamination clean-up are important to implement. A number of remedial actions presented in the previous sections, there is still factors which hinder the development of large scale of cleanup technology.

i. Characteristic of organic, inorganic and synthetic chemical elements is well documented in literature. Research on the synthesis of new green chemical elements for industry application is important to replace the original materials in order to protect the sustainable environment.
ii. Although most of the bioremediation studies for surface water and soil contamination were proven to be effective in removing or reducing contaminant, such studies were done at lab scale. Thus, there is a need to investigate on large scale to better understand the suitable condition for optimization remediation process.
iii. The groundwater is a major important water resource for some countries for irrigation, drinking and bathing. However, waste dump brought adverse impact to the groundwater quality. Hence, an effective corrective action is necessary to improve the water quality, especially the metal contamination, which may lead to serious illness and death. Study on the in situ bioremediation by the indigenous bacteria could be carried in depth for cleanup process.

## 9. Reference


Abbas Alkarkhi, F., Ismail, N., & Easa, A. M. (2008). Assessment of arsenic and heavy metal contents in cockles (Anadara granosa) using multivariate statistical techniques. *Journal of hazardous materials,* Vol. 150, No. 3, pp 783-789, ISSN 0304-3894

Acevedo-Aguilar, F. J., Espino-Saldaña, A. E., Leon-Rodriguez, I. L., Rivera-Cano, M. E., Avila-Rodriguez, M., Wrobel, K., et al. (2006). Hexavalent chromium removal in vitro and from industrial wastes, using chromate-resistant strains of filamentous fungi indigenous to contaminated wastes. *Canadian journal of microbiology,* Vol. 52, No. 9, pp 809-815, ISSN 0008-4166

Ahmad, M., Bajahlan, A. S., & Hammad, W. S. (2008). Industrial effluent quality, pollution monitoring and environmental management. *Environmental Monitoring and Assessment,* Vol. 147, No. 1, pp 297-306, ISSN 0167-6369

Ahmed, G., Miah, M. A., Anawar, H. M., Chowdhury, D. A., & Ahmad, J. U. (2011). Influence of multi-industrial activities on trace metal contamination: an approach towards surface water body in the vicinity of Dhaka Export Processing Zone (DEPZ). *Environmental Monitoring and Assessment,* Vol., No., pp 1-10, ISSN 0167-6369

Al-Shami, S. A., Rawi, C. S. M., HassanAhmad, A., & Nor, S. A. M. (2010). Distribution of Chironomidae (Insecta: Diptera) in polluted rivers of the Juru River Basin, Penang, Malaysia. *Journal of Environmental Sciences,* Vol. 22, No. 11, pp 1718-1727, ISSN 1001-0742

Ansari, A., Singh, I., & Tobschall, H. (1999). Status of anthropogenically induced metal pollution in the Kanpur-Unnao industrial region of the Ganga Plain, India. *Environmental Geology,* Vol. 38, No. 1, pp 25-33, ISSN 0943-0105

Babel, S., & Kurniawan, T. A. (2003). Low-cost adsorbents for heavy metals uptake from contaminated water: a review. *Journal of hazardous materials,* Vol. 97, No. 1-3, pp 219-243, ISSN 0304-3894

Bennicelli, R., Stpniewska, Z., Banach, A., Szajnocha, K., & Ostrowski, J. (2004). The ability of Azolla caroliniana to remove heavy metals (Hg (II), Cr (III), Cr (VI)) from municipal waste water. *Chemosphere,* Vol. 55, No. 1, pp 141-146, ISSN 0045-6535

Bobmanuel, N., Gabriel, U., & Ekweozor, I. (2006). Direct toxic assessment of treated fertilizer effluents to Oreochromis niloticus, Clarias gariepinus and catfish hybrid(Heterobranchus bidorsalis x Clarias gariepinus). *African Journal of Biotechnology,* Vol. 5, No. 8, pp 635-642, ISSN 1684-5315

Boopathy, R. (2000). Factors limiting bioremediation technologies. *Bioresource Technology,* Vol. 74, No. 1, pp 63-67, ISSN 0960-8524

Botalova, O., & Schwarzbauer, J. (2011). Geochemical Characterization of Organic Pollutants in Effluents Discharged from Various Industrial Sources to Riverine Systems. *Water, Air, & Soil Pollution,* Vol., No., pp 1-22, ISSN 0049-6979

Brix, H. (1997). Do macrophytes play a role in constructed treatment wetlands? *Water science and technology,* Vol. 35, No. 5, pp 11-18, ISSN 0273-1223

Camargo, J. A. (1992). Macroinvertebrate responses along the recovery gradient of a regulated river (Spain) receiving an industrial effluent. *Archives of Environmental Contamination and Toxicology,* Vol. 23, No. 3, pp 324-332, ISSN 0090-4341

Chu, K., Hashim, M., Phang, S., & Samuel, V. (1997). Biosorption of cadmium by algal biomass: adsorption and desorption characteristics. *Water science and technology,* Vol. 35, No. 7, pp 115-122, ISSN 0273-1223

Devi, N. L., Yadav, I. C., Shihua, Q., Singh, S., & Belagali, S. (2010). Physicochemical characteristics of paper industry effluents—a case study of South India Paper Mill (SIPM). *Environmental Monitoring and Assessment,* Vol., No., pp 1-11, ISSN 0167-6369

Din, Z., & Ahamad, A. (1995). Changes in the scope for growth of blood cockles (Anadara granosa) exposed to industrial discharge. *Marine pollution bulletin,* Vol. 31, No. 4-12, pp 406-410, ISSN 0025-326X

Dwivedi, S., Mishra, A., Kumar, A., Tripathi, P., Dave, R., Dixit, G., et al. (2011). Bioremediation potential of genus <i>Portulaca</i> L. collected from industrial areas in Vadodara, Gujarat, India. *Clean Technologies and Environmental Policy,* Vol., No., pp 1-6, ISSN 1618-954X

Dwivedi, S., Srivastava, S., Mishra, S., Dixit, B., Kumar, A., & Tripathi, R. (2008). Screening of native plants and algae growing on fly-ash affected areas near National Thermal Power Corporation, Tanda, Uttar Pradesh, India for accumulation of toxic heavy metals. *Journal of hazardous materials,* Vol. 158, No. 2-3, pp 359-365, ISSN 0304-3894

Dwivedi, S., Srivastava, S., Mishra, S., Kumar, A., Tripathi, R., Rai, U., et al. (2010). Characterization of native microalgal strains for their chromium bioaccumulation potential: Phytoplankton response in polluted habitats. *Journal of hazardous materials,* Vol. 173, No. 1-3, pp 95-101, ISSN 0304-3894

Ekweozor, I., Bobmanuel, N., & Gabriel, U. (2010). Sublethal Effects of Ammoniacal Fertilizer Effluents on three Commercial Fish Species from Niger Delta Area, Nigeria. *Journal of Applied Sciences and Environmental Management,* Vol. 5, No. 1, ISSN 1119-8362

Esmaeili Taheri, H., Hatamipour, M., Emtiazi, G., & Beheshti, M. (2008). Bioremediation of DSO contaminated soil. *Process Safety and Environmental Protection,* Vol. 86, No. 3, pp 208-212, ISSN 0957-5820

Esteves, A., Valdman, E., & Leite, S. (2000). Repeated removal of cadmium and zinc from an industrial effluent by waste biomass Sargassum sp. *Biotechnology letters,* Vol. 22, No. 6, pp 499-502, ISSN 0141-5492

Ezeronye, O., & Okerentugba, P. (1999). Performance and efficiency of a yeast biofilter for the treatment of a Nigerian fertilizer plant effluent. *World Journal of Microbiology and Biotechnology,* Vol. 15, No. 4, pp 515-516, ISSN 0959-3993

Faisal, M., & Hasnain, S. (2005). Comparative study of Cr (VI) uptake and reduction in industrial effluent by Ochrobactrum intermedium and Brevibacterium sp. *Biotechnology letters,* Vol. 26, No. 21, pp 1623-1628, ISSN 0141-5492

Ferrara G., B. G., Senesi N., Mondelli D. and LA Gheza V. (2003). Total and potentially phytotoxic trace Metals in southeastern Italian Soil. *JFAE,* Vol. 1, No., pp 279-286.

Garcia-Serna, J., Perez-Barrigon, L., & Cocero, M. (2007). New trends for design towards sustainability in chemical engineering: Green engineering. *Chemical Engineering Journal,* Vol. 133, No. 1-3, pp 7-30, ISSN 1385-8947

Girvin, D. C., & Scott, A. J. (1997). Polychlorinated biphenyl sorption by soils: Measurement of soil-water partition coefficients at equilibrium. *Chemosphere,* Vol. 35, No. 9, pp 2007-2025, ISSN 0045-6535

Gouda, M. (2000). Studies on chromate reduction by three Aspergillus species. *Fresenius Environmental Bulletin,* Vol. 9, No. 11/12, pp 799-808, ISSN 1018-4619

Govil, P., Sorlie, J., Murthy, N., Sujatha, D., Reddy, G. L. N., Rudolph-Lund, K., et al. (2008). Soil contamination of heavy metals in the Katedan industrial development area, Hyderabad, India. *Environmental Monitoring and Assessment,* Vol. 140, No. 1, pp 313-323, ISSN 0167-6369

Gratão, P. L., Prasad, M. N. V., Cardoso, P. F., Lea, P. J., & Azevedo, R. A. (2005). Phytoremediation: green technology for the clean up of toxic metals in the environment. *Brazilian Journal of Plant Physiology,* Vol. 17, No. 1, pp 53-64, ISSN 1677-0420

Hale KL, M. S., Lombi E, Stack SM, Terry N, Pickering IJ, George GN, Pilon-Smits EAH. (2001). Molybdenum sequestration in Brassica species, a role for anthocyanins? *Plant Physiol,* Vol. 126, No., pp 1391–1402.

Haq, R., Qazi, J., & Shakoori, A. (1998). Growth and survival of protozoa isolated from a tannery effluent. *Folia microbiologica,* Vol. 43, No. 1, pp 109-112, ISSN 0015-5632

Hiller, E., Zemanová, L., Sirotiak, M., & Jurkovi , L. (2011). Concentrations, distributions, and sources of polychlorinated biphenyls and polycyclic aromatic hydrocarbons in bed sediments of the water reservoirs in Slovakia. *Environmental Monitoring and Assessment,* Vol. 173, No. 1, pp 883-897, ISSN 0167-6369

Honda, R., Tsuritani, I., Ishizaki, M., & Yamada, Y. (1997). Zinc and Copper Levels in Ribs of Cadmium-Exposed Persons with Special Reference to Osteomalacia* 1,* 2. *Environmental research,* Vol. 75, No. 1, pp 41-48, ISSN 0013-9351

Horvat, T., Vidakovic-Cifrek, Z., Orescanin, V., Tkalec, M., & Pevalek-Kozlina, B. (2007). Toxicity assessment of heavy metal mixtures by Lemna minor L. *Science of the total environment,* Vol. 384, No. 1-3, pp 229-238, ISSN 0048-9697

Jain, C., & Ali, I. (2000). Adsorption of cadmium on riverine sediments: Quantitative treatment of the large particles. *Hydrological processes,* Vol. 14, No. 2, pp 261-270, ISSN 1099-1085

Jain, C., & Sharma, M. (2002). Adsorption of cadmium on bed sediments of river Hindon: Adsorption models and kinetics. *Water, Air, & Soil Pollution,* Vol. 137, No. 1, pp 1-19, ISSN 0049-6979

Jain, C., Singhal, D., & Sharma, M. (2004). Adsorption of zinc on bed sediment of River Hindon: adsorption models and kinetics. *Journal of hazardous materials,* Vol. 114, No. 1-3, pp 231-239, ISSN 0304-3894

Jain, C., Singhal, D., & Sharma, M. (2005). Metal pollution assessment of sediment and water in the river Hindon, India. *Environmental Monitoring and Assessment,* Vol. 105, No. 1, pp 193-207, ISSN 0167-6369

Janardhana Raju, N., Ram, P., & Dey, S. (2009). Groundwater quality in the lower Varuna River basin, Varanasi district, Uttar Pradesh. *Journal of the Geological Society of India,* Vol. 73, No. 2, pp 178-192, ISSN 0016-7622

Jonathan, M., Srinivasalu, S., Thangadurai, N., Ayyamperumal, T., Armstrong-Altrin, J., & Ram-Mohan, V. (2008). Contamination of Uppanar River and coastal waters off Cuddalore, Southeast coast of India. *Environmental Geology,* Vol. 53, No. 7, pp 1391-1404, ISSN 0943-0105

Jordao, C., Pereira, M., & Pereira, J. (2002). Metal contamination of river waters and sediments from effluents of kaolin processing in Brazil. *Water, Air, & Soil Pollution,* Vol. 140, No. 1, pp 119-138, ISSN 0049-6979

Kamal, M., Ghaly, A., Mahmoud, N., & Cote, R. (2004). Phytoaccumulation of heavy metals by aquatic plants. *Environment international,* Vol. 29, No. 8, pp 1029-1039, ISSN 0160-4120

Khérici-Bousnoubra, H., Khérici, N., Derradji, E., Rousset, C., & Caruba, R. (2009). Behaviour of chromium VI in a multilayer aquifer in the industrial zone of Annaba, Algeria. *Environmental Geology,* Vol. 57, No. 7, pp 1619-1624, ISSN 0943-0105

Kirchhoff, M. M. (2005). Promoting sustainability through green chemistry. *Resources, conservation and recycling,* Vol. 44, No. 3, pp 237-243, ISSN 0921-3449

Kisku, G., Barman, S., & Bhargava, S. (2000). Contamination of soil and plants with potentially toxic elements irrigated with mixed industrial effluent and its impact on the environment. *Water, Air, & Soil Pollution,* Vol. 120, No. 1, pp 121-137, ISSN 0049-6979

Kivaisi, A. K. (2001). The potential for constructed wetlands for wastewater treatment and reuse in developing countries: a review. *Ecological Engineering,* Vol. 16, No. 4, pp 545-560, ISSN 0925-8574

Kocan, A., Petrik, J., Jursa, S., Chovancova, J., & Drobna, B. (2001). Environmental contamination with polychlorinated biphenyls in the area of their former manufacture in Slovakia. *Chemosphere,* Vol. 43, No. 4-7, pp 595-600, ISSN 0045-6535

Kramer, U. (2005). Phytoremediation: novel approaches to cleaning up polluted soils. *Current opinion in Biotechnology,* Vol. 16, No. 2, pp 133-141, ISSN 0958-1669

Kratochvil, D., & Volesky, B. (1998). Advances in the biosorption of heavy metals. *Trends in Biotechnology,* Vol. 16, No. 7, pp 291-300, ISSN 0167-7799

Kratochvil, D., Volesky, B., & Demopoulos, G. (1997). Optimizing Cu removal/recovery in a biosorption column. *Water Research,* Vol. 31, No. 9, pp 2327-2339, ISSN 0043-1354

Küpper, H., Küpper, F., & Spiller, M. (1996). Environmental relevance of heavy metal-substituted chlorophylls using the example of water plants. *Journal of Experimental Botany,* Vol. 47, No. 2, pp 259, ISSN 0022-0957

Kuppusamy, M., & Giridhar, V. (2006). Factor analysis of water quality characteristics including trace metal speciation in the coastal environmental system of Chennai Ennore. *Environment international,* Vol. 32, No. 2, pp 174-179, ISSN 0160-4120

Lesage, E., Mundia, C., Rousseau, D. P. L., Moortel, A., Laing, G. D., Tack, F. M. G., et al. (2008). Removal of heavy metals from industrial effluents by the submerged aquatic plant Myriophyllum spicatum L. *Wastewater Treatment, Plant Dynamics and Management in Constructed and Natural Wetlands,* Vol., No., pp 211-221.

Machado, M. D., Soares, H. M. V. M., & Soares, E. V. (2010). Removal of Chromium, Copper, and Nickel from an Electroplating Effluent Using a Flocculent Brewer's Yeast Strain of Saccharomyces cerevisiae. *Water, Air, & Soil Pollution,* Vol. 212, No. 1, pp 199-204, ISSN 0049-6979

MacNaughton, S. J., Stephen, J. R., Venosa, A. D., Davis, G. A., Chang, Y. J., & White, D. C. (1999). Microbial population changes during bioremediation of an experimental oil spill. *Applied and Environmental Microbiology,* Vol. 65, No. 8, pp 3566, ISSN 0099-2240

Maine, M. A., Duarte, M. V., & Su é, N. L. (2001). Cadmium uptake by floating macrophytes. *Water Research,* Vol. 35, No. 11, pp 2629-2634, ISSN 0043-1354

Maine, M. A., Su é, N. L., & Lagger, S. C. (2004). Chromium bioaccumulation: comparison of the capacity of two floating aquatic macrophytes. *Water Research,* Vol. 38, No. 6, pp 1494-1501, ISSN 0043-1354

Malaysia(DOE), D. o. E. (1998). *Classification of Malaysian Rivers, Juru River.,* Department of Environment, Ministry of Science, Technology and the Environment Malaysia.

Marčiulionienė D., M. D., Kiponas D., Lukšienė B., and Butkus D. (2004). Toxicity to Tradescantia of technogenic Radionuclides and their Mixture with heavy Metals. *Environ. Toxicology in vitro,* Vol. 19, No., pp 346-350.

Megharaj, M., Avudainayagam, S., & Naidu, R. (2003). Toxicity of hexavalent chromium and its reduction by bacteria isolated from soil contaminated with tannery waste. *Current microbiology,* Vol. 47, No. 1, pp 51-54, ISSN 0343-8651

Mishra, S., Srivastava, S., Tripathi, R. D., Kumar, R., Seth, C. S., & Gupta, D. K. (2006). Lead detoxification by coontail (Ceratophyllum demersum L.) involves induction of phytochelatins and antioxidant system in response to its accumulation. *Chemosphere,* Vol. 65, No. 6, pp 1027-1039, ISSN 0045-6535

Mitrowska, K., & Posyniak, A. (2004). Determination of malachite green and its metabolite, leucomalachite green, in fish muscle by liquid chromatography. *Bulletin-Veterinary Institute in Pulawy,* Vol. 48, No. 2, pp 173-176, ISSN 0042-4870

Mohan, B., & Hosetti, B. (1999). Aquatic plants for toxicity assessment. *Environmental research,* Vol. 81, No. 4, pp 259-274, ISSN 0013-9351

Montvydienė, D., Marčiulionienė, D., Karlavičienė, V., & Hogland, W. (2008). Phytotoxicity Assessment Of Effluent Waters, Surface Water And Sediments Dangerous Pollutants (Xenobiotics) in Urban Water Cycle. In P. Hlavinek, O. Bonacci, J. Marsalek & I. Mahrikova (Eds.), (pp. 171-180): Springer Netherlands.

Morales-Barrera, L., & Cristiani-Urbina, E. (2008). Hexavalent chromium removal by a Trichoderma inhamatum fungal strain isolated from tannery effluent. *Water, Air, & Soil Pollution,* Vol. 187, No. 1, pp 327-336, ISSN 0049-6979

O'Brien, T. J., Ceryak, S., & Patierno, S. R. (2003). Complexities of chromium carcinogenesis: role of cellular response, repair and recovery mechanisms. *Mutation Research/Fundamental and Molecular Mechanisms of Mutagenesis,* Vol. 533, No. 1-2, pp 3-36, ISSN 0027-5107

Olajire, A., & Imeokparia, F. (2001). Water quality assessment of Osun River: studies on inorganic nutrients. *Environmental Monitoring and Assessment,* Vol. 69, No. 1, pp 17-28, ISSN 0167-6369

Pehlivan, E., & Arslan, G. (2007). Removal of metal ions using lignite in aqueous solution--Low cost biosorbents. *Fuel processing technology,* Vol. 88, No. 1, pp 99-106, ISSN 0378-3820

Perfus Barbeoch, L., Leonhardt, N., Vavasseur, A., & Forestier, C. (2002). Heavy metal toxicity: cadmium permeates through calcium channels and disturbs the plant water status. *The Plant Journal,* Vol. 32, No. 4, pp 539-548, ISSN 1365-313X

Petrik, J., Drobna, B., Kocan, A., Chovancova, J., & Pavuk, M. (2001). Polychlorinated biphenyls in human milk from Slovak mothers. *Fresenius Environmental Bulletin,* Vol. 10, No. 4, pp 342-348, ISSN 1018-4619

Pickering IJ, W. C., Bubner B, Ellis D, Persans MW, Yu EY,. (2003). Chemical form and distribution of selenium and sulfur in the selenium hyperaccumulator Astragalus bisculatus. *Plant Physiol* Vol. 131, No., pp 1460–1467, ISSN

Purnomo, A. S., Mori, T., Takagi, K., & Kondo, R. (2011). Bioremediation of DDT contaminated soil using brown-rot fungi. *International Biodeterioration & Biodegradation,* Vol., No., ISSN 0964-8305

Purushotham, D., Narsing Rao, A., Ravi Prakash, M., Ahmed, S., & Ashok Babu, G. (2011). Environmental impact on groundwater of Maheshwaram Watershed, Ranga Reddy district, Andhra Pradesh. *Journal of the Geological Society of India,* Vol. 77, No. 6, pp 539-548, ISSN 0016-7622

Radi , S., Stipani ev, D., Cvjetko, P., Mikeli , I. L., Raj i , M. M., Širac, S., et al. (2010). Ecotoxicological assessment of industrial effluent using duckweed (Lemna minor L.) as a test organism. *Ecotoxicology,* Vol. 19, No. 1, pp 216-222, ISSN 0963-9292

Radić, S., Stipaničev, D., Cvjetko, P., Mikelić, I., Rajčić, M., Širac, S., et al. (2010). Ecotoxicological assessment of industrial effluent using duckweed (Lemna minor L.) as a test organism. *Ecotoxicology,* Vol. 19, No. 1, pp 216-222, ISSN 0963-9292

Rai, P. (2011). An eco-sustainable green approach for heavy metals management: two case studies of developing industrial region. *Environmental Monitoring and Assessment,* Vol., No., pp 1-28, ISSN 0167-6369

Rai, P., & Tripathi, B. (2007). Heavy metals adsorption characteristics of free floating aquatic macrophyte Spirodela poyrhhiza. *Journal of Environmental Research and Development,* Vol.148, No. 1-4, pp 75-84.

Rai, P. K. (2008). Heavy metal pollution in aquatic ecosystems and its phytoremediation using wetland plants: An ecosustainable approach. *International journal of phytoremediation,* Vol. 10, No. 2, pp 133-160, ISSN 1522-6514

Rai, P. K. (2009b). Heavy metal phytoremediation from aquatic ecosystems with special reference to macrophytes. *Critical Reviews in Environmental Science and Technology,* Vol. 39, No. 9, pp 697-753, ISSN 1064-3389

Rai, P. K., & Tripathi, B. (2009a). Comparative assessment of Azolla pinnata and Vallisneria spiralis in Hg removal from GB Pant Sagar of Singrauli Industrial region, India. *Environmental Monitoring and Assessment,* Vol. 148, No. 1, pp 75-84, ISSN 0167-6369

Rehman, A., & Sohail Anjum, M. (2010). Cadmium Uptake by Yeast, Candida tropicalis, Isolated from Industrial Effluents and Its Potential Use in Wastewater Clean-Up Operations. *Water, Air, & Soil Pollution,* Vol. 205, No. 1, pp 149-159, ISSN 0049-6979

Rooney, C. P., Zhao, F. J., & McGrath, S. P. (2007). Phytotoxicity of nickel in a range of European soils: Influence of soil properties, Ni solubility and speciation. *Environmental pollution,* Vol. 145, No. 2, pp 596-605, ISSN 0269-7491

Roosens, N., Verbruggen, N., Meerts, P., XIMÉNEZ EMBÚN, P., & Smith, J. (2003). Natural variation in cadmium tolerance and its relationship to metal hyperaccumulation for seven populations of Thlaspi caerulescens from western Europe. *Plant, Cell & Environment,* Vol. 26, No. 10, pp 1657-1672, ISSN 1365-3040

Saifullah, S., Khan, S., & Ismail, S. (2002). Distribution of nickel in a polluted mangrove habitat of the indus delta. *Marine pollution bulletin,* Vol. 44, No. 6, pp 551-576, ISSN 0025-326X

Sani, R., Azmi, W., & Banerjee, U. (1998). Comparison of static and shake culture in the decolorization of textile dyes and dye effluents byPhanerochœte chrysosporium. *Folia microbiologica,* Vol. 43, No. 1, pp 85-88, ISSN 0015-5632

Scholz, M. (2006). *Wetland systems to control urban runoff,* Elsevier Science,0444527346,

Seung-Mok, L., & Diwakar, T. (2009). Application of ferrate (VI) in the treatment of industrial wastes containing metal-complexed cyanides: A green treatment. *Journal of Environmental Sciences,* Vol. 21, No. 10, pp 1347-1352, ISSN 1001-0742

Shakeri, A., & Moore, F. (2010). The impact of an industrial complex on freshly deposited sediments, Chener Rahdar river case study, Shiraz, Iran. *Environmental Monitoring and Assessment,* Vol. 169, No. 1, pp 321-334, ISSN 0167-6369

Shedbalkar, U., & Jadhav, J. P. (2011). Detoxification of malachite green and textile industrial effluent by Penicillium ochrochloron. *Biotechnology and Bioprocess Engineering,* Vol. 16, No. 1, pp 196-204, ISSN 1226-8372

Shiozawa, T., Suyama, K., Nakano, K., Nukaya, H., Sawanishi, H., Oguri, A., et al. (1999). Mutagenic activity of 2-phenylbenzotriazole derivatives related to a mutagen, PBTA-1, in river water. *Mutation Research/Genetic Toxicology and Environmental Mutagenesis,* Vol. 442, No. 2, pp 105-111, ISSN 1383-5718

Singh, K. P., Mohan, D., Singh, V. K., & Malik, A. (2005). Studies on distribution and fractionation of heavy metals in Gomti river sediments--a tributary of the Ganges, India. *Journal of hydrology,* Vol. 312, No. 1-4, pp 14-27, ISSN 0022-1694

Singh, N., Sharma, B., & Bohra, P. (2000). Impact assessment of industrial effluent of arid soils by using satellite imageries. *Journal of the Indian Society of Remote Sensing,* Vol. 28, No. 2, pp 79-92, ISSN 0255-660X

Singh, R., Sengupta, B., Bali, R., Shukla, B., Gurunadharao, V. V. S., & Srivatstava, R. (2009). Identification and mapping of chromium (VI) plume in groundwater for remediation: A case study at Kanpur, Uttar Pradesh. *Journal of the Geological Society of India,* Vol. 74, No. 1, pp 49-57, ISSN 0016-7622

Srinivasa Gowd, S., & Kotaiah, B. (2000). Groundwater pollution by Cystine manufacturing industrial effluent around the factory. *Environmental Geology,* Vol. 39, No. 6, pp 679-682, ISSN 0943-0105

Stottmeister, U., Wießner, A., Kuschk, P., Kappelmeyer, U., Kästner, M., Bederski, O., et al. (2003). Effects of plants and microorganisms in constructed wetlands for wastewater treatment. *Biotechnology Advances,* Vol. 22, No. 1-2, pp 93-117, ISSN 0734-9750

Tiwari, D., Kim, H. U., Choi, B. J., Lee, S. M., Kwon, O. H., Choi, K. M., et al. (2007). Ferrate (VI): A green chemical for the oxidation of cyanide in aqueous/waste solutions. Vol., No., ISSN 1093-4529

Tiwari, D., Yang, J., & Lee, S. (2005). Applications of ferrate (VI) in the treatment of wastewaters. *ENVIRONMENTAL ENGINEERING RESEARCH-SEOUL-,* Vol. 10, No. 6, pp 269, ISSN 1226-1025

Tiwari, D., Yang, J. K., Chang, Y. Y., & Lee, S. M. (2008). Application of Ferrate (VI) on the decomplexation of Cu (II) EDTA. *Environmental Engineering Research,* Vol. 13, No., pp 131–135.

Tiwari, K., Dwivedi, S., Mishra, S., Srivastava, S., Tripathi, R., Singh, N., et al. (2008). Phytoremediation efficiency of Portulaca tuberosa rox and Portulaca oleracea L. naturally growing in an industrial effluent irrigated area in Vadodra, Gujrat, India. *Environmental Monitoring and Assessment,* Vol. 147, No. 1, pp 15-22, ISSN 0167-6369

Valarmathi, S., & Azariah, J. (2003). Effect of copper chloride on the enzyme activities of the crab Sesarma quadratum (Fabricius). *Turkish Journal of Zoology,* Vol. 27, No. 3, pp 253-256, ISSN 1300-0179

Valdman, E., & Leite, S. (2000). Biosorption of Cd, Zn and Cu by Sargassum sp. waste biomass. *Bioprocess and Biosystems Engineering,* Vol. 22, No. 2, pp 171-173, ISSN 1615-7591

Vasanthavigar, M. Prasanna, K. S., M. V. (2011). Evaluation of groundwater suitability for domestic, irrigational, and industrial purposes: a case study from Thirumanimuttar river basin, Tamilnadu, India. *Environ Monit Assess* DOI: 10.1007/s10661-011-1977-y

Venkata Mohan, S., Rama Krishna, M., Muralikrishna, P., Shailaja, S., & Sarma, P. (2007). Solid phase bioremediation of pendimethalin in contaminated soil and evaluation of leaching potential. *Bioresource Technology,* Vol. 98, No. 15, pp 2905-2910, ISSN 0960-8524

Vymazal, J. (2002). The use of sub-surface constructed wetlands for wastewater treatment in the Czech Republic: 10 years experience. *Ecological Engineering,* Vol. 18, No. 5, pp 633-646, ISSN 0925-8574

Vymazal, J., Svehla, J., Kropfelova, L., & Chrastny, V. (2007). Trace metals in Phragmites australis and Phalaris arundinacea growing in constructed and natural wetlands. *Science of the total environment,* Vol. 380, No. 1-3, pp 154-162, ISSN 0048-9697

Wang, J., & Chen, C. (2006). Biosorption of heavy metals by Saccharomyces cerevisiae: A review. *Biotechnology Advances,* Vol. 24, No. 5, pp 427-451, ISSN 0734-9750

Wang, Q., Kim, D., Dionysiou, D. D., Sorial, G. A., & Timberlake, D. (2004). Sources and remediation for mercury contamination in aquatic systems--a literature review. *Environmental pollution,* Vol. 131, No. 2, pp 323-336, ISSN 0269-7491

Watanabe, K. (2001). Microorganisms relevant to bioremediation. *Current opinion in Biotechnology,* Vol. 12, No. 3, pp 237-241, ISSN 0958-1669

WHO. (2009). Potassium in drinking-water. Background Document for Preparation of WHO Guidelines for Drinking-Water Quality, *WHO/HSE/WSH/09.01/7.*

Yadav, A., Gopesh, A., S. Pandey, R., Rai, D. K., & Sharma, B. (2007). Fertilizer Industry Effluent Induced Biochemical Changes in Fresh water Teleost, Channa striatus (Bloch). *Bulletin of environmental contamination and toxicology,* Vol. 79, No. 6, pp 588-595, ISSN 0007-4861

Yadav, A., Neraliya, S., & Singh, R. (2005). Effect of fertilizer industrial effluent on the behavior and morphology of fresh water catfish, Heteropeneustes fossilis (Bloch). *Proc Nat Acad Sci India,* Vol. 75, No. 111, pp 191–195, ISSN

Yap, C., Ismail, A., Tan, S., & Omar, H. (2002). Concentrations of Cu and Pb in the offshore and intertidal sediments of the west coast of Peninsular Malaysia. *Environment international,* Vol. 28, No. 6, pp 467-479, ISSN 0160-4120

Yavuz, M., Gode, F., Pehlivan, E., Ozmert, S., & Sharma, Y. C. (2008). An economic removal of Cu2+ and Cr3+ on the new adsorbents: Pumice and polyacrylonitrile/pumice composite. *Chemical Engineering Journal,* Vol. 137, No. 3, pp 453-461, ISSN 1385-8947

Zafar, S., Aqil, F., & Ahmad, I. (2007). Metal tolerance and biosorption potential of filamentous fungi isolated from metal contaminated agricultural soil. *Bioresource Technology,* Vol. 98, No. 13, pp 2557-2561, ISSN 0960-8524

Zahn, T., & Braunbeck, T. (1995). Cytotoxic effects of sublethal concentrations of malachite green in isolated hepatocytes from rainbow trout (Oncorhynchus mykiss). *Toxicology in vitro,* Vol. 9, No. 5, pp 729-741, ISSN 0887-2333

Zhu, Y., Zayed, A., Qian, J. H., De Souza, M., & Terry, N. (1999). Phytoaccumulation of trace elements by wetland plants: II. Water hyacinth. *Journal of Environmental Quality*, Vol. 28, No. 1, pp 339-344, ISSN 0047-2425

# 2

# Anodic Materials with High Energy Efficiency for Electrochemical Oxidation of Toxic Organics in Waste Water

Wang Yun-Hai, Chen Qing-Yun*, Li Guo and Li Xiang-Lin
*State Key Laboratory of Multiphase Flow, Xi'an Jiaotong University, China*

## 1. Introduction

With the development of human society, discharge of toxic organics in industrial waste water increased sharply. It is difficult to treat these types of waste water efficiently by traditional bio-process to meet the more critical discharge standard. Advanced oxidation processes (AOPs) attracted more and more attention as efficient methods to remove the toxic organics from waste water. Compared with other AOPs, electrochemical oxidation process was considered as an effective and environmental friendly process due to its simplicity in operation, robustness in system configuration, strong oxidizing ability, reliable performance for a wide variety of toxic organics and chemical reagents free.

The electrochemical system is basically composed of anode, cathode, electrolyte and cell. Oxidation occurs on the anode while reduction occurs on the cathode simultaneously. For the electrochemical oxidation process, the toxic organics oxidation can be performed in several different ways, including direct and indirect oxidation, which are shown in Fig. 1. The oxidation mechanisms are generally observed to be influenced by the electrode material, the electrolyte composition and experimental conditions.

In direct electrolysis, the toxic organics are oxidized directly on the anode surface after its adsorption without involvement of any other substances except the electrons, as shown in Fig. 1(a).

Direct electrochemical oxidation is theoretically possible at low potentials, before oxygen evolution occurs. But the reaction rate is usually low and the anodic potential has to be controlled at a constant lower value to avoid the oxygen evolution reaction. At this low potential, the formation of polymer layer on the anode surface will be accelerated and the anode will lose its activity. This deactivation actually depends on the adsorption properties of the anode surface and the properties of the organic substrates. Aromatic compounds such as phenol (Foti et al., 1997; Gattrell & Kirk, 1993), chlorophenols (Rodgers et al., 1999; Rodrigo et al. 2001), naphthol (Panizza & Cerisola, 2003; Panizza & Cerisola, 2004) and pyridine (Iniesta et al., 2001a) were reported to form polymer layer on anode surface easily.

* Corresponding author

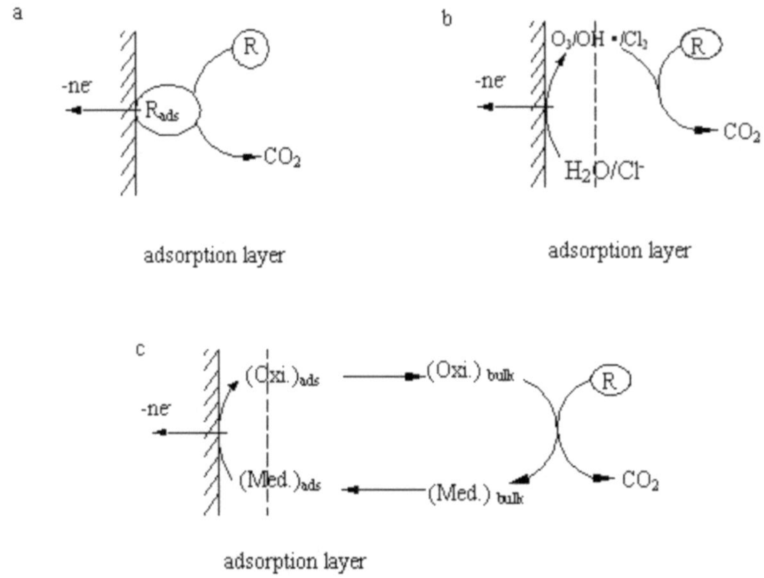


Fig. 1. Scheme of the electrochemical processes for the oxidation of organics: (a) direct electrolysis; (b) irreversible indirect electrolysis; (c) reversible indirect electrolysis

In order to accelerate the organics oxidation rate, in practical applications, higher oxidation potential on anode is needed and other side reactions like oxygen evolution occur. Thus another important mechanism usually accompanied with the organics electrochemical oxidation.

In indirect oxidation, the organic substrates do not donate electrons directly to the anode, but they would react with the active oxidants generated from the anode as showed in Fig. 1 b (irreversible) and c (reversible). In reversible process, the redox reagents are turned over many times and recycled. The redox couples can be metal ions such as $Ag^{+}/Ag^{2+}$ (Farmer et al., 1992), $Co^{2+}/Co^{3+}$ (Leffrang et al., 1995), $Fe^{2+}/Fe^{3+}$ (Dhooge & Park, 1983) or inorganic ions such as $Cl^{-}/ClO^{-}$ (Comninellis & Nerini, 1995; Szpyrkowicz et al., 1995), $Br^{-}/BrO^{-}$ (Martinez-Huitle et al., 2005). These redox couples can be added to or present in the waste water. In irreversible process, strong oxidants like ozone (Chen et al., 2010; Wang et al., 2006), chlorine (Panizza & Cerisola, 2003), hydrogen peroxide (Brillas et al., 1996; Brillas et al., 1995; Do & Chen, 1993) and hydroxyl free radicals (Johnson et al., 1999) etc. are generated and in-situ applied to mineralize the organic pollutants.

In the electrochemical system, the most important component is the anodic material. Different anodic materials show diverse effectiveness for toxic organics oxidation and the organics oxidation mechanisms are also different. According to the model developed by Comninellis (1994), anode materials are classified into two types as active and nonactive anodes. The active anodes such as carbon, graphite, platinum, iridium oxide and ruthenium oxide electrodes have lower oxygen evolution overpotential and are good electrocatalysts for oxygen evolution reaction. The nonactive anodes such as tin dioxide, lead dioxide, boron doped diamond electrodes have higher oxygen evolution overpotential and are poor electrocatalysts for oxygen evolution reactions. On active anodes, organics usually are partially oxidized while organics can be mineralized to carbon dioxide completely on nonactive anodes. The electrochemical oxidation process usually consumes large amount of electrical energy. In order to increase the energy efficiency and decrease the energy consumption, people have paid much attention to investigate more effective anodic materials.

Till now, lead dioxide, tin dioxide and diamond as anodic materials have relative higher energy efficiency for electrochemical oxidation of toxic organics in waste water due to their high oxygen evolution potentials which can hinder the oxygen evolution side reaction in electrochemical oxidation process. These anodic materials have attracted much research interest and many research papers have been published. Also different preparation methods can affect the electrodes performance because the preparation methods have effect on the electrode morphology, crystal structure and other characteristics. Except the anodic materials mentioned above, $Ta_2O_5$-$IrO_2$, $Nb_2O_5$-$IrO_2$ and Pt-$Ta_2O_5$ etc. also are nonactive anodes and show good electrochemical properties. They may also have the potential to be used in electrochemical oxidation process. From the above mentioned considerations, it is necessary to summarize the efficient anodic materials for electrochemical oxidation of toxic organics.

In this work, the development in last decade of the anodic materials including lead dioxide, tin dioxide and diamond materials will be reviewed in detail from their preparation, structure characterization and performance. A few other anodic materials will also be introduced briefly.

## 2. Lead dioxide

Lead dioxide has a long history of use as anode for the oxidation of organics and ozone generation because of its good conductivity and large overpotential for oxygen evolution. Lead dioxide anodes also have a lower cost compared to those based on precious metals. Lead dioxide anode can be prepared with different phase structures, surface morphologies with different doping and preparation conditions. Thus this can contribute to the fundamental understanding of the relationship between the structure and catalytic properties, which is very important to all catalysis fields (Li et al., 2011). Pure lead dioxide is an odorless dark-brown crystalline powder which is nearly insoluble in water. It has two major polymorphs, alpha and beta, which occur naturally as rare minerals scrutinyite and plattnerite, respectively. The alpha form has orthorhombic symmetry, space group Pbcn (No. 60), Pearson symbol oP12, lattice constants a = 0.497 nm, b = 0.596 nm, c = 0.544 nm, Z = 4 (four formula units per unit cell). The symmetry of the beta form is tetragonal, space group P42/mnm (No. 136), Pearson symbol tP6, lattice constants a = 0.491 nm, c = 0.3385 nm, Z = 2. The crystal structures of alpha and beta form are shown in Fig. 2.

### 2.1 The preparation of lead dioxide

Lead dioxide is usually prepared by anodic deposition method. Basically, the lead source exists as lead(II) salt in the aqueous electrolyte, while electrode substrate as anode, under an anodic current, lead dioxide will deposit on the anode. The mechanism for lead dioxide formation on anode is still not clear. Velichenko (1996) proposed a two steps mechanism. The first step is the formation of adsorbed hydroxyl free radicals and the second step is the hydroxyl free radicals reacting with lead(II) ions to form $Pb(OH)^{2+}$ as intermediate. And soon the $Pb(OH)^{2+}$ is further oxidized to $PbO_2$.

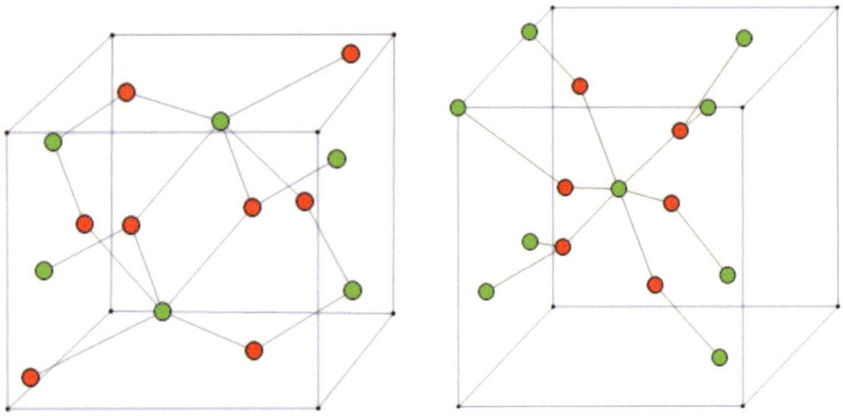

Fig. 2. The crystal structures for α-$PbO_2$ (left) and β-$PbO_2$(right) (green dots represent Pb while red dots represent O)

The literature on the anodic deposition of lead dioxide is very extensive. And it is clear that the adhesion, surface morphology and ratio of alpha/beta form and catalytic activity can be influenced by substrate and its pretreatment process, dopant and its concentration, additives, depositing pH, depositing current density and potential, lead source and concentration and temperature etc. In order to improve lead dioxide anode performance including the activity and lifetime, people tried many different dopants, substrates and other techniques for the anodic deposition of lead dioxide. The earlier lead dioxide preparation was based on in situ oxidizing lead or lead alloy. This type of anode always suffered from continuous corrosion of the underlying lead substrate. It is only more recently that the stable lead dioxide can be anodic deposited on inert substrates such as titanium and carbon.

Though some academic publications employed Pt, Au (Chang & Dennis, 1991; Velichenko et al., 1995; Yeo et al., 1992), graphite (Munichandraiah & Sathyanarayana, 1987), tantalum (Tahar & Savall, 1999) or glassy carbon (González-García et al., 2002; Sáez et al., 2011a; Velayutham & Noel, 1991) as substrate for lead dioxide anode electro-deposition, titanium is mostly used as the lead dioxide electrode substrate because titanium has relative low cost, strong mechanical strength and good adhesion to the lead dioxide layer. Titanium as electrode substrate is usually pretreated in order to remove the existing titanium oxide layer, grease and other contaminants on the surface. The usually used pretreatment process may include sandblasting, alkaline degrease followed by etching in boiling oxalic acid or hydrochloric acid. After this pretreatment, the substrate is relative clean and rough and suitable for anodic depositing lead dioxide. With the clean and rough surface, the deposited lead dioxide layer may be attached strongly. In order to further improve the electrode, some interlayer is also applied to titanium substrate before lead dioxide deposition. This interlayer may be helpful to improve the electrode conductivity, stability and activity. The interlayers proposed include platinum (Devilliers et al., 2003), $TiO_2/Ta_2O_5$ (Ueda et al., 1995) and $TiO_2/RuO_2$ (Hine et al., 1984). Also some people proposed multi interlayers before catalyst layer applied. Ueda et al. (1995) employed a thermally formed oxide layer $TiO_2/Ta_2O_5$ attached to the substrate, which could improve the layers adhesion to the substrate. After the oxide interlayer, Ueda et al. employed another alpha lead dioxide layer between the oxide interlayer and catalyst layer because the alpha lead dioxide layer is stress free and can prevent the corrosive electrolyte penetrating to the substrate to destroy the anode. Finally an active beta lead dioxide and tantalum oxide composite layer was employed as catalyst layer and a stable and active lead dioxide anode was achieved. Feng and Johnson (1991) and Mohd and Pletcher (2006) proposed that F doped lead dioxide interlayer could give good adhesion and stability, while Bi or Fe doped lead dioxide catalyst layer could give the desired electrocatalytic activity. Except employing the interlayers to enhance the adhesion and control the inside stress of the lead dioxide anode, some soluble organic additives were also applied in the electroplating bath. Munichandraiah and Sathyanarayana (1987) found that Teepol could improve the adherence while creating a higher surface area. Ghaemi et al. (2006) and Wen et al. (1990) found that by adding Triton X-100 to the electroplating bath, the adhesion between the substrate and catalyst layer was improved, so as the mechanical strength and oxygen evolution potential. Adding gelatin and sodium dodecylsulfonate to the electroplating bath could modify the electrode morphology and decrease the beta lead dioxide content (Wen et al., 1990). Polyvinyl pyridine could be used to control both the morphology and crystallite size (Ghasemi et al., 2007). Other additives such as tetraalkylammonium ions were also applied in lead dioxide anode preparation process (Ghanasekaran et al., 1976; Pletcher & Wills, 2004). Low et al. (2009) reported that a highly reflective lead dioxide electrode could be prepared by adding hexadecyltrimethylammonium chloride or bromide to the electroplating bath.

In last decades, people tried different dopants to improve the lead dioxide coated anode characteristics and large improvements were achieved without any doubt. Generally, the doped lead dioxide anode was prepared by simply adding doping elements into the electroplating bath. And the effective dopants are generally $F^-$, $Bi^{3+}$, $Fe^{3+}$, $Co^{2+}$ etc. $F^-$ and $Fe^{3+}$ doping as mentioned above, could enhance the adhesion and mechanical strength (Feng & Johnson, 1991; Mohd & Pletcher, 2006). While $Bi^{3+}$ (Feng & Johnson, 1991; Iniesta et al., 2001c; Liu et al., 2008b; Mondal et al., 2001; Mohd & Pletcher, 2006; Yeo et al., 1989) and $Co^{2+}$ (Andrade et al., 2008; Velichenko et al.,2002) were reported to enhance the organics oxidation and reduce the electrode fouling by organics. During the organics oxidation, it was suggested that the dopants promote the formation of adsorbed OH radicals by increasing the oxygen vacancies concentration in lead dioxide lattice. In addition, preparation and properties of the rare earth doped lead dioxide were seldom reported. Kong et al. (2007) reported that the crystal structure of the $PbO_2$ electrodes was influenced by doping with different rare earth oxides. The presence of $Er_2O_3$ and $La_2O_3$ in the $PbO_2$ films could enhance the direct anodic oxidation, which was helpful to mineralize organic wastes. Yang et al. (2010) reported that Nd, Sm, Gd, Ce doping did not change the crystal form of $PbO_2$ electrodes but improve the crystal purity of beta-$PbO_2$. SEM results indicated that the morphology of the electrode surface had changed at different degrees. The lifetime and activity were improved by rare earth doping.

Nanostructured material is a hot topic in modern research because nano-material usually has the particles size in nano meter scale and relative large surface area. So people also paid effort to prepare lead dioxide anode with nano-structure. The nano-structured lead dioxide anode usually was prepared by employing nano-structured template as deposition substrate. Inguanta et al. (2008) and Perret et al. (2009) proposed electrodepositing lead dioxide onto commercial anodic alumina membranes with mean pore diameter of c.a. 210 nm. The electroplating was conducted in a lead nitrate solution bath and after the deposition, the alumina was dissolved in sodium hydroxide to achieve free standing lead dioxide nanowire clusters. Bartlett et al. (2002) deposited lead dioxide onto self-assembled polystyrene microspheres and then the polystyrene was dissolved out by an organic solvent to get a highly ordered lead dioxide nano-structure. Tan et al. (2011) prepared nano lead dioxide by employing self-organized $TiO_2$ nanotube arrays as substrate.

In general, composites in metal electroplating have been used to improve the strength, wear resistance, corrosion resistance, and catalytic activity. Thus people also utilized this composites technology to prepare lead dioxide anode. PTFE and polypyrrole have good hydrophobic property and they are applied to introduce hydrophobicity and assist the gases release, aid the interaction between the organic molecules and electrode surface (Hwang & Lee, 1996; Ho & Hwang, 1994; Tong et al., 2008; Yoshiyama et al., 1994; Zhao et al., 2010). The PTFE and polypyrrole may also have the function of reducing the coating films inside stress thus to improve the electrode stability. Fu et al. (2010) reported that $PbO_2$/PVA composites electrode also had excellent electrocatalytic activity and longer life time and higher corrosion resistance due to the addition of PVA deducing the lead dioxide particle size and increasing the conductivity. Binh et al. (2011) very recently tried to prepared $PbO_2$/PANI composite anode. People also used composites materials like $PbO_2/RuO_2$ (Musiani et al., 1999; Bertoncello et al., 1999), $PbO_2/Co_3O_4$ (Musiani, 1996; Bertoncello et al., 1999), $PbO_2/IrO_2$ (Musiani, 2000) to get a more stable and active anode for oxygen evolution. More recently, the $PbO_2/TiO_2$ composites have attracted considerable attention due to its potential application in photo electrocatalytic oxidation of organic waste in wastewater (Velichenko et al., 2009). Li et al. (2006a; 2006b) reported the electrochemically assisted photocatalytic degradation of toxic organic waste on composites $PbO_2/TiO_2$ photoelectrode.

Except the above mentioned, anodic deposition of lead dioxide was attempted to be assisted by ultrasound waves (González-García et al., 2002; Sáez et al., 2011b; Ghasemi et al., 2008). The results indicated that the ultrasound could improve the prepared lead dioxide stability, adhesion and mechanical strength.

### 2.2 Lead dioxide anode application in waste water treatment

Lead dioxide anode has wide applications in oxygen evolution, ozone generation, lead acid batteries, manufacturing fine chemicals and water and waste water treatment. For waste water treatment, due to its relative high cost compared to the most favorable biological process, the electro oxidation process was usually used to oxidize organic wastes which are toxic and resistant to biological treatment. These organic compounds usually include phenol (Andrade et al., 2008; Tahar & Savall, 1999), aniline (Hmani et al., 2009), benzoquinone, chlorinated phenol (Cao et al., 2009; Tan et al.,2011), nitrophenol (Liu et al., 2008a), naphthol (Panizza & Cerisola, 2003; Panizza & Cerisola, 2004), cyanide (Hine et al., 1986), benzene (Hamza et al., 2011), cresols (Flox et al., 2009), chloranilic acid, indoles, tannic acid, 1,2-dichloroethane, herbicides (Panizza et al., 2008), pesticides (Youssef et al., 2010), surfactants (Weiss et al., 2006) and dyes.

Borras et al. (2003) studied the initial stages of oxidation of aqueous solutions of p-chlorophenol (p-CP) and p-nitrophenol (p-NP) on Bi-doped $PbO_2$ electrodes. Benzoquinone and aliphatic acids were identified as the primary oxidation intermediates. Oxidation of benzoquinone was found to be the slow step during the early stages of the electrochemical combustion process. The effect of competing adsorption of p-CP and p-NP on Bi–$PbO_2$ was also examined, and the presence of p-NP in solution was found to inhibit the p-CP oxidation during concurrent oxidation of both phenols.

Liu et al. (2008a; 2008b; 2009) studied the electrocatalytic oxidation of o-nitrophenol (o-NP), m-nitrophenol (m-NP) and p-nitrophenol (p-NP), dinitro-phenol and trinitro-phenol at Bi-doped lead dioxide anode in acid medium by cyclic voltammetry and bulk electrolysis. The results of voltammetric studies indicated that these nitrophenol isomers were indirectly oxidized by hydroxyl radicals in the solutions. Molecular configuration including the electron character and hydrogen bonds of NPs significantly influenced the electrocatalytic oxidation of these isomers. The efficiency for electrocatalytic oxidation of NPs lay in the order of p-NP > m-NP > o-NP, while 2,6-dinitrophenol >2,5-dinitrophenol > 2,4-dinitrophenol > 2,4,6-trinitrophenol. Hydroquinone, catechol, resorcinol, benzoquinone, aminophenols, glutaconic acid and maleic acid and oxalic acid had been detected as soluble products during the electrolysis of NPs.

Awad and Abo (2005) investigated the electrocatalytic degradation of Acid Blue and Basic Brown dyes from artificial wastewater on lead dioxide anode in different conductive electrolytes. It was shown that complete degradation of these dyes was dependent primarily on type and concentration of the conductive electrolyte. The highest electrocatalytic activity was achieved in the presence of NaCl. The possibility of electrode poisoning as a result of growth of adherent film on the anode surface or production of stable intermediates not easily further oxidized by direct electrolysis in $H_2SO_4$ might be accountable for the poor performance observed in this conductive electrolyte. Optimizing the conditions that ensure effective electrochemical degradation of Acid Blue and Basic Brown dyes on lead dioxide electrode necessitated the control of all the operating factors.

The above mentioned studies, however, were conducted in lab with controlled pH and low current density, which were seldom used in real waste water treatment plants. Only few papers dealt with real waste water such as landfill leachate (Cossu et al., 1998), tannery waste, dye plant effluents etc (Ciriaco et al., 2009). However, due to the complex composition of the real waste water, the mechanism for organic compounds oxidation was difficult to discuss. Before the technology could be practically use, a lot of research work including the bench scale and pilot scale test should be done.

## 3. Tin dioxide

Tin dioxide is a well known n-type semiconductor with a wide band gap (>3.7eV). It crystallizes with the rutile structure, also called cassiterite structure, which belongs to the space group D14 (P42/mnm), with a unit cell containing two tin and four oxygen atoms (Fig. 3).

In Fig. 3, the oxygen atoms are placed approximately at the corner of regular octahedron, and tin atoms are placed approximately at the corners of regular octahedron, and tin atoms are located approximately at the corners of an equivalent triangle. The lattice parameters are a=b=0.4737 nm, c=0.3185 nm. The d-spacing for $SnO_2$(110) and $SnO_2$(101) is 0.3350 nm and 0.2643 nm, respectively. Due to its advantageous electrical, electrochemical and optical properties, $SnO_2$ has been widely used in various fields such as gas sensors, solar cells and electrode material for electrolysis and lithium ion batteries.

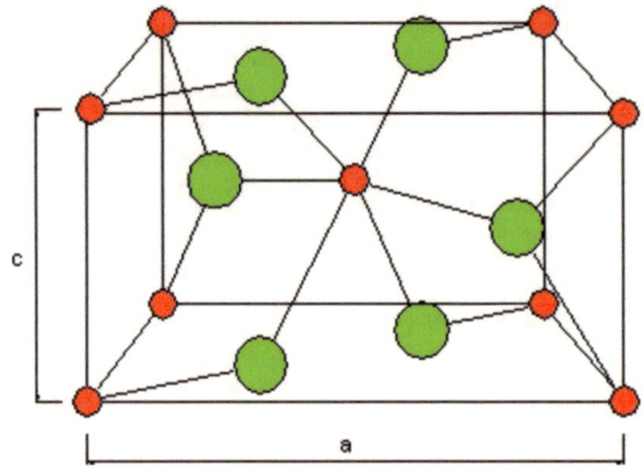


Fig. 3. The cassiterite crystal structure for tin dioxide lattice (small dots are Sn atoms and larger dots are O atoms)

### 3.1 The preparation of tin dioxide

Tin dioxide as an anode material is usually prepared on a substrate, which can support the tin dioxide active layer. The mostly used substrate is titanium, the same as lead dioxide anode used. Also the titanium shall be pretreated before active layer coating. The pretreatment process is also the same as described in section 2.1, that is sandblasting, degreasing by alkaline or acetone and boiling in acid. However, tin dioxide layer can not be prepared by anodic deposition. It is prepared usually by thermal decomposition method, in which tin salt can be oxidized to tin oxide in air at high temperature. According to the coating solution preparation method, the thermal decomposition method can be classified into dip-coating pyrolysis and sol-gel dip coating method.

#### 3.1.1 The dip-coating pyrolysis method

In this method, the coating solution is directly prepared from metal salt. That is to dissolve the metal chloride into alcohol solvent. The alcohol solvent can be absolute ethanol, isopropanol, butanol or the mixture of them. If any dopants needed, the doping elements shall also be dissolved into the mixture. In order to get a uniform solution, a few drops of hydrochloric acid are usually added to avoid the hydrolysis of the metal salts. Then the pretreated titanium substrate can be dipped into the coating solution and then a liquid film formed on the titanium surface. Then the substrate is moved to an oven to dry at 80-100 ℃ for about 10 min to evaporate the solvent. After then, the titanium substrate is moved to a furnace, in there, the coated tin salt is transformed to tin oxide under high temperature. This procedure, as illustrated in Fig.4, is repeated for 15~30 times until a required tin oxide film thickness achieved.

The dip-coating pyrolysis procedure is relative simple, easy operating and cost effective. However, this method may also introduce non-uniform tin oxide layer with cracks and islands structure due to the inside stress deduced from the thermal treatment. The typical antimony doped tin dioxide electrode surface morphology is shown in Fig. 5. This structure may contribute to the poor stability.

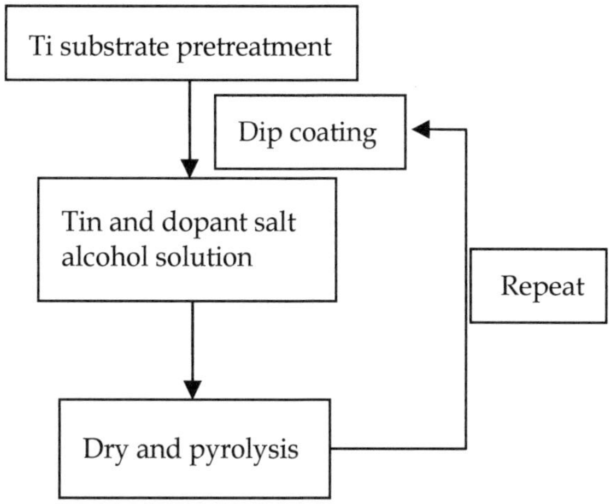


Fig. 4. The illustration for the tin oxide electrode preparation procedure

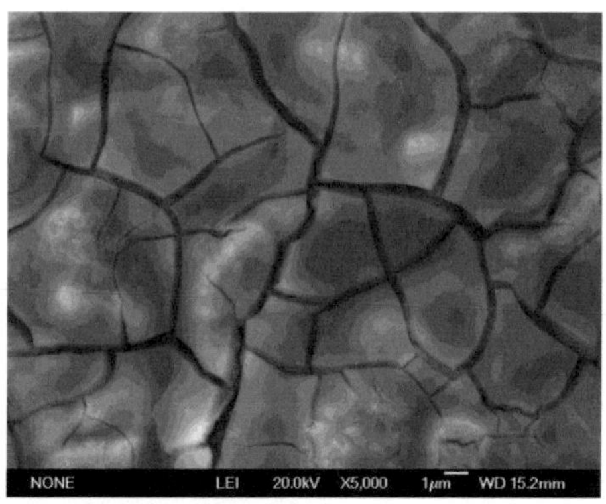


Fig. 5. Typical SEM image of antimony doped tin dioxide electrode prepared by dip-coating pyrolysis

In this procedure, the coating solution composition and pyrolysis temperature are key factors to affect the electrode activity. From the extensive report on doped tin dioxide electrode preparation, the optimized pyrolysis temperature usually lies in the range of 500-600 ℃ (Comninellis & Chen, 2010). The antimony is the mostly used dopant because its higher valence (+5) can introduce more oxygen vacancies in the tin dioxide lattice, which can improve the electrode conductivity and introduce more active sites for electrochemical oxidation. Also $Sb^{5+}$ has the similar ionic radius, which can facilitate higher dopant concentration in the lattice. However, the optimized Sb doping concentration varied greatly (in the range between 1.6%-10% atomic) among different research groups (Vicent et al., 1998; Montilla et al., 2002; Montilla et al., 2004a; Montilla et al., 2004b). This was possible on one hand because people usually used the atomic percent in the coating solution, the final Sb concentration in the electrode was not determined accurately. On the other hand, the details of the preparation procedure may be different, such as the drying time and temperature, heating times and durations etc. which would affect the electrode performance even with the similar doping levels.

### 3.1.2 Sol-gel dip coating method

In order to overcome the shortages derived by direct pyrolysis method, sol-gel dip coating method was developed to prepare tin dioxide electrode. Actually, the sol-gel dip coating method is not a new concept in material chemistry. It has wide application in panel display manufacturing, nano materials preparation etc. In this technique for tin dioxide electrode preparation, the similar Ti substrate pretreatment is needed, while the coating solution is colloidal rather than an alcoholic solution.

The colloidal can be achieved both from inorganic metal salt or organic metal compounds. In a typical colloidal preparation procedure using inorganic metal salt as precursor, ammonia is added to the inorganic salt solution to facilitate the hydrolysis of metal salts. After hydrolysis, the colloidal is washed to remove the chloride ions, which is believed to be harmful to the electrode stability. Then the colloidal with water as solvent is heated up to 80 ℃ to finish the transformation from metal hydroxide to metal oxide and facilitate the doping ions into the tin oxide lattice. On the completion of hydrolysis and doping, the colloidal's color is becoming deep from yellow (lower Sb concentration) to grey (higher Sb concentration). In order to get a more uniform sol, some additives such as oxalic acid or polyethylene glycol can be added into the colloidal solution.

By coating the present colloidal solution on titanium followed high temperature heating for many times, the prepared electrode showed relative smooth and uniform surface structure. A typical SEM image of antimony doped tin dioxide electrode prepared by sol-gel dip coating method is showed in Fig. 6.

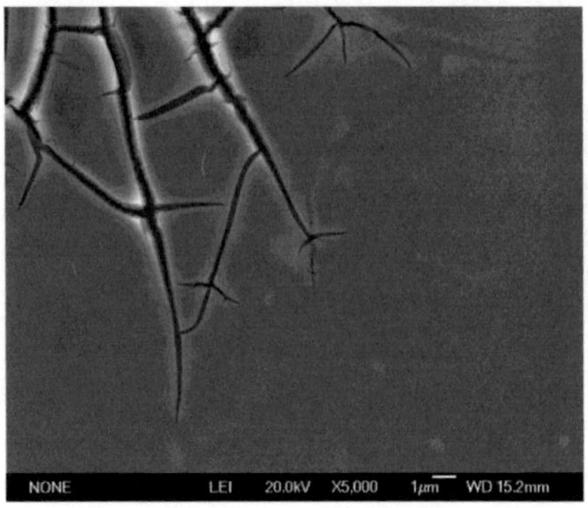


Fig. 6. SEM image of antimony doped tin dioxide electrode prepared by sol-gel dip coating method

From Fig. 6, the relative smooth surface and fewer cracks may prevent the corrosive electrolyte penetrating inside the catalyst layer, thus enhance the electrode stability. In another hand, the free contamination from chloride ions in the electrode may retard the hydration of the oxide layer and also contribute to a longer lifetime. Also in sol-gel dip coating method, the dopants concentration is easily controlled. Thus this technique attracts more and more attention in tin dioxide electrode preparation, though it suffers from the preparation of colloidal solution.

Except the above mentioned preparation method, people also attempted to prepare tin oxide electrode by electroplating tin metal or tin alloy onto titanium, followed by an oxidation process in oven with air. In a typical electroplating procedure for antimony doped tin dioxide electrodes, the cleaned Ti substrate was first cathodic deposited in acidic solution containing $Sn^{2+}$ and followed in citric acid solution containing $Sb^{3+}$ with the same current density at room temperature. The metal ions ($Sn^{2+}$, $Sb^{3+}$) were reduced to metal alloy (SbSn) on the cathode. Repeat the electroplating procedure 3-5 times to get a thick metal alloy layer. Finally, the electrodes were calcined at high temperature for 3 h to obtain the Sb-$SnO_2$/Ti electrodes.

Except the preparation techniques, people also tried a lot of dopants to improve the tin oxide electrode activity. Except the above mentioned antimony dopant, Pt/Sb co-doped tin dioxide electrode was prepared and reported to have longer lifetime (Vicent et al., 1998). Berenguer et al. (2009) found that Ru and Pt co-doped tin dioxide electrode had better electrocatalytic activity for oxygen evolution. It was reported that with trace amount of nickel and antimony doped tin dioxide electrode showed promising applications for anodic oxidation and the current efficiency for ozone generation could reach upto 50%, which was much higher than that on lead dioxide (Christensen et al., 2009; Wang et al., 2005). In order to improve the adhesion and stability, people also tried to apply interlayer between the substrate and the top active layer. Debiemme-Chouvy et al. (2011) prepared conductive $SnO_2$ films deposited on a glass substrate by the spray pyrolysis technique. In order to improve the electrochemical behaviour of the $SnO_2$ films, conductive F-doped layer was first deposited followed by the deposition of a Sb-doped layer to form a bi-layer film. Ding et al. (2010) introduced a Sb doped tin dioxide inter-layer onto a titanium using electrodeposition method followed by coating its surface using thermo-decomposition procedures.

### 3.2 Tin dioxide anode application in waste water treatment

Extensive research indicates that tin dioxide and doped tin dioxide electrode has good activity toward organics degradation (Comninellis & Vercesi, 1991; Comninellis & Pulgarin, 1993; Houk et al., 1998; Stuki et al., 1991; Steve et al., 1999; Polcaro et al., 1999; Vicent et al., 1998). Generally, the organic waste is similar to those being treated by lead dioxide, which are toxic and resistant to biological treatment.

Houk et al. (1998) evaluated electrochemical incineration of benzoquinone in aqueous media using a quaternary metal oxide electrode, which is a Ti or Pt anode coated with a film of the oxides of Ti, Ru, Sn and Sb. They concluded that the quaternary metal oxide films applied to Ti or Pt substrates exhibited high and persistent activity as anode materials for the electrochemical incineration of benzoquinone.

Feng et al. (2008) doped Gd into Ti-based Sb-$SnO_2$ anodes to investigate its performance on electrochemical destruction of phenol. They found that doping the anode with rare earth gadolinium (Gd) could improve the electrochemical oxidation rate of phenol. Cui et al. (2009a) prepared rare earth Ce, Eu, Gd and Dy doped Ti/Sb–$SnO_2$ electrodes by thermal decomposition and the performance of electrodes for the electro-catalytic decomposition of a model pollutant phenol was investigated. They also investigated the bisphenol A degradation on tin dioxide anode (Cui et al., 2009b).

Cossu et al. (1998) investigated landfill leachate oxidation on $SnO_2$ and $PbO_2$ anodes, respectively. Tin dioxide electrode showed similar performance as the $PbO_2$. Polcaro et al. (1999) studied electrochemical oxidation of 2-chlorophenol at Ti/$PbO_2$ and Ti/$SnO_2$ anodes. Results showed that, although similar average faradaic yields were obtained using Ti/$PbO_2$

or $Ti/SnO_2$ anodes, the latter material is preferred because of its better ability to oxidize toxic compounds.

Adams et al. (2009) fabricated four different $SnO_2$-based electrodes ($SnO_2$–$Sb_2O_5$/Ti, $SnO_2$–$Sb_2O_5$–$PtO_x$/Ti, $SnO_2$–$Sb_2O_5$–$RuO_2$/Ti and $SnO_2$–$Sb_2O_5$–$IrO_2$/Ti) using thermal decomposition method and systemically studied their stability and electrocatalytic activity towards the degradation of 2-nitrophenol (2-NPh), 3-nitrophenol (3-NPh) and 4-nitrophenol (4-NPh). It was found that by incorporating different metal oxides into the Sb-doped $SnO_2$ coating, the lifetime was significantly increased.

Wang et al. (2006) and Chen et al. (2010) found out that introducing trace amount of nikel into antimony doped tin oxide electrode could enhance its activity toward 4-chlorophenol and phenol degradation. The possible reason may be due to the hybrid effect of electro-oxidation and ozone oxidation because ozone could be in-situ generated on this type anode during electrolysis.

The doped tin dioxide electrode showed promising activity toward organics degradation, the activity in some cases is higher than lead dioxide, however its lifetime is shorter. Their stability must be improved further before its practical application in real wastewater treatment.

## 4. Diamond

Diamond is a fascinating material and exhibits many unique technologically important properties, including high thermal conductivity, wide band gap, high electron and hole mobility, high breakdown electric field, hardness, optical transparency and chemical inertness. Diamond has a cubic lattice constructed from $sp^3$-hybridized tetrahedral arranged carbon atoms with each carbon atom bonded to four other carbon atoms. The stacking sequence is ABCABC with every third layer plane identical as showed in Fig. 7. This structure is fundamentally different from that of graphite. Impurities in diamond can make it an insulator with a resistivity of $>10^6$ Ohm m and a band gap of 5.5 eV.

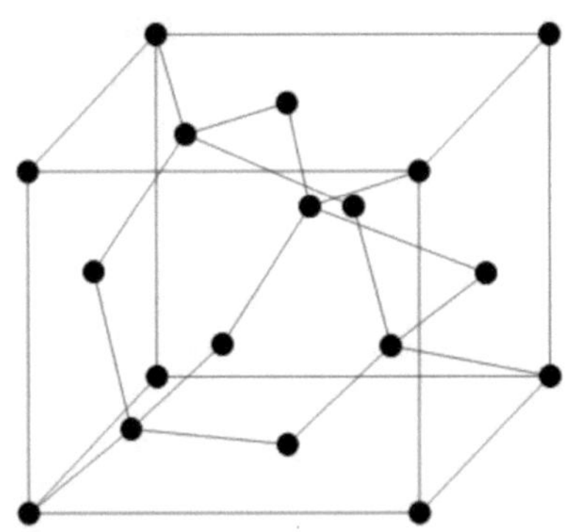

Fig. 7. Diamond lattice structure

### 4.1 Diamond preparation

Diamond films are usually synthesized by chemical vapor deposition (CVD). Early CVD of diamond was carried out by thermal decomposition of carbon containing gases under high temperature. However, this technique may suffer low growth rate, which will retard its commercial application. In addition, the electrode may be contaminated with non-diamond carbon and required frequent interruptions to remove the accumulated graphite by hydrogen etching at high temperature.

In order to improve the diamond conductivity, boron was usually added to the diamond film as dopant. Boron to carbon ratio from ca. 0.02 to ca. $10^{-6}$ was widely accepted by different research groups. The standard method used for CVD boron doped diamond (BDD) was initially thermal diffusion and ion implantation (Prawer, 1995). These processes were made ex situ, after the film growth, and the no contamination of the growth reactor (Martin, et al., 1996.) is the main advantage. However, alternative techniques have shown that highly doped films could be prepared using in situ process from a solid source of boron inside the reactor and by introducing of $B_2O_3$ in a methanol-acetone mixture (Okano et al., 1990). It was observed better results evidenced by more homogeneity in films bulk, also showing a linear relation between the doping levels and boron concentration in the precursor source. Boron doping using other boron sources was also reported such as $B_2H_6$ by Mort et al. (1989) and Fujimori et al. (1990), while $B(OCH_3)_3$ was used as boron source by Ran et al. (1993).

### 4.2 Boron doped diamond (BDD) electrode application in waste water treatment

BDD film was found to be the most active anodic material for degradation of refractory or priority pollutants such as isopropanol and acetic acid (Foti et al., 1999), different carboxylic acids (Gandini et al., 2000), 4-chlorophenol (Rodrigo et al., 2001; Gherardini et al., 2001), phenol (Iniesta et al., 2001b), 3-methylpyridine (Iniesta et al., 2001a), benzoic acid (Montilla et al., 2002), polyacrylates (Bellagamba et al., 2002), 4-chlorophenoxyacetic acid (Boye et al., 2002) and an amaranth dyestuff (Hattori et al., 2003). All these investigations could confirm very high current efficiencies for COD removal, usually higher than 90%. Migliorinia et al. (2011) investigated the electrochemical degradation of Reactive Orange (RO) 16 Dye on BDD/Ti electrode.

BDD electrode as a developing anodic material showed very promising applications in electrochemical oxidation for waste water treatment due to its high energy efficiency and stability. However, its high cost and difficulty to find a proper substrate are its major drawbacks. In fact, only silicon, tantalum, niobium and tungsten as substrate can obtain stable diamond film. Silicon is brittle and its conductivity is poor. Tantalum, niobium and tungsten are too expensive. Titanium is a good candidate with low cost, strong mechanical strength and good electrical conductivity. However, diamond film deposited on titanium is not stable enough. During long term electrolysis, the diamond film can be detached from the titanium substrate. Thus there are still a lot of work to do before this promising anode material commercial available.

## 5. Other anodic materials

According to the model developed by Comninellis (1994), the $Ta_2O_5$ and $Nb_2O_5$ are also non-active anodes. Tantalum oxide and niobium oxide show very high electrochemical stability. They have powerful oxidizing ability and can be used to oxidize water to ozone efficiently (Santana et al., 2004; Da Silva et al., 2004; Kaneda et al., 2005; Awad et al. 2006; Kitsuka et al., 2007). Their preparation is usually followed a thermal decomposition method and the procedure is similar to that of tin dioxide electrode except the coating solution used. Though they are non-active eletrodes for oxygen evolution and may show high efficiency for toxic organics degradation, their application in waste water treatment is seldom reported except a few reports on ozone generation. Perhaps it is due to the high cost to prepare these types electrodes because precious metal is usually used.

## 6. Summary and prospect

In this section, the non-active anode materials including lead dioxide, tin dioxide, diamond, tantalum oxide and niobium oxide electrode were briefly reviewed from their fundamental concept, preparation techniques and application in waste water treatment by electrochemical oxidation. Electrochemical oxidation technology showed promising application in oxidation of toxic and non-biodegradable organic wastes. However, from the view point of anodic material, which is one of the key factors of the electrochemical oxidation system, there is still a lot of work to do before the electrochemical oxidation technology can be widely applied in real waste water treatment plants. For example, the cost of diamond shall be further decreased, while the stability of doped tin dioxide shall be improved.

## 7. Acknowledgments

The present work is finicially supported by Research Fund for the Doctoral Program of Higher Education of China (20090201120010) and the Fundamental Research Funds for the Central Universities in China.

## 8. References

Adams B., Tian M. and Chen A. C. 2009, Design and electrochemical study of $SnO_2$-based mixed oxide electrodes. *Electrochim. Acta*, 54, 1491–1498

Andrade L. S., Rocha-Filho R. C., Bocchi N, Biaggio S. R., Iniesta J., Garcia V. and Montiel V. 2008, Degradation of phenol using Co- and Co,F-doped $PbO_2$ anodes in electrochemical filter-press cells. *J. Hazard. Mater.*, 153, 1-2, 252-260

Awad H. S. and Abo N. 2005, Electrochemical degradation of Acid Blue and Basic Brown dyes on $Pb/PbO_2$ electrode in the presence of different conductive electrolyte and effect of various operating factors. *Chemosphere*, 61, 9, 1327-1335

Awad M. I., Sata S., and Kaneda K. 2006, Ozone electrogeneration at a high current efficiency using a tantalum oxide-platinum composite electrode. *Electrochem. Commun.*, 8,1263-1269

Bartlett P. N., Dunford T. and Ghanem M. A. 2002, Templated electrochemical deposition of nanostructured macroporous $PbO_2$. *J. Mater. Chem.*, 12, 3130-3135

Bellagamba R., Michaud P. A., Comninellis C. and Vatistas N. 2002, Electro-combustion of polyacrylates with boron-doped diamond anodes. *Electrochem. Commun.* 4, 171-175

Berenguer R., Quijada C. and Morallón E. 2009, Electrochemical characterization of $SnO_2$ electrodes doped with Ru and Pt. *Electrochim. Acta*, 54, 5230–5238

Bertoncello R., Furlanetto F., Guerriero P. and Musiani M. 1999, Electrodeposited composite electrode materials: effect of the concentration of the electrocatalytic dispersed phase on the electrode activity. *Electrochim. Acta*, 44, 4061-4068

Binh P. T., Mai T. T., Truong N. X. and Yen T. H. 2011, Synthesis of hybrid nanocomposite based on $PbO_2$ and polyaniline coated onto stainless steel by cyclic voltammetry. *Asian J. Chem.*, 23, 8, 3445-3448

Borras C., Laredo T. and Scharifker B. R. 2003, Competitive electrochemical oxidation of p-chlorophenol and p-nitrophenol on Bi-doped $PbO_2$. *Electrochim. Acta,* 48, 19, 2775-2780

Boye B., Michaud P. A., Marselli B., Dieng M. M., Brillas E. and Comninellis C. 2002, Anodic oxidation of 4-chlorophenoxyacetic acid on synthetic boron-doped diamond electrodes. *New Diamond Front. Carbon Technol.*, 12, 63-68

Brillas E., Bastida R. M., Llosa E. and Casado J. 1995, Electrochemical destruction of aniline and chloroaniline for waste water treatment using a carbon PTFE $O_2$-fed cathod. *J. Electrochem. Soc.*, 142, 1733-1741

Brillas E., Mur E. and Casado J. 1996, Iron(II) catalysis of the mineralization of aniline using a carbon-PTFE $O_2$-fed cathode. *J. Electrochem. Soc.*, 143, L49-L53

Cao J. L., Zhao H. Y., Cao F. H., Zhang J. Q. and Cao C. A. 2009, Electrocatalytic degradation of 4-chlorophenol on F-doped $PbO_2$ anodes. *Electrochim. Acta,* 54, 9, 2595-2602

Chang H. and Dennis C. J. 1991, Voltammetric response of dimethyl sulphoxide DMSO at gold electrodes modified by thin films of bismuth-doped lead dioxide. *Anal. Chim. Acta,* 248, 85-94

Chen Q. Y., Shi D. D., Zhang Y. J. and Wang Y. H. 2010, Electrochemical degradation of phenol on a novel antimony and nickel doped tin dioxide electrode. *Water Sci. Technol.,* 62,9,62-65

Christensen P. A., Lin W. F., Christensen H., Imkum A., Jin J. M., Li G. and Dyson C. 2009, Room temperature, electrochemical generation of ozone with 50% current efficiency in 0.5M sulfuric acid at cell voltages <3V. *Ozone Sci. Eng.*, 31,287-293

Ciriaco L., Anjo C., Pacheco M. J., Lopes A. and Correia J, 2009, Electrochemical degradation of Ibuprofen on Ti/Pt/$PbO_2$ and Si/BDD electrodes. *Electrochim. Acta* ,54, 5, 1464-1472

Comninellis C. 1994, Electrocatalysis in the electrochemical conversion/combustion of organic pollutants for wastewater treatment. *Electrochim. Acta,* 39, 1857-1862

Comninellis C. and Chen G. H., 2010, *Electrochemistry for the environment*, Springer, ISBN978-0387-36922-8, Berlin

Comninellis C. and Nerini A., 1995, Anodic oxidation of phenol in the presence of NaCl for wastewater treatment. *J. Appl. Electrochem.,* 25, 23-28

Comninellis C. and Pulgarin C. 1993, Electrochemical oxidation of phenol for wastewater treatment using $SnO_2$ anodes. *J. Appl. Electrochem.*, 23, 108-112

Comninellis C. and Vercesi G. P., 1991, characterization of DSA type oxygen evolving electrodes, choice of coating. *J. Appl. Electrochem.*, 21, 4, 335-340

Cossu R., Polcaro A. M., Lavagnolo M. C., Mascia M., Palmas S. and Renoldi F. 1998, Electrochemical treatment of landfill leachate: oxidation at Ti/$PbO_2$ and Ti/$SnO_2$ anodes. *Environ. Sci. Technol.*, 32, 22, 3570-3573

Cui Y. H., Feng Y. J. and Liu Z. Q. 2009a, Influence of rare earths doping on the structure and electro-catalytic performance of Ti/Sb– $SnO_2$ electrodes. *Electrochim. Acta,* 54, 4903–4909

Cui Y. H., Li X. Y. and Chen G. H. 2009b, Electrochemical degradation of bisphenol A on different anodes. *Water Res.*, 43, 1968–1976

Da-Silva L. M., Ftanco D. V. and De-Faria L. A. 2004, Surface, kinetics and electrocatalytic properties of Ti/($IrO_2$+$Ta_2O_5$)electrodes, prepared using controlled cooling rate, for ozone production. *Electrochim. Acta,* 49, 3977-3988

Debiemme-Chouvy C., Hua Y., Hui F., Duval J. L. and Cachet H. 2011, Electrochemical treatments using tin oxide anode to prevent biofouling. *Electrochim. Acta,* 56, 28, 10364-10370

Devilliers D., Thi M. T. D., Mahe E. and Xuan Q. L. 2003, Cr(III) oxidation with lead dioxide-based anodes. *Electrochim. Acta*, 48,4301-4309

Dhooge P. M. and Park S. M. 1983, Electrochemistry of coal slurries-2. studies on various experimental parameters affecting oxidation of coal slurries. *J. Electrochem. Soc.*, 130, 1029-1036

Ding E., Feng Y. J. and Lu J. W. 2010, Study on the Service life and deactivation mechanism of Ti/$SnO_2$-Sb electrode by physical and electrochemical methods. *Russ. J. Electrochem.*, 46, 1, 72–76

Do J. S. and Chen C. P.,1993, In-situ oxidative degradation of formaldehyde with electro-generated hydrogen peroxide. *J. Electrochem. Soc.*, 140, 1632-1637

Farmer J. C., Wang F. T., Hawley-Fedder R. A. Lewis P. R., Summers L. J. and Foiles L. 1992, Electrochemical treatment of mixed and hazardous wastes: Oxidation of ethylene glycole and benzene by silver(II). *J. Electrochem. Soc.*, 139, 654-662

Feng J. and Johnson D. C. 1991, Electrocatalysis of Anodic oxygen-transfer reactions: titanium substrates for pure and doped lead dioxide films. *J. Electrochem. Soc.*, 138, 3328-3337

Feng Y. J., Cui Y. H., Logan B. and Liu Z. Q. 2008, Performance of Gd-doped Ti-based Sb-$SnO_2$ anodes for electrochemical destruction of phenol. *Chemosphere* 70, 1629–1636

Flox C., Arias C., Brillas E., Savall A. and Groenen-Serrano K. 2009, Electrochemical incineration of cresols: A comparative study between $PbO_2$ and boron-doped diamond anodes. *Chemosphere*, 74, 10, 1340-1347

Foti, G., Gandini, D. and Comninellis C. 1997, Anodic oxidation of organics on thermally prepared oxide electrodes. *Curr. Top. Electrichem.*, 5, 71-91

Foti G., Gandini D., Comninellis C., Perret A. and Haenni W. 1999, Oxidation of organics by intermediates of water discharge on $IrO_2$ and synthetic diamond anodes. *Electrochem. Solid State Lett.* 2, 228-232

Fu F., Yang W.H. and Wang H.H. 2010, Preparation and performance of Ti-based lead dioxide electrode modified with polyethylene glycol. *J. Inorg. Mater.* 25, 653-658

Fujimori N., Nakahata H. and Imai T. 1990, Properties of boron doped epitaxial diamond films. *Jpn. J. Appl. Phys.*, 29, 824-827

Gandini D., Mahe E, Michaud P. A., Haenni W., Perret A. and Comninellis C. 2000, Oxydation of carboxylic acid at boron-doped diamond electrodes. *J. Appl. Electrochem.* 30, 1345-1349

Gattrell M. and Kirk D. 1993, A study of oxidation of phenol at platinum and preoxidized platinum surfaces. *J. Electrochem. Soc.*, 140, 1534-1540

Ghaemi M., Ghafouri E. and Neshati J. 2006, Influenceof the nonionic surfactant Triton X-100 on electro crystallization and electrochemical performance of lead dioxide electrode. *J. Power Sources*, 157,550-562

Ghanasekaran K. S. A., Narasimham K. C. and Udupa H. V. K. 1976, The effect of the additive cetyl trimethyl ammonium bromide on the electrodeposition of lead dioxide. *J. Appl. Electrochem.*, 6,189-198

Ghasemi S., Mousavi M. F. and Shamsipur M., 2007, Electrochemical deposition of lead dioxide in the presence of polyvinylpyrrolidone A morphological study. *Electrochim. Acta*, 53, 459-467

Ghasemi S., Mousavi M. F., Shamsipur M. and Karami H. 2008, Sonochemical-assisted synthesis of nano-structured lead dioxide. *Ultrason. Sonochem.*, 15, 4, 448-455

Gherardini L., Michaud P. A., Panizza M., Comninellis C. and Vatistas N. 2001, Electrochemical oxidation of 4-chlorophenol for wastewater treatment. Definition of normalized current efficiency. *J. Electrochem. Soc.* 148, D78-D82

González-García J., Sáez V., Iniesta I., Montiel V. and Aldaz A. 2002, Electrodeposition of $PbO_2$ on glassy carbon electrodes: influence of ultrasound power. *Electrochem. Commun.*, 4-5, 370-373

Hamza M., Ammar S. and Abdelhédi R. 2011, Electrochemical oxidation of 1,3,5-trimethoxybenzene in aqueous solutions at gold oxide and lead dioxide electrodes. *Electrochim. Acta,* 56, 11, 3785-3789

Hattori S., Doi M., Takahashi E., Kurosu T., Nara M., Nakamatsu S., Nishiki Y., Furuta T. and Iida M. 2003, Electrolytic decomposition of amaranth dyestuff using diamond electrodes. *J. Appl. Electrochem.* 33, 85-89

Hine F., Yasuda M., Iida T. and Ogata Y. 1986, On the oxidation of cyanide solutions with lead dioxide coated anode. *Electrochim. Acta,* 31, 11, 1389-1395

Hine F., Yasuda M., Iida T., Ogata Y. and Hara K. 1984, On the $RuO_2$-$TiO_2$ interlayer of $PbO_2$ electrodeposited Ti anode. *Electrochim. Acta,* 29, 1447-1452

Hmani E., Elaoud S. C., Samet Y. and Abdelhédi R. 2009, Electrochemical degradation of waters containing O-Toluidine on $PbO_2$ and BDD anodes. *J. Hazard. Mater.*, 170, 2-3, 928-933

Ho C. N. and Hwang B. J. 1994, Effect of hydrophobicity on the hydrophobic-modified polytetrafluoroethylene/$PbO_2$ electrode towards oxygen evolution. *J. Electroanal. Chem.*, 377, 177-190

Houk L. L, Johnson S. K., Feng J., Houk R. S. and Johnson D. C. 1998, Electrochemical incineration of benzoquinone in aqueous media using a quaternary metal oxide electrode in the absence of a soluble supporting electrolyte. *J. Appl. Electrochem*, 28, 1167-1177

Hwang B. J. and Lee K. L. 1996, Electrocatalytic oxidation of 2-chlorophenol on a composite $PbO_2$/polypyrrole electrode in aqueous solution. *J. Appl. Electrochem.*, 26, 153-159

Inguanta R. Piazza S and Sunseri C. 2008, Growth and characterization of ordered $PbO_2$ nanowire arrays. *J. Electrochem. Soc.*, 155, K205-K210

Iniesta J., Michaud P. A., Panizza M. and Comninellis C. 2001a, Electrochemical oxidation of 3-methylpyridine at a boron doped diamond electrode: Application to electroorganic synthesis and waste wastewater treatment. *Electrochem. Commun.*, 3,346-351

Iniesta J., Michaud P. A., Panizza M., Cerisola G., Aladaz A. and Comninellis C. 2001b, Electrochemical oxidation of phenol at boron-doped diamond electrode. *Electrochim. Acta,* 46, 3573-3577

Iniesta J., Gonzalez-Garcia J., Exposito E., Montiel V. and Aldaz A. 2001c, Influence of chloride ion on electrochemical degradation of phenol in alkaline medium using bismuth doped and pure $PbO_2$ anodes. *Water Res.*, 35, 3291-3300

Johnson S. K., Houk L. L., Feng J., Houk R. S. and Johnson D. C. 1999, Electrochemical incineration of 4-chlorophenol and the identification of products and intermediates by mass spectrometry. *Environ. Sci. Technol.*, 33, 2638-2644

Kaneda K., Ikematsu M., and Koizumi Y. 2005, Ozone generation by a TaOx and Pt composite insulator-coated Ti electrode. *Electrochem. Solid Stat. Lett.*, 8, 6, J13-J16

Kitsuka K., Kaneda K., and Ikematsu M. 2007, A spin-coated Si/TiOx/Pt/TaOx electrode for electrochemical ozone generation. *Chem. Lett.*, 36, 6, 806-807

Kong J., Shi S., Kong L. and Zhu X. 2007, Preparation and characterization of $PbO_2$ electrodes doped with different rare earth oxides. *Electrochim. Acta,* 53, 4, 2048-2054

Leffrang U., Flory K., Galla U. and Schmeider H. 1995, Organic waste destruction by indirect electrooxidation. *Sep. Purif. Technol.*, 30, 1883-1899

Li G., Qu J., Zhang X. and Ge J. 2006a, Electrochemically assisted photocatalytic degradation of Acid Orange 7 with b-$PbO_2$ electrodes modified by $TiO_2$. *Water Res.*, 40,213-220

Li G., Qu J., Zhang X., Liu H. and Liu H. 2006b, Electrochemicallya ssisted photocatalytic degradationof Orange II: Influence of initial pH values. *J. Mol. Catal. A: Chem.*, 259, 238-244

Li X. H., Pletcher D. and Walsh F. C. 2011, Electrodeposited lead dioxide coatings. *Chem. Soc. Rev.*, 40, 3879-3894

Liu H., Liu Y., Zhang C, and Shen R. 2008a, Electrocatalytic oxidation of nitrophenols in aqueous solution using modified $PbO_2$ electrodes. *J. Appl. Electrochem.*, 38,101-108

Liu Y., Liu H. and Li Y. 2008b, Comparative study of the electrocatalytic oxidation and mechanism of nitrophenols at Bi-doped lead dioxide anodes. *Appl. Catal.B, 84*, 1-2, 297-302

Liu Y., Liu H., Ma J. and Wang X. 2009, Comparison of degradation mechanism of electrochemical oxidation of di- and tri-nitrophenols on Bi-doped lead dioxide electrode: Effect of the molecular structure. *Appl. Catal.B*,91, 1-2, 284-299

Low C. T. J., Pletcher D. and Walsh F. C. 2009, The electrodeposition of highly reflective lead dioxide coatings. *Electrochem. Commun.*, 11, 6, 1301-1304

Martin H. B., Argoitia A., Landau U., Anderson A. B. and Angus J. C. 1996, Hydrogen and oxygen evolution on boron-doped diamond electrodes. *J. Electrochem. Soc.*, 143, 6, L133-L136.

Martinez-Huitle C. A., Ferro S. and De-Battisti A. 2005, Electrochemical incineration in the presence of halides. *Electrochem. Solid State Lett.*, 8, 35-39

Migliorinia F. L., Bragab N. A. and Alvesc S. A. 2011, Anodic oxidation of wastewater containing the Reactive Orange 16 Dye using heavily boron-doped diamond electrodes. *J. Hazard. Mater.*, 192, 3, 1683-1689

Mohd Y. and Pletcher D. 2006, The fabrication of lead dioxide layers on a titanium substrate. *Electrochim. Acta*, 52,786-793

Mondal K., Mandich N. V. and Lalvani S. B. 2001, Regeneration of hexavalent chromium using a Bi-doped $PbO_2$ anode. *J. Appl. Electrochem.*, 31,165-173

Montilla F., Michaud P. A., Morallon E., Vazquez J. L. and Comninellis C. 2002, Electrochemical oxidation of benzoic acid at boron-doped diamond. *Electrochim. Acta*, 47, 3509-3514

Montilla F., Morallón E., De-Battisti A. and Vázquez J. L. 2004a, Preparation and characterization of antimony-doped tin dioxide electrodes. Part 1. Electrochemical characterization, *J. Phys. Chem. B*, 108, 5036–5043

Montilla F., Morallón E., De-Battisti A., Barison S., Daolio S. and Vázquez J. L. 2004b, Preparation and characterization of antimony-doped tin dioxide electrodes. 3. XPS and SIMS characterization, *J.Phys.Chem.B*, 108, 15976–15981

Mort J., Kuhman D., Machonkin M., Morgan M., Jansen F., Okumura K., Legrice Y. M. and Nemanich R. J. 1989, Boron doping of diamond thin films. *Appl. Phys. Lett.*, 55, 11,1121-1123

Munichandraiah N. and Sathyanarayana S. 1987, Insoluble anode of porous lead dioxide for electrosynthesis: preparation and characterization. *J. Appl. Electrochem.*, 17, 22-32

Musiani M. 1996, Anodic deposition of $PbO_2/Co_3O_4$ composites and their use as electrodes for oxygen evolution reaction. *Chem. Commun.* 21, 2403-2404

Musiani M. 2000, Electrodeposition of composites: an expanding subject in electrochemical materials science. *Electrochim. Acta.*, 45, 3397-3402

Musiani M., Furlanetto F. and Bertoncello R. 1999, Electrodeposited $PbO_2$+$RuO_2$: a composite anode for oxygen evolution from sulphuric acid solution. *J. Electroanal. Chem.*, 465,160-167

Okano K., Akiba Y., Kurosu T., Lida M. and Nakamura T. J. 1990, Synthesis of B-doped diamond film. *Cryst. Growth*, 99, 1192-1195

Panizza M. and Cerisola G. 2003, Electrochemical oxidation of 2-naphthol with in situ electro-generated active chlorine. *Electrochim. Acta*, 48,1515-1519

Panizza M. and Cerisola G. 2004, Influence of anode material on the electrochemical oxidation of 2-naphthol: Part 2. Bulk electrolysis experiments. *Electrochim. Acta*, 49, 19, 3221-3226

Panizza M., Sires I. and Cerisola G. 2008, Anodic oxidation of mecoprop herbicide at lead dioxide. *J. Appl. Electrochem.*, 38, 7, 923-929

Perret P., Brousse T., Belanger D. and Guay D. 2009, Electrochemical template synthesis of ordered lead dioxide nanowires. *J. Electrochem. Soc.*, 156, A645-A651

Pletcher D. and Wills R. G. A. 2004, A novel flow battery: A lead acid battery based on an electrolyte with soluble lead(II), Part II. Flow cell studies. *Phys. Chem. Chem. Phys.*, 6, 1779-1785

Polcaro A. M., Palmas S. and Renoldi F. 1999, On the performance of Ti/$SnO_2$ and Ti/$PbO_2$ anodes in electrochemical degradation of 2-chlorophenol for wastewater treatment. *J. Appl. Electrochem.* 29, 2, 147-151

Prawer, S. 1995, Ion implantation of diamond and diamond films. *Diamond Relat. Mater.*, 4, 862-865

Ran J. G., Zheng C. Q., Ren J. and Hong S. M. 1993, Properties and texture of B-doped diamond films as thermal sensor. *Diamond Relat. Mater.*, 2, 793-796

Rodgers J. D., Jedral W. and Bunce N. J. 1999, Electrochemical oxidation of chlorinated phenols, *Environ. Sci. Technol.* 33, 1453-1457

Rodrigo M. A., Michaud P. A., Duo I., Panizza M., Cerisola G. and Comninellis C. 2001, Oxidation of 4-chlorophenol at boron-doped diamond electrodes for wastewater treatment. *J. Electrochem. Soc.*, 148, D60-D64

Sáez V., Esclapez M. D., Frías-Ferrer A. J., Bonete P., Tudela I., Díez-García M. I. and González-García J. 2011a, Lead dioxide film sonoelectrodeposition in acidic media: Preparation and performance of stable practical anodes. *Ultrason. Sonochem.*, 18, 4, 873-880

Sáez E., Marchante M. I., Díez M. D., Esclapez P., Bonete T., Lana-Villarreal J., González G. and Mostany J. 2011b, A study of the lead dioxide electrocrystallization mechanism on glassy carbon electrodes. Part I: Experimental conditions for kinetic control. *Mater. Chem. Phys.*, 125, 1-2, 46-54

Santana M. H. P. › De-Faria L. A. and Boodts J. F. C. 2004, Investigation of the properties of Ti/$IrO_2$-$Nb_2O_5$ electrodes for simultaneous oxygen evolution and electmchemical ozone production, EOP. *Electrochim. Acta*, 49, 12, 1925-1935

Steve K. J., Linda L. H., Feng J., Houk R. S. and Johnson D. C. 1999, Electro-chemical incineration of 4-chlorophenol and the identification of products and intermediates by mass spectrometry. *Environ. Sci. Technol.*, 33, 2638-2644

Stuki S., Kotz R. and Carcer B. 1991, Electrochemical waste water treatment using high overvoltage anodes. Part II, anode performance and applications. *J. Appl. Electrochem.* 21, 2, 99-104

Szpyrkowicz L., Naumczyk J. and Zilio-Grandi F. 1995, Electrochemical treatment of tannery wastewater using Ti/Pt and Ti/Pt/Ir electrodes. *Water Res.* 29, 517-524

Tahar N. B. and Savall A. 1999, A comparison of different lead dioxide coated electrodes for the electrochemical destruction of phenol. *J. New Mater. Electrochem. Syst.*, 2, 1, 19-26

Tan C., Xiang B., Li Y. J., Fang J. W. and Huang M. 2011, Preparation and characteristics of a nano-$PbO_2$ anode for organic wastewater treatment. *Chem. Eng. J.*, 166, 1, 15-21

Tong S. P., Ma C. A., Feng H. 2008, A novel $PbO_2$ electrode preparation and its application in organic degradation. *Electrochim. Acta*, 53, 6, 3002-3006

Ueda M., Watanabe A., Kameyama T., Matsumoto Y., Sekimoto M. and Shimamune T. 1995, Performance-characteristics of a new-type of lead dioxide-coated titanium anode. *J. Appl. ELectrochem.*, 25,817-822

Velayutham D. and Noel M. 1991, The influence of electrolyte media on the deposition-dissolution behaviour of lead dioxide on glassy carbon electrode, *Electrochim. Acta*, 36, 13, 2031-2035

Velichenko A. B. 1996, Mechanism of lead dioxide electrodeposition, *J. Electroanal. Chem.*, 45, 127-132

Velichenko A. B., Amadelli R., Baranova E. A., Girenko D. V. and Danilov F. I. 2002, Electrodeposition of Co-doped lead dioxide and its physicochemical properties. *J. Electroanal. Chem.*, 527, 1-2, 56-64

Velichenko A. B., Amadelli R., Knysh V. A., Lukyanenko T. V. and Danilov F. I. 2009, Kinetics of lead dioxide electrodeposition from nitrate solutions containing colloidal $TiO_2$. *J. Electroanal. Chem.*, 632, 192-196

Velichenko A. B., Girenko D. V. and Danilov F. I. 1995, Electrodeposition of lead dioxide at an Au electrode, *Electrochim. Acta*, 40, 17, 2803-2807

Vicent F., Morallón E., Quijada C., Vázquez J. L., Aldaz A. and Cases F. 1998, Characterization and stability of doped $SnO_2$ anodes. *J. Appl. Electrochem.*, 28, 607–612

Wang Y. H., Cheng S. A., Chan K. Y. and Li X. Y. 2005, Electrolytic generation of ozone on antimony and nickel doped tin oxide electrode. *J. Electrochem. Soc.*, 152, 11, D197-D200

Wang Y. H., Chan K. Y., Li X. Y. and So S. K. 2006, Electrochemical treatment of 4-chlorophenol on novel nickel antimony doped tin dioxide, *Chemosphere*, 65,8, 1087-1093

Weiss E., Groenen-Serrano K. and Savall A. 2006, Electrochemical degradation of sodium dodecylbenzene sulfonate on boron doped diamond and lead dioxide anodes. *J. New Mater. Electrochem. Syst.*, 9, 3, 249-256

Wen T. C., Wei M. G. and Lin K. L. 1990, Electrocrystallization of $PbO_2$ deposits in the presence of additives. *J. Electrochem. Soc.*, 137, 2700-2702

Yang W. H., Wang H. H. and Fu F. 2010, Preparation and performance of Ti/Sb-$SnO_2$/beta-$PbO_2$ electrode modified with rare earth, *Rare Met. Mater. Eng.*, 39, 1215-1218

Yeo I. H., Lee Y. S. and Johnson D. C. 1992, Growth of lead dioxide on a gold electrode in the presence of foreign ions, *Electrochim.Acta*, 37, 10, 1811-1815

Yeo I. H., Kim S., Jacobson R. and Johnson D. C. 1989, Electrocatalysis of anodic oxygen transfer reactions: comparison of structural data with electrocatalytic phenomena for bismuth-doped lead dioxide. *J. Electrochem. Soc.*, 136,1395-1401

Yoshiyama A., Nonaka T., Sekimoto M. and Nishiki Y. 1994, Preparation of PTFE composite-plated hydrophobic β-$PbO_2$ electrodes. *Chem. Lett.*, 1565-1568

Youssef S., Lamia A. and Ridha A. 2010, Anodic oxidation of chlorpyrifos in aqueous solution at lead dioxide electrodes. *J. Electroanal. Chem.*, 650, 1, 152-158

Zhao G. H., Zhang Y. G., Lei Y. Z., Lv B. Y., Gao J. X., Zhang Y. A. and Li D. M. 2010, Fabrication and electrochemical treatment application of a novel lead dioxide anode with superhydrophohic surfaces, high oxygen evolution potential, and oxidation capability. *Environ. Sci. Technol.*, 44, 5, 1754-1759

# 3

# Impact of Sewage and Industrial Effluents on Soil-Plant Health

Rajinder Singh Antil
*Department of Soil Science, CCS Haryana Agricultural University, Hisar, India*

## 1. Introduction

Effluent irrigation has been practiced for centuries throughout the world (Shuval *et al.*, 1986; Tripathi *et al.*, 2011). It provides farmers with a nutrient enriched water supply and society with a reliable and inexpensive system for wastewater treatment and disposal (Feigin *et al.*, 1991). In India also being a cheap source of irrigation farmers are applying this water to their fields. Rapid industrialization, population explosion and more urbanization in India have created enormous problems of environmental pollution in terms of generating the variable quantity and quality of solid and liquid wastes. In developing countries, there has not been much emphasis on the installation of sewage treatment plants and all the industrial effluents are generally discharged in to the sewage system. The sewage waters are used as potential source irrigation for raising vegetables and fodder crops around the sewage disposal sites which are directly or indirectly consumed by human beings. Soil contamination by sewage and industrial effluents has affected adversely both soil health and crop productivity. Sewage and industrial effluents are the rich sources of both beneficial as well as harmful elements. Since some of these effluents are a rich source of plant nutrients, therefore soil provides the logical sink for their disposal. But many untreated and contaminated sewage and industrial effluents may have high concentration of several heavy metals such Cd, Ni, Pb and Cr (Arora *et al.*, 1985; Narwal *et al.*, 1993). Their continuous disposal on agricultural soils has resulted in soil sickness (Narwal *et al.*, 1988) and accumulation of some of the toxic metals in soil (Adhikari *et al.*, 1993; Antil & Narwal, 2005, 2008; Antil *et al.*, 2004, 2007; Gupta *et al.*, 1986, 1994, 1998; Kharche *et al.*, 2011; Narwal *et. al.*, 1993) which may pose serious human and animal health. Ground water in Punjab has been contaminated by Hg and Pb to such an extent that it is causing DNA of the people, who drink it, to mutate (Bajwa, 2008). The present chapter, therefore, discusses the composition of sewage and industrial effluents in India and their possible effect on soil-plant health.

## 2. Sewage water

Raw sewage water available from cities is a mixture of domestic, commercial and industrial activities. Currently more than 450 cities in India generate more than 17 million cubic meters of raw sewage water per day (Bijay-Singh, 2002). Since the raw sewage water is rich in organic matter and essential nutrients, sewage farming is quite common in all urban areas.

In the country as a whole, about 200 sewage farms, covering as area of about 50,000 ha, are utilizing sewage waters to supplement the nutrients and water supply. Some city sewage waters where industrial effluent is discharged in to sewer system may contain toxic metals in high amounts. Thus the composition of domestic sewage may be changed with the type of industries discharging their effluents.

### 2.1 Composition of sewage water

The composition of sewage water is quite variable depending upon the contributing source, mode of collection and treatment provided. Although a large proportions of these sewage waters is organic in nature and contains essential plant nutrients but sometimes toxic metals are also present in appreciable amounts. The sewage water generated in India contains more than 90% water. The solid portion contains 40-50% organics, 30-40% inert materials, 10-15% bio-resistant organics and 5-8% miscellaneous substances on oven dry weight basis (Antil & Narwal, 2008).

The chemical composition of sewage waters varied from site to site which was in accordance with the type of industries discharging their effluents. Some city sewage waters where industrial effluent is discharged in to sewage system may contain toxic metals in high amounts. Many investigations have found variations in pH, electrical conductivity (EC), suspended solids, organic C, $CO_3$, $HCO_3$, $Ca^{++}$, $Mg^{++}$ and other essential and toxic elements in the sewage water from Indian cities. The pH of the sewage water of different cities ranged from 7.2 to 8.3 (Table 1) and it was within normal range and irrigation with these waters in not going to cause any significant change in the soil pH due to high buffering capacity of the soils. The EC of sewage waters collected from different cities varied from 1.1 to 3.8 dS $m^{-1}$ and their continuous use in the agricultural fields may cause increase in salinity of the soils and ultimately restricting the plant growth. Organic C content of sewage waters of different cities ranged from 59 to 480 mg $L^{-1}$ being lowest in sewage water from Bhatinda and highest in sewage water of Gurgaon. Since organic C has a direct relationship with biochemical oxygen demand (BOD), their harmful effects may be due to high BOD values. The sodium absorption ratio (SAR) ranged from 0.8 to 10.4 m $mol^{-½}L^{-½}$ and some of these waters are not suitable for irrigation. Apart from these properties, the sewage waters contained appreciable amount of N, P and K. However, their content in sewage water varied from city to city. The amounts of N, P and K in sewage waters ranged from 8 to 106, 4.2 to 53 and 19 to 2500 mg $L^{-1}$, respectively.

Sewage waters of many cities also contained appreciable amounts of micro-nutrients and toxic metals. In general the content of heavy metals was higher in cities of Haryana and West Bengal than cities of Punjab (Adhikari *et al*., 1997; Gupta & Mitra 2002; Gupta *et al*., 1986; Narwal *et al*,. 1993; Singh & Kansal, 1985a). The Zn, Mn, Fe, Cu, Pb, Cd and Ni of different cities of Haryana varied from 2.8 to 11.0, 1.6 to 15.0, 5.6 to 205, 0.6 to 1.9, 1.5 to 40, 0.15 to 5.80 and 1.0 to 6.4 mg $L^{-1}$, respectively (Table 2). The wide variation in metal content of sewage water of different cities is a reflection of variability of sources of metals entering in the sewage system. Arora *et al.* (1985) and Tiwana *et al.* (1987) also reported wide variation in the concentration of heavy metals in sewage water collected from different outlets of Ludhiana city. The concentration of Cr in sewage water of Ludhiana city was found one thousand times higher than that of New York City.

| Location | pH | EC($dSm^{-1}$) | Organic C ($mgL^{-1}$) | SAR (m mol-½L-½) | N (mg $L^{-1}$) | P (mg $L^{-1}$) | K (mg $L^{-1}$) |
|---|---|---|---|---|---|---|---|
| Haryana | | | | | | | |
| Faridabad | 7.5 | 3.8 | 180 | 10.0 | 35 | 20 | 110 |
| Gurgaon | 7.4 | 3.8 | 480 | 10.1 | 50 | 35 | 275 |
| Bahadurgarh | 7.2 | 3.6 | 90 | 4.6 | 8 | 53 | 388 |
| Hisar | 7.4 | 2.1 | 105 | 10.4 | 106 | 25 | 294 |
| Panipat | 8.2 | 1.6 | 15 | 0.8 | 61 | 18 | 2500 |
| Sonepat | 7.7 | 2.6 | 30 | 0.9 | 71 | 17 | 2125 |
| Punjab | | | | | | | |
| Abohar | 8.3 | 1.4 | 200 | ND | 29 | 4.2 | 46 |
| Bhatinda | 7.8 | 3.0 | 59 | ND | 17 | 5.6 | 60 |
| Jalandhar | 7.9 | 1.1 | 117 | ND | 38 | 7.6 | 35 |
| Amritsar | 7.2 | 1.7 | 108 | ND | 34 | 8.4 | 42 |
| West Bengal | | | | | | | |
| Calcutta | 8.0 | 2.0 | ND | 2.9 | ND | 7.0 | 20 |
| Dhapa | 7.8 | 1.2 | ND | 2.9 | ND | ND | 19 |

ND - Not determined
Source (Adhikari *et al.*, 1997; Narwal *et al.*, 1993, 2002; Singh & Kansal, 1985a)

Table 1. Chemical composition of sewage water of some cities of India.

| Location | Zn | Mn | Fe | Cu | Pb | Cd | Ni |
|---|---|---|---|---|---|---|---|
| Haryana | | | | | | | |
| Faridabad | 4.2 | 1.6 | 5.6 | 1.0 | 3.0 | 0.48 | 2.0 |
| Gurgaon | 2.8 | 3.2 | 25 | 0.6 | 2.5 | 0.15 | 1.0 |
| Bahadurgarh | 5.0 | 5.6 | 34 | 0.8 | 1.5 | 0.15 | 1.0 |
| Hisar | 10.0 | 15.0 | 145 | 1.9 | 38.0 | 5.80 | 2.8 |
| Panipat | 11.0 | 14.0 | 205 | 1.7 | 38.0 | 5.60 | 3.4 |
| Sonepat | 10.0 | 14.0 | 66 | 1.7 | 40.0 | 5.60 | 6.4 |
| Punjab | | | | | | | |
| Abohar | 0.14 | 0.10 | 1.6 | 0.09 | 0.03 | 0.01 | ND |
| Bhatinda | 0.20 | 0.14 | 2.9 | 0.35 | 0.04 | 0.01 | ND |
| Jalandhar | 0.38 | 0.23 | 5.3 | 0.37 | 0.09 | 0.02 | ND |
| Amritsar | 0.31 | 0.30 | 4.1 | 0.18 | 0.06 | 0.01 | ND |
| West Bengal | | | | | | | |
| Calcutta | 0.30 | 0.66 | 656 | 0.07 | 2.9 | 1.3 | 58 |
| Dhapalock | ND | 2.3 | ND | 3.35 | 18.0 | 3.8 | ND |

ND - Not determined
Source (Adhikari *et al.*, 1997; Gupta & Mitra, 2002; Gupta *et al*,. 1986; Narwal *et al.*, 1993; Singh & Kansal, 1985a).

Table 2. Heavy metal content (mg $L^{-1}$) in sewage water of some cities of India.

The composition of sewage waters is not constant and changes within the year due to several factors. The extent of the contaminants will be least during the rainy seasons. During the same year, the pH of the sewage water of one location of Hisar varied from 7.3 to 8.7 and EC from 0.52 to 1.55 $dSm^{-1}$ (Antil & Narwal, 2008). The variation in the composition of sewage water within season has also been reported by Singh & Kansal (1985b) for different cities of Punjab. The concentration of metals in sewage water is season specific (Adhikari *et al.*, 1994). For example, Pb content varied from 8.1 to 30 mg $L^{-1}$ during monsoon and 11.0 to 41.3 mg $L^{-1}$ during summer. Similarly, Cd ranged from 4.0 to 22.0 mg $L^{-1}$ in summer season while it was not detected during monsoon and Cr content in raw sewage during summer was 7.2 mg $L^{-1}$ while in monsoon it was 1.6 mg $L^{-1}$.

Kharche *et al.* (2011) found that the pH of sewage waters of Ahmednagar city of Maharashtra was neutral to slightly alkaline (7.4 to 8.4) which was within safe limits of 6.0 to 8.5 as suggested by United States Salinity Laboratory staff (USSL) (1954). The EC (0.97 to 1.77 dS $m^{-1}$), BOD (100-210 mg $L^{-1}$), chemical oxygen demand (COD) (700-940 mg $L^{-1}$), total dissolved solids (400-1200 mg $L^{-1}$) values of effluents were higher than the recommended limits prescribed by FAO (1985). The concentration of Fe (4.81-7.26 mg $L^{-1}$), Mn (0.45-1.17 mg $L^{-1}$), Zn (0.63-2.00 mg $L^{-1}$), Cu (0.024-1.18 mg $L^{-1}$) and B (2.11-5.75 mg $L^{-1}$) was higher in sewage effluents indicating that they are good source of plant micronutrients which may help in mitigating emerging problems of their deficiency that are otherwise overcome by application of costly chemical fertilizers (Kharache *et al.*, 2011). However, their concentrations in sewage effluents was higher than the recommended maximum concentration suggested by Indian Standards (1982) and FAO (1985) (Table 3) indicating that they can lead to contamination of soil and may be potentially toxic for plants, human and animal health.

| Heavy metal | *Recommended maximum concentration |
|---|---|
| Fe | 5.0 |
| Mn | 0.2 |
| Zn | 2.0 |
| Cu | 0.2 |
| B | 2.0 |
| Cd | 0.01 |
| Cr | 0.1 |
| Ni | 1.2 |

Source (*FAO, 1985; Indian Standards, 1982)

Table 3. Recommended maximum concentration (mg $L^{-1}$) of heavy metals for use of effluents in agriculture.

### 2.2 Composition of sewage sludge

The agricultural use of sewage sludge has been a common practice in waste disposal in the recent decades. Since it contains plant nutrients and organic matter, sewage sludge may be beneficial to soils and their productivity. However, depending on its source, sewage sludge often contains considerable amount of toxic metals and organic toxicants. The presence of toxic metals in sewage sludges collected from different treatment plants of some cities of India has also been reported by Adhikari *et al.* (1993) and Jurwarkar *et al.* (1991). The contents of Cd, Cr, Ni and Pb varied from 1.5 to 8.3, 17.3 to 176.2, 11.7 to 191.5 and 14.3 to 200 mg $L^{-1}$, respectively (Table 4).

The Central Pollution Control Board (CPCB) formulated a Ganga project in 1984 to clean the water of Ganga river. This project aimed at the installation of sewage treatment plants at various sites along Ganga river in the important cities and towns. Cities selected for the installation of sewage treatment plants include 26 cities in Uttar Pradesh, 15 in Bihar and 59 in West Bengal have been identified as the worst polluting cities because they contribute 84% of the total pollutants of the Ganga. Similarly under Yamuna Action Plan, sewage treatment plants have been installed at Yamunanagar, Karnal, Panipat, Gurgaon and Faridabad in Haryana. The chemical composition of sewage sludge indicates that that it contains appreciable amount of useful plant nutrients but at the same time also contain toxic metals (Table 5). So that the continuous use of sewage sludge may result in accumulation of toxic metals in the soils. Thus, it should be used carefully and soil must be monitored continuously to prevent the entry of toxic metals in food chain.

| Name of city | Cd | Cr | Ni | Pb |
|---|---|---|---|---|
| Allahabad | 3.5 | 60.4 | 32.3 | 76.8 |
| Delhi | 1.5 | 82.0 | 191.5 | 41.7 |
| Nagpur | 1.5 | 49.2 | 14.8 | 24.3 |
| Madras | 8.3 | 38.5 | 60.5 | 16.5 |
| Kanpur | 0.6 | 17.3 | 11.7 | 14.3 |
| Jaipur | 7.3 | 176.2 | 37.5 | 66.0 |
| Pagladanga | 4.0 | 101.0 | ND | 200.0 |
| Topsia | 2.0 | 101.3 | ND | 185.0 |

Source (Adhikari *et al.*, 1993; Jurwarkar *et al.*, 1991).

Table 4. Heavy metal (mg $kg^{-1}$) content of sewage sludge of some cities in India.

| Element | Range | Mean | Element | Range | Mean |
|---|---|---|---|---|---|
| N (%) | 0.75-1.32 | 1.06 | Mn (mg $kg^{-1}$) | 154-721 | 259 |
| P (%) | 0.31-2.31 | 0.68 | Fe (mg $kg^{-1}$) | 3263-12606 | 8139 |
| K (%) | 0.22-1.67 | 0.50 | Cu (mg $kg^{-1}$) | 24-2050 | 460 |
| S (%) | 0.29-1.94 | 1.00 | Pb (mg $kg^{-1}$) | 14.7-139.7 | 73.2 |
| Ca (%) | 1.00-4.33 | 2.48 | Ni (mg $kg^{-1}$) | 4.1-624.2 | 123.7 |
| Mg (%) | 0.39-0.78 | 0.59 | Cr (mg $kg^{-1}$) | 10-3796 | 715 |
| Zn (mg $kg^{-1}$) | 100-4835 | 1376 | Cd (mg $kg^{-1}$) | 1.02-6.88 | 3.56 |
| Cr (mg $kg^{-1}$) | 10-3796 | 715 | Cr (mg $kg^{-1}$) | 10-3796 | 715 |
| Cd (mg $kg^{-1}$) | 1.02-6.88 | 3.56 | Cd (mg $kg^{-1}$) | 1.02-6.88 | 3.56 |

Source (Antil & Narwal, 2005)

Table 5. Composition of sewage sludge from various locations in Haryana (n=8).

## 2.3 Effect on soil properties

In India, land around the cities receiving sewage waters containing both the domestic and the industrial wastes. These wastes are suitable for crop production provided the content of major plant nutrients is high and that of toxic elements is low. Its long-term application would affect the physical, chemical and biological properties of soils (Antil *et al.*, 2007).

### 2.3.1 Physical properties

The bulk density of soils irrigated with sewage water was low (1.2 to 1.39 Mg $m^{-3}$) as compared to those for the well-irrigated soils (Table 6). The hydraulic conductivity was also higher (1.10 to 1.33 cm $h^{-1}$) for sewage irrigated soils.

| Soil depth (cm) | Soil physical property | | | |
|---|---|---|---|---|
| | Bulk density (Mg $m^{-3}$) | Hydraulic conductivity (cm $hr^{-1}$) | Water retention (%) | |
| | | | 33 kPa | 1500 kPa |
| | Sewage-irrigated soils | | | |
| 0-30 | 1.28 (1.20-1.36) | 1.22 (1.12-1.33) | 43.2 (40.0-46.4) | 15.4 (13.2-17.2) |
| 30-60 | 1.30 (1.22-1.39) | 1.19 (1.10-1.28) | 45.5 (42.0-49.4) | 18.5 (17.3-19.8) |
| | Well-irrigated soils | | | |
| 0-30 | 1.40 (1.39-1.41) | 1.12 (1.10-1.15) | 38.0 (38.0-38.0) | 13.4 (12.8-14.0) |
| 30-60 | 1.38 (1.34-1.41) | 0.94 (0.90-0.98 | 42.6 (40.8-44.4) | 15.7 (14.6-16.8) |

Source (Kharche *et al.*, 2011)

Table 6. Mean values of physical properties of sewage- and well- irrigated soils.

Mathan (1994) recorded significantly lower bulk density and increased hydraulic conductivity in sewage farm soils with sewage irrigation for 15 years. This can be attributed to improvement in total porosity and aggregate stability in the sewage-irrigated soils due to addition of organic matter which plays an important role in improving soil physical environment. Otis (1984), however, reported that the application of sewage reduced the hydraulic conductivity of soils due to pore clogging by suspended solids. This can be justified as the organic suspended solids may impede water transmission initially by temporarily plugging soil surface and by clogging of pores; however, the effect of organic matter addition through sewage on aggregation improves soil structure and enhances water transmission. The soil moisture retention of sewage-irrigated soils was also slightly higher as compared to that of sewage-free soils which can be ascribed to addition of organic matter through sewage. Rattan *et al.* (2001) observed enhanced available water content in the soils due to continuous application of sewage waters.

### 2.3.2 Chemical properties

With the continuous long-term application of sewage water to field crops some of the heavy metals may get accumulated in the soil profile and may limit plant growth. Even domestic sewage water from household and dairy effluents application for longer times drastically reduced the hydraulic conductivity of the soil, increased the pH, EC, DTPA extractable and total Zn, Mn, Fe, Cu, Cd, Ni and Pb (Narwal *et al.*, 1988). Singh & Kansal (1985b) also postulated the problems like sickness, salinity and alkalinity in Punjab where farmers irrigate their fields too often with sewage water especially during summer season. Several studies conducted in India revealed that irrigation with sewage water had a profound effect on chemical properties of soils (Table 7). The pH of the soil decreased with sewage irrigation at all the locations. The decrease of about one unit in pH with sewage water irrigation was observed in soils of Haryana (Narwal *et al.*, 1993), but such large decrease in pH of the soils in Punjab was not seen. Higher reduction in pH of Haryana soils was due to the low pH of sewage water applied. The EC of the soils increased at all the locations because of high EC of sewage waters. The EC of Bahadurgarh (9.74 $dSm^{-1}$) and Faridabad (4.26 $dSm^{-1}$) were higher than the soils of other cities and it may be due to presence of high salt content of sewage water of these cities. This indicated that long-term use of sewage waters may develop the salinity problem and ultimately will render the soils unproductive due to high amounts of salt accumulation.

The organic C of the soil increased with sewage irrigation owing to the presence of C in sewage waters at all the location. However, highest increase in C content of soil was observed at Faridabad. Loehr *et al.*, (1979) and Narwal *et al.* (1988) reported that the long-term application of sewage water or dairy effluents of high organic matter resulted in soil sickness due to poor aeration and higher BOD. These results indicated that unwise use of sewage waters may deteriorate the soil physical environment due to accumulation of salts and poor soil aeration. The concentrations of heavy metals at all the locations increased with the sewage water irrigation. Gupta *et al.* (1986) and Narwal *et al.* (1993) found that the application of sewage water on agricultural land for long period increased the total content of Zn, Mn, Cu, Pb, Ni and Cd in soils of Haryana (Table 8). The extent of increase in metal content was higher in Sonipat soil which received sewage irrigation for 15 years. The results indicate that amounts of Cd increased four times and that of Ni three times in sewage irrigated soils. This increase was mainly due to mixing of Cycles industry effluent in the sewage system.

| Location | pH | | EC ($dS\ m^{-1}$) | | Organic C(%) | |
|---|---|---|---|---|---|---|
| | TW | SW | TW | SW | TW | SW |
| Haryana | | | | | | |
| Faridabad | 8.22 | 7.20 | 0.46 | 4.26 | 0.57 | 2.52 |
| Gurgaon | 8.10 | 6.95 | 0.33 | 3.04 | 0.42 | 0.53 |
| Bahadurgarh | 8.10 | 7.35 | 2.21 | 9.74 | 0.30 | 0.36 |
| Hisar | 8.26 | 8.37 | 0.19 | 0.23 | 1.02 | 1.50 |
| Panipat | 8.45 | 8.55 | 0.19 | 0.32 | 0.38 | 0.87 |
| Sonepat | 8.70 | 7.94 | 0.06 | 0.65 | 0.18 | 0.39 |
| Punjab | | | | | | |
| Abohar | 8.47 | 8.19 | 0.66 | 0.78 | 0.37 | 0.72 |
| Bhatinda | 8.38 | 8.45 | 0.57 | 0.66 | 0.23 | 0.60 |
| Jalandhar | 8.35 | 7.59 | 0.19 | 0.31 | 0.31 | 1.09 |
| Amritsar | 8.24 | 7.94 | 0.23 | 0.97 | 0.45 | 1.16 |

TW - Tube well-irrigated soil, SW - Sewage-irrigated soil
Source (Brar *et al.*, 2000; Narwal *et al.*, 1993, 2002; Singh & Kansal, 1985b)

Table 7. Effect of sewage irrigation on chemical characteristics of soils.

| Location | Duration of sewage irrigation (year) | Micronutrients | | | Toxic metals | | |
|---|---|---|---|---|---|---|---|
| | | Zn | Mn | Cu | Pb | Ni | Cd |
| Faridabad | 0 | 85 | 350 | 25 | 37 | 33 | 1.9 |
| | 10 | 158 | 370 | 68 | 50 | 34 | 1.9 |
| Gurgaon | 0 | 83 | 340 | 25 | 25 | 42 | 1.9 |
| | 10 | 83 | 340 | 25 | 25 | 42 | 1.9 |
| Bahadurgarh | 0 | 65 | 310 | 16 | 25 | 33 | 1.9 |
| | 15 | 75 | 325 | 25 | 38 | 50 | 1.9 |
| Hisar | 0 | 8.26 | 8.37 | 0.19 | 0.23 | 1.02 | 1.50 |
| | 20 | 205 | 609 | 71 | 51 | 54 | 2.9 |
| Panipat | 0 | 140 | 656 | 37 | 40 | 48 | 3.4 |
| | 10 | 175 | 656 | 51 | 48 | 46 | 2.4 |
| Sonepat | 0 | 91 | 281 | 14 | 20 | 26 | 0.8 |
| | 15 | 183 | 406 | 52 | 32 | 72 | 3.2 |

Source (Gupta *et al.*, 1986, 1998; Narwal *et al.*, 1993)

Table 8. Effect of sewage effluent irrigation on total metal content (mg $kg^{-1}$) in soils of Haryana.

Build up of heavy metals in sewage water irrigated soils at other cities has also been reported. Mitra & Gupta (1997) has reported 143 times more accumulation of Cd in sewage effluent irrigated soils around Calcutta over non-sewage irrigated soils, followed by Zn (47 times), Pb (18.5 times), Cr (5.6 times), Co (3.9 times), Cu (3.6 times), Fe (2.4 times) and Ni (2.3 times). In soils of cultivated fields of Durgapur Industrial Belt (DIB) of West Bengal (comprising of steel plants, chemical factories, pharmaceuticals, fertilizer factory and hundreds of auxiliary factories) total Zn, Cu, Cd and Pb contents varied from 200 to 570, 24 to 69, 3 to 9 and 53 to 324 mg $kg^{-1}$, respectively (Antil & Narwal, 2008). In agricultural lands around Ludhiana city, increase in total Cd, Ni and Co contents of sewage effluent irrigated soils was 36, 86 and 46% over tube well irrigated soils (Azad *et al.*, 1986).

The effect of long-term irrigation with sewage effluents on available metal status of agricultural soils has also been reported at various places. Singh & Kansal (1983) reported that the use of sewage water for irrigation substantially increases the accumulation of DTPA extractable Zn, Cu, Pb and Cd in soils of different cities of Punjab. The accumulation was higher in soils receiving sewage water of industrial cities (Table 9).

Under Keshopur effluent irrigation system of Delhi, the DTPA extractable Zn, Cu, Fe and Ni has increased by 253, 202, 237 and 153%, respectively in effluent irrigated soils as compared to that of tube well irrigated soils in two decades (Rattan *et al.*, 2001). Kharche *et al.* (2011) found that the total content of Fe, Mn, Zn, Cu, Cd, Cr and Ni in the soils irrigated with sewage water for more than three decades was 1.05, 1.24, 3.98, 1.51, 2.10, 1.62, and 1.24 times higher as compared to their content in the well-irrigated soils. Although, the concentration of heavy metals in the sewage-irrigated soils is below the maximum permissible limits as given by Department of Environment (1989) and Kabata & Pendias (1992), but it can be observed that these soils expected to approach these critical concentration in few years from now and may become toxic on accumulation of these metals

| Location | Zn | | Cu | | Pb | | Cd | |
|---|---|---|---|---|---|---|---|---|
| | TW | SW | TW | SW | TW | SW | TW | SW |
| Abohar | 0.95 | 3.69 | 0.40 | 1.53 | 0.35 | 0.76 | 0.05 | 0.06 |
| Bhantinda | 0.98 | 2.75 | 0.40 | 1.67 | 0.20 | 0.69 | 0.02 | 0.04 |
| Jalandhar | 1.82 | 5.66 | 0.90 | 1.98 | 0.85 | 0.89 | 0.05 | 0.17 |
| Amritsar | 0.95 | 7.30 | 0.95 | 5.49 | 0.50 | 1.69 | 0.05 | 0.17 |
| Ludhiana | 2.13 | 4.13 | 0.62 | 3.44 | 0.41 | 2.57 | 0.06 | 0.20 |

TW - Tube well-irrigated soil, SW - Sewage-irrigated soil
Source (Singh & Kansal, 1983)

Table 9. Effect of sewage irrigation on DTPA extractable metals (mg $kg^{-1}$) in soils of Punjab.

by continuous use of sewage (Table 10). In addition, if rather strict criteria of Ontario Canada (Seto & Deangelis, 1978) are considered for categorization soils in polluted category, most of the soils fall under contaminated category with respect to these heavy metals.

| Element | Kabata and Pendias (1992) | Department of Environment (1989) | Ontario Canada by Seto and Deangelis (1978) |
|---|---|---|---|
| Cd | 3-8 | 3 | 3 |
| Ni | 100 | 110 | 22 |
| Cu | 60-125 | 200 | 100 |
| Mn | 1500-3000 | - | 1500 |
| Zn | 70-400 | 450 | 216 |
| Cr | - | 400 | - |

Table 10. Suggested critical soil concentration values (mg $kg^{-1}$) for phyto-toxicity of heavy metals.

### 2.3.3 Biological properties

The microbial count in sewage-irrigated soils was higher for bacteria, fungi and actinomycetes which was about 1.34, 1.52 and 1.18 times (for 0-30 cm) higher as compared to that in normal soils, respectively (Table 11). This may be due to the suspended organic material added to soil through sewage which serves as a source of energy for microbial population (Joshi & Yadav, 2005; Seeker & Sopper, 1988).

| Soil depth (cm) | Soil microbial count | | |
|---|---|---|---|
| | Bacteria ($10^6$ x $g^{-1}$) | Fungi ($10^5$ x $g^{-1}$) | Actinimycetes ($10^4$ x $g^{-1}$) |
| | Sewage-irrigated soils | | |
| 0-30 | 20.8 (15-26) | 10.7 (8-12) | 5.33 (4-6) |
| 30-60 | 19.2 (13-24) | 9.3 (7-11) | 4.0 (3-6) |
| | Well-irrigated soils | | |
| 0-30 | 15.5 (15-16) | 7.0 (6-8) | 4.5 (4-5) |
| 30-60 | 11.5 (11-12) | 6.0 (5-7) | 3.5 (3-4) |

Source (Kharche *et al.*, 2011)

Table 11. Mean soil microbial counts of sewage- and well-irrigated soils.

## 2.4 Effect on plants

Several investigators have reported positive effect of sewage irrigation on crop yield. Mahida (1981) obtained higher yields of vegetable crops irrigated with untreated sewage water compared to irrigation with canal water (Table 12). Experiments conducted at National Environmental Engineering Research Institute, Nagpur revealed that the continuous use of untreated sewage for irrigation significantly reduced the yield of wheat, cotton and paddy. However, the use of primary treated sewage proved to be beneficial for both wheat and cotton (Juwarker *et al.*, 1991). Application of sewage water resulted remarkable increase on the mean plant height (3.4%), number of tiller/plant (31.8%), length of ear (18.8%) in wheat crop. As a result, straw as well as grain yields also were increased significantly by 43.1 and 34.3%, respectively due to application of sewage water (ISSS Annual Report, 2006-07).

| Crop | Source of irrigation | |
|---|---|---|
| | Canal water | Sewage water |
| Beet root | 8.75 | 16.27 |
| Carrot | 9.71 | 11.75 |
| Radish | 7.26 | 8.33 |
| Potato | 6.12 | 9.33 |
| Ginger | 6.04 | 9.80 |
| Knolkhol | 9.70 | 16.57 |
| Cabbage | 9.27 | 12.13 |
| Cauliflower | 6.96 | 9.09 |
| French beans | 6.63 | 8.06 |
| Tomato | 10.01 | 13.38 |
| Tobacco | 1.12 | 1.25 |
| Groundnut | 2.88 | 2.90 |

Source (Mahida, 1981)

Table 12. Effect of sewage irrigation on yield (t $ha^{-1}$) of vegetable crops.

The accumulation of some of the usually prevalent heavy metals (Ni, Cd, Cr, Pb etc.) in sewage irrigated soils in different plant parts have been reported. Mitra & Gupta (1999) has reported that the contents of heavy metals in sewage irrigated radish, gourd, spinach and cauliflower around Calcutta were comparatively 2 to 40 times higher than the non-sewage irrigated vegetables. Brar *et al.* (2000) also reported higher accumulation of metals in leaves and tubers of potato grown on sewage irrigated soils as compared with ground water irrigated soils. Leafy vegetables and root crops are known to accumulate higher amounts of heavy metals than grain crops. Other factors like the composition, duration and rate of sewage irrigation, soil types, soil reaction and interaction among metals also affect their uptake. Kansal *et. al.* (1996) collected the plant samples of different crops from sewage irrigated soils and found that these plants contained higher amount of Zn, Mn, Fe, Cu, Cd and Pb than plants from tube well irrigated soils (Table 13). The leafy crops accumulated higher amounts of heavy metals than other crops. The maximum concentration of Mn Cu, Cd and Pb was found in spinach followed by berseem and least was absorbed by wheat. This indicate that selection of a crop which absorbed the lowest amount of heavy metals may prevent the entry of toxic metals from contaminated soils into food chain.

| Crop | Source of irrigation | Zn | Mn | Fe | Cu | Cd | Pb |
|---|---|---|---|---|---|---|---|
| Maize | TW | 38 | 26 | 128 | 8 | 0.85 | 1.98 |
| | SW | 53 | 37 | 170 | 12 | 1.74 | 3.82 |
| Berseem | TW | 25 | 25 | 293 | 9 | 0.69 | 1.93 |
| | SW | 46 | 38 | 356 | 15 | 1.67 | 4.48 |
| Cauliflower heads | TW | 22 | 11 | 80 | 3 | 0.48 | 1.27 |
| | SW | 43 | 15 | 117 | 7 | 1.80 | 1.60 |
| Cauliflower leaves | TW | 22 | 24 | 91 | 4 | 0.80 | 2.69 |
| | SW | 33 | 35 | 170 | 7 | 2.24 | 5.23 |
| Spinach | TW | 28 | 31 | 270 | 10 | 0.50 | 3.29 |
| | SW | 44 | 48 | 332 | 15 | 2.59 | 6.08 |
| Fenugreek | TW | 40 | 25 | 29 | 12 | 0.64 | ND |
| | SW | 48 | 35 | 71 | 17 | 1.25 | ND |
| Coriander | TW | 37 | 27 | 47 | 8 | 0.85 | ND |
| | SW | 62 | 29 | 64 | 14 | 1.64 | ND |
| Wheat | TW | 34 | 15 | 10 | 8 | 0.45 | ND |
| | SW | 43 | 24 | 46 | 10 | .82 | ND |

TW - Tube well water, SW - Sewage water, ND - Not determined
Source (Kansal *et al.*, 1996)

Table 13. Heavy metal contents (mg $kg^{-1}$) of various crops grown on tube well and sewage water irrigated soils.

Kharche *et al.* (2011) reported higher concentration of heavy metals in the cabbage plant grown on the sewage-irrigated soils (Table 14). The mean concentration of Fe, Mn, Zn, Cu, Cd, Cr and Ni in cabbage grown on sewage-irrigated soils was about 1.11, 7.51, 1.72, 7.66, 4.36, 1.26 and 1.91 times than their content in well-irrigated soils, respectively. The concentration of heavy metals in cabbage is higher as compared to the suggested permissible tolerance levels (Table 17) as suggested by Council for Agricultural Science and Technology (1976); Melsted (1973) and Naidu *et al.* (1996).

| Heavy metal | Concentration (mg $kg^{-1}$) | | Suggested tolrent level * (mg $kg^{-1}$) |
|---|---|---|---|
| | Sewage-irrigated | Well-irrigated | |
| Fe | 821.7 (524-1361) | 736.5 (702-771) | 750 |
| Mn | 220.3 (186-263) | 29.3 (28.2-30.5) | 300 |
| Zn | 124.9 (101-159) | 72.4 (70.2-74.7) | 300 |
| Cu | 127.3 (118-138) | 16.6 (15.7-17.5) | 150 |
| Cd | 2.75 (2.25-3.25) | 0.63 (0.50-0.75) | 3 |
| Cr | 2.09 (1.62-2.47) | 1.65 (1.55-1.75) | 2 |
| Ni | 134.7 (123-142) | 70.5 (65-76) | 50 |

Source (Kharche *et al.*, 2011)

Table 14. Mean heavy metal contents in cabbage plant grown on sewage and well irrigated soils

## 3. Industrial effluents

With the industrialization of the country a large volume of liquid and solid wastes are generated daily. The quality of these wastes depends upon the nature of the industry and type of treatments given to these waters before their release from factory premises. The use of agricultural for the disposal of industrial effluents is becoming a widespread practice. Such materials may contain various toxic metals that could accumulate in excessive quantities in soils. Also, soil pollution by heavy metals is one of the major environmental problems associated with the application of effluents from industries involved in metal processing (Antil & Narwal, 2008).

### 3.1 Composition

In India, seventeen categories of heavily polluting industries have been identified by CPCB. They are cement, thermal power plant, distilleries, sugar, fertilizer, integrated iron and steel, oil refineries, pulp and paper, petrochemicals, pesticides, tanneries, basic drugs and pharmaceuticals, dye and dye intermediates, caustic soda, zinc smelter, copper smelter and aluminum smelter. Generally, these industries discharge their effluents into city sewage system, nearby water bodies or adjoining agricultural lands which cause environmental problem. The composition of some of the effluents from different type of industries is quite variable (Table 15). Effluents of Zn smelter and paper mills are acidic in nature having pH 3.5 and 4.8, respectively. The effluents of oil refinery, paper mill, distillery and sugar mill are rich in organic C ranging from 820 to 28350 mg $L^{-1}$. When such effluents of high BOD are disposed on soils they may develop anaerobic conditions and the soil becomes sick. Highest amount of toxic metals were observed in case of cycle industry effluents compared to the other industrial effluents. Thus, the composition of the industrial effluents varied according to the nature of process for manufacturing and the raw material used. Narwal *et al.*, (1992) found that effluents from cycle industry at Sonipat, Haryana, had high amounts of metals particularly Cu (64 mg $L^{-1}$) and Ni (30 mg $L^{-1}$).

| Characteristic | Industrial effluent | | | | | |
|---|---|---|---|---|---|---|
| | Oil refinery | Paper mill | Distillery | Cycle industry | Spent wash | Zn smelter |
| pH | 6.9 | 4.8 | 7.1 | 7.1 | 7.2 | 3.5 |
| EC (dS $m^{-1}$) | 0.5 | 1.2 | 13.0 | 4.8 | 29.0 | 7.7 |
| C (mg $L^{-1}$) | 820 | 1350 | 28350 | 23 | 2225 | ND |
| N (mg $L^{-1}$) | 140 | 168 | 42 | 84 | 1200 | ND |
| K (mg $L^{-1}$) | 0.1 | 0.5 | 1576 | 163 | 6681 | ND |
| Fe (mg $L^{-1}$) | 80 | 240 | 459 | 92 | 61 | 0.7 |
| Mn (mg $L^{-1}$) | 7.8 | 16.4 | 37.9 | 15 | 4 | ND |
| Cu (mg $L^{-1}$) | 4.0 | 10.9 | 19.3 | 64.0 | 0.8 | 13.0 |
| Zn (mg $L^{-1}$) | 30.0 | 6.4 | 37.9 | 2.4 | 1.2 | ND |
| Pb (mg $L^{-1}$) | 1.0 | 3.8 | 8.7 | ND | 0.7 | ND |
| Ni (mg $L^{-1}$) | 3.0 | 4.2 | 6.6 | 30.0 | 0.7 | ND |
| Cd (mg $L^{-1}$) | 0.1 | 0.2 | 0.3 | 5.4 | 0.06 | 0.02 |

ND - not determined
Source (Antil *et al.*, 2004; Gupta *et. al.*, 1994; Narwal et al. , 1992; Totawat, 1993; Zalawadia *et al.* 1997)

Table 15. Composition of some industrial effluents

Effluents of Zn smelter near Udaipur, Rajasthan, contained the high amount of Zn (13 mg $L^{-1}$) than the permissible limit (Totawat, 1993). Leather tanning contaminates the soil with Cr, As and B. In Punjab, tannery effluents contained as high as 9.2-13.8 mg $L^{-1}$ of hexavalent Cr. The tannery water in Tamil Nadu had deteriorated the quality of surface and ground water making it unfit for drinking and agriculture. Effluents of electroplating units in Punjab are highly acidic (pH 2.1) and contain high amounts of pollutants like Ni (2.4 to 52 mg $L^{-1}$), cyanides (0.4 to 4.4 mg $L^{-1}$). Zalawadia *et al.* (1997) surveyed several industries effluents of Gujarat and also found higher amounts of heavy metals. The release of such effluents on agricultural lands will adversely affect the quality of crops grown on these soils, making unfit for consumption for animals and human beings.

### 3.2 Effect on soil properties

Industrial effluents when released in the open or on agricultural land contaminate the soil with heavy metals and organic pollutants. The total Pb, Ni, Cd and Cr were higher in soils irrigated with lead battery and distillery industries effluent compared to soils irrigated either with canal or tube well water (Table 16). In case of distillery effluent irrigated soil, about 14 times increase in organic C was observed. It may be because of presence of higher C amounts in distillery effluent samples (Antil *et. al.*, 1999). Lead content was almost 11.5 times more in lead battery effluent irrigated soils as compared to canal or tube well irrigated soils (Mahata & Antil, 2004). The EC, organic C and toxic metal content (Pb, Ni and Cd) of soil increased due to irrigation with cycle industrial effluent (Antil *et al.*, 2004) as compared to tube well irrigated soils (Table 17). Excessive use of cycle industry effluent had converted the productive land to unproductive land and soil became rust colored and fluffy. In this soil, very high amounts of toxic metals were accumulated.

| Characteristic | Source of irrigation | |
|---|---|---|
| | Lead battery cell effluent | Distillery effluent |
| pH (1:2 soil:water) | 8.5 (8.3) | 8.0 (7.2) |
| EC (1:2) dS $m^{-1}$ | 0.31 (0.34) | 0.9 (1.1) |
| Organic C (%) | 0.45 (0.40) | 0.2 (2.8) |
| Total Pb, mg $kg^{-1}$ | 52.1 (589.9) | 0.5* (3.2) |
| Total Cd, mg $kg^{-1}$ | 1.0 (5.0) | 0.4* (1.2) |
| Total Ni, mg $kg^{-1}$ | 41.9 (48.3) | 0.6* (0.9) |
| Total Cr, mg $kg^{-1}$ | 1.1 (4.4) | ND |

Values in parentheses denote the effluents irrigated soils; * - DTPA extractable toxic metals
Source (Antil & Narwal, 2008; Mahata & Antil, 2004).

Table 16. Soil characteristics and toxic metal content (mg $kg^{-1}$) in soils irrigated with lead battery cell and distillery effluents

### 3.3 Effect on plants

During the survey of sewage and industrial effluents composition, we found a site at Sonipat which had turned barren due to excessive irrigation with cycle industry effluent. Crops grown on this soil indicated the high accumulation of toxic metal. The health hazard problems due to Ni absorption by crops grown on metal polluted soil was more in carrot

| Characteristic | Source of irrigation | |
|---|---|---|
| | Tube-well / canal water | Cycle industry effluent |
| pH | 8.7 | 7.5 |
| EC (dS $m^{-1}$) | 0.06 | 0.16 |
| Organic C (%) | 0.18 | 0.48 |
| Zn | 91 | 12188 |
| Mn | 281 | 281 |
| Fe | 15675 | 72600 |
| Cu | 14 | 1199 |
| Pb | 20 | 280 |
| Cd | 0.8 | 5.4 |
| Ni | 26 | 6000 |

Source (Antil *et al.* , 2004; Narwal *et al.*, 1992)

Table 17. Soil characteristics and toxic metal content (mg $kg^{-1}$) in soils irrigated with cycle industry effluent.

followed by spinach, fenugreek and wheat (Table 18). The accumulation of high amounts of heavy metal in plants may influence the consumer's health. The high intake of metals by human affects the body system and may deteriorate the health.

| Toxic metal | Shoot | | | Grain |
|---|---|---|---|---|
| | Carrot | Fenugreek | Spinach | Wheat |
| Ni | 434 | 167 | 300 | 65 |
| Cd | 13.4 | 5.5 | 4.5 | 4.7 |

Source (Narwal *et al.*, 1992)

Table 18. Toxic metal content (mg $kg^{-1}$) of different crops grown on polluted soil.

Application of agro-based industrial effluent (Jam and pickle industry) to agricultural lands adversely influenced the germination and growth of different crops (Verma & Kumar, 2004). A significant variation in percent germination of wheat, maize and mustard with respect to different concentration of effluents was observed (Table 19). The germination percentage

| Effluent concentration (%) | Germination (%) | | | Average height of seedling (cm) | | | Average length of root (cm) | | |
|---|---|---|---|---|---|---|---|---|---|
| | Wheat | Maize | Mustard | Wheat | Maize | Mustard | Wheat | Maize | Mustard |
| 0 (control) | 100 | 92 | 100 | 27.8 | 55 | 20.7 | 12 | 25 | 9.5 |
| 25 | 89 | 80 | 90 | 27.6 | 49.5 | 17.4 | 12.8 | 20.5 | 8 |
| 50 | 84 | 68 | 69 | 26.0 | 47 | 16.8 | 10.5 | 20.1 | 7.7 |
| 75 | 52 | 28 | 38 | 20.6 | 45.0 | 15.0 | 8.3 | 19.4 | 6.8 |
| 100 | 24 | 23 | 27 | 21 | 40 | 14.2 | 4.3 | 18.9 | 6.3 |

Source (Verma & Kumar 2004)

Table 19. Effect of agro-based industrial effluent on germination and development of wheat, maize and mustard crops.

decreased in all the crop plants and inhibition of germination was significant. The high acidic nature of effluent might have decreased the germination. The morphometric analysis showed that the height of seedling and length of root was also reduced with the application of effluent. Thus, effluents must be diluted either with canal water or tube well water to avoid their adverse effect on plant growth.

## 4. Possible solutions of problems associated with the sewage and industrial effluents

To exploit the sewage waters as a potential source of irrigation and maintain environment, the sewage waters must be diluted either with canal or underground waters to avoid the excessive accumulation of soluble salts in the soils. It will help in maintaining the productivity of agricultural crops without any harmful effect on soil properties.

Entry of heavy metals into food chain can be reduced by adopting soil and crop management practices, which immobilize these metals in soils and reduce their uptake by plants.

Heavy phosphate application and also the application of kaolin/zeolite to soils can reduce the availability of heavy metals.

Application of organic manures can mitigate the adverse effect of the toxic metals on crops. Thus in the soils contaminated with high amount of toxic metals, application of organic manures is recommended to boost the yield potentials as well as decrease the metal availability to plants.

Raising hyper accumulator plants (mustard/trees) in toxic metals contaminated soils is recommended to avoid the entry of toxic metals in the food chain.

The sewage/industrial effluents, sludge and the soils must be monitored continuously to avoid the excessive accumulation of toxic metals in the soils and then transfer in the food chain.

There should be strict Government legislation that only those sewage and industrial effluents be used in the fields which are cleaned through sewage and effluent treatment plants.

Highest priorities should be given to proper disposal of solid and liquid effluents from industries for proper land management.

## 5. Conclusion

The major environmental concern is an urbanizing India relate to high levels of water pollution due to poor waste disposal, inadequate sewerage and drainage, and improper disposal of industrial effluents. The sewage and industrial effluents contain essential nutrients or possess properties which can easily be utilized for irrigating the field crops. But the sewage water of many cities where industrial effluent is mixed in the sewage system contained toxic metals. Continuous use of sewage and industrial effluents irrigation recorded improvement in water retention, hydraulic conductivity, organic C and build-up of available N, P, K and micronutrient status and soil microbial count. The EC although increased due to sewage irrigation, it was within the tolerance limit to cause any soil salinity hazard. The toxic metals like Cd, Cr and Ni were found to be accumulated in the soil and plant due to long-term use of sewage irrigation. Hence, Cd, Cr and Ni are more likely to be the elements that may become health hazards for consumers of the crops grown in sewage-irrigated soils. The concentration of these metals was greater in leafy vegetables than in grain crops. This warrants the potential hazard to soil and plant health suggesting necessity of their safe use after pretreatment to safe guard soil health and reduces the risk of animal and human health hazard. To avoid or delay such problems, continuous monitoring of quality of sewage and industrial effluents available in the country and their impact on soil-plant health is required in order to make use of sewage waters as a cheap potential alternative source of plant nutrients in agriculture. Based on these results the farmers around cities have been alerted about the adverse effects of unwise long-term application of sewage water to crops.

## 6. Future research needs

Research should be done to study the long-term effects of sewage and industrial effluents on salt and toxic metal accumulation in soils and their effect on soil biological health and crop productivity.

Effect of sewage/industrial effluents and heavy metal pollution in soils should be studied on fixed sites.

Bio-transpiration of the contaminates through farm forestry and the critical concentrations of toxic metals in soil and plants for better animal and human health needs to be initiated.

To develop eco-friendly technology for the use of sewage and industrial effluents to improve crop productivity and soil quality and to protect of quality of farm produce and environment from degradation.

## 7. References

Adhikari, S.; Gupta, S.K. & Banerjee, S.K. (1993). Heavy metals content of city sewage and sludge. *Journal of Indian Society of Soil Science,* 41: 160-172.

Adhikari, S.; Gupta, S.K. & Banerjee, S.K. (1997). Long-term effect of raw sewage application on the chemical composition of ground water. *Journal of Indian Society of Soil Science,* 45: 392-394.

Antil, R.S. & Narwal, R.P. (2005). Problems and prospectus of utilization of sewer water in Haryana. In: *Management of Organic Wastes for Crop Production,* K.K. Kapoor, P.K. Sharma, S.S. Dudeja & B.S. Kundu, (Ed.), 159-168, Department of Microbiology, CCS Haryana Agricultural University, Hisar, India.

Antil, R.S. & Narwal, R.P. (2008). Influence of sewer water and industrial effluents on soil and plant health. In: *Groundwater resources: Conservation and management,* V.D. Puranik, V.K. Garg, A. Kaushik, C.P. Kaushik, S.K. Sahu, A.G. Hegde, T.V. Ramachandarn, I.V. Saradhi & P. Prathibha, (Ed.), 37-46, Department of

Environmental Science and Engineering, GJU Science and Technology Hisar, India.

Antil, R.S.; Arora, U. & Kuhad, M.S. (1999). Leaching and transformation of urea in soils treated with sewage water and distillery effluent. *Proceedings of International Conference on Contaminants in Soil Environment in Australasia Pacific Region*, pp. 464-466, New Delhi, India, Dec. 12-17, 1999.

Antil, R.S.; Dinesh & Dahiya, S.S. (2007). Utilization of sewer water and its significance in INM. *Proceedings of ICAR sponsored Winter School on Integrated Nutrient Management*, pp 79-83, Department of Soil Science and Directorate of Human Resource Management, CCS Haryana Agricultural University, Hisar, India, Dec. 4-24, 2007.

Antil, R.S.; Kumar, V., Kethpal, T.S., Narwal, R.P., Sharma, S.K., Mittal, S.B., Singh, J. & Kuhad, M.S. (2004). Extent of land degradation in different agro-climatic zones of Haryana. *Fertilizer News*, 49: 47-59.

Arora, B.R., Azad, A.S., Singh, B. & Sekhon, G.S. (1985). Pollution potential of municipal waste waters of Ludhiana, Punjab. *Indian Journal of Ecology*, 12: 1-7.

Azad, A.S.; Sekhon, G.S. & Arora, B.R. (1986). Distribution of cadmium, nickel and cobalt in sewage water irrigated soils. *Journal of Indian Society of Soil Science*, 34: 619-622.

Bajwa, H. (2008). Water contamination causing DNA mutation. *The Indian Express*, pp. 5, Feb. 16, 2008, Chandigarh edition, India.

Bijay-Singh (2002). Soil pollution and its control. In: *Fundamentals of Soil Science, Indian Society of Soil science, 499-514*, Indian Agricultural Research Institute, New Delhi.

Brar, M.S.; Mahli, S.S., Singh, A.P., Arora, C.L. & Gill, K.S. (2000). Sewage water irrigation effects on some potentially toxic trace elements in soils and potato plants in North Western India. *Canadian Journal of Soil Science*, 80: 465-71.

Council for Agricultural Science and Technology (1976). *Application of sewage sludge to crop land: Appraisal of potential hazards of the heavy metals to plants and animals*. Report No. 64, Office of Water Programs, U.S. Environmental Protection Agency EPA-429/9-76-013.

Department of Environment (1989). The use of sewage sludge in agriculture. *A National Code of Practice*, HSMO, London, UK.

FAO (1985). *Water Quality for Agriculture*, R.S. Ayers & D.W. Westcot (Ed.), Irrigation and drainage paper 29 Rev. 1, pp 1-174, FAO Rome.

Feigin, A.; Ravina, I. & Shalhevet, J. (1991). Irrigation with treated sewage effluent. Management for Environmental Protection. *Advanced Series in Agricultural Sciences* 17, pp 224, Springer-Verla.

Gupta, A.P.; Antil, R.S. & Singh, A. (1986). *Proceedings of National Seminar on Environmental Pollution Control and Monitoring*, 419-425, CSIO Chandigarh, India, October 22-26, 1986.

Gupta, A.P.; Narwal, R.P. & Antil, R.S. (1998). Sewer waters composition and their effect on soil properties. *Bioresource Technology*, 65: 171-173.

Gupta, A.P.; Narwal, R.P., Antil, R.S., Singh, A. & Poonia, S.R. (1994). Impact of sewage water irrigation on soil health. In: *Impact of Modern Agriculture on Environment*, R.K. Behl, S.K. Arora & P. Tauro (Ed.), 109-117, CCS Haryana Agricultural University, Hisar and Max Muller Bhawan, New Delhi.

Gupta, S.K. & Mitra, A. (2002). In: *Advances in Land Resource Management for 21st Century*, 446-469, Soil Conservation Society of India, New Delhi.

Indian Standards (IS) (1982). Tolerance limits for inland surface water subject to pollution, *Indian Standards IS*, 2296.

IISS Annual Report (2006-07). Recycling and rational use of different waste in agriculture and remediation of contaminated soils. *Annual Report of Indian Institute of Soil science (ISSS)*, pp 72-83, Bhopal, India.

Joshi, P.K. & Yadav, R.K. (2005). Effect of sewage on microbiological and chemical properties and crop growth in reclaimed alkali soil. *Proceedings of the International Conference on Soil, water and Environment Quality, Issues and Stratigies*, Jan. 28 – Feb. 1, 2005, New Delhi.

Jurwakar, A.S.; Jurwakar, Asha, Deshbharatar, P.B. & Bal, A.S. (1991). *Asian Experiences in Integrated Plant Nutrition*, 178-201, RAPA. FAO, Bangkok.

Kabata, P. and Pendias, H. (1992). *Trace Elements in Soils and Plants*, CRC Press Inc. Boca Raton, pp. 1-356, Florida, USA.

Kansal, B.D.; Kumar, R. & Sokka, R. (1996). The influence of municipal wastes and soil properties on the accumulation of heavy metals in plants. *In: Heavy metals in the Environment*, 413-416, CEC Consultants Ltd. Edinburg, UK.

Kharche, V.K.; Desai, V.N. & Pharande, A.L. (2011). Effect of sewage irrigation on soil properties, essential nutrients and pollutant element status of soils and plants in a vegetable growing area around Ahmednagar city in Maharashtra. *Journal of Indian Society of Soil Science*, 59: 177-184.

Loehr, R.C., Jowell, W.J., Novak, J.D., Charkson, W.W. & Friedman, G.S. (1979). *Land application of wastes*, Vol. II Van Noruend Keinhold Co., New York.

Mahata, M.K. & Antil, R.S. (2004). Effect of organic matter and levels of organic carbon on urease activity of selected surface soil contamination with lead. *Enviornment and Ecology*, 22: 314-318.

Mahida, U.N. (1981). Influence of sewage irrigation on vegetable crops. In: *Water pollution and disposal of waste water on land*, Tata Mcgrew Hill Pub., New Delhi.

Mathan, K.K. (1994). Studies on the influence of long-term municipal sewage effluent irrigation on soil properties. *Bioresource Technology*, 48: 275-276.

Melsted, S.W. (1973). *Soil-plant Relationship*, Recycling municipal sludges and effluents on land, pp. 121-128. National Association of State University and Land-Grant Colleges, Washington, D.C.

Mitra, A. & Gupta, S.K. (1997). Assessment of ground water quality from sewage farming area of east Calcutta. *Indian Journal of Environment Protection*, 17: 442.

Mitra, A. & Gupta, S.K. (1999). Effect of sewage water irrigation on essential plant nutrient and element status in vegetable growing areas around Calcutta. *Journal of Indian Society of Soil Science*, 47: 99-105.

Naidu, R., Kookuna, R.S., Oliver, D.P., Rogers, S. & Mc Laughlin, M.J. (1996). *Contaminants and the Soil Environment in the Australasia Pacific Region*, Kluwer Academic Publ., London.

Narwal, R.P., Gupta, A.P., & Dahiya, R.S. (2002). Effect of long term sewage water irrigation on soil health. *Proceedings 11th National Symposium on Environment*, pp. 249-251, Rajasthan Agricultural College, Udaipur, June 5-7, 2002.

Narwal, R.P.; Antil, R.S. & Gupta, A.P. (1992). Soil pollution through industrial effluent and its management. *Journal of Soil Contamination*, 1: 265-272.

Narwal, R.P.; Gupta, A.P., Singh, A. & Karwasra, S.P.S. (1993). Composition of some city waste waters and their effect on soil characteristics. *Annals of Biology*, 9: 239-245.

Narwal, R.P.; Singh, M. & Gupta, A.P. (1988). Effect of different sources of irrigation on the physico-chemical properties of soil. *Indian Journal of Environment and Agriculture*, 3: 27-34.

Otis, R.J. (1984). In onsite treatment. 4th *National Symposium on Individual and small community sewage systems*, New Orleans, Lousiana, December 10-11, 1984.

Rattan, R.K., Datta, S.P., Singh, A.K., Chonkar, P.K. & Suribau, K. (2001). Effect of long-term application of sewage effluents on available nutrient and available water status in soils under Keshopur effluent irrigation scheme in Delhi. *Journal of water Management* 9: 21-26.

Seaker, E.M. and Sopper, W.E. (1988). Municipal sludge for minespoil reclamation. 1. Effects of microbial population and activity. *Journal of Environment Quality* 17: 591-597.

Seto, P. and Dengelis, P. (1978). Concept of sludge utilization on agricultural land. In: Sludge utilization and disposal conference. In: *Proceedings No. 6 Environment*, pp. 138-155, Ottawa, Canada.

Shuval, H.I.; Adin, A., Fattal, B., Rawitz, E. & Yekutiel, P. (1986). Wastewater irrigation in developing countries, *Health effects and technical solutions*, pp 325, World Bank Technical Report 51.

Singh, J. & Kansal, B.D. (1985a). Amount of heavy metal in the waste water of different towns of Punjab and its evaluation for irrigation. *Journal of Research Punjab Agricultural University*, 22: 17-24.

Singh, J. & Kansal, B.D. (1985b). Effect of long term application of municipal waste water on some chemical properties of soil. *Journal of Research Punjab Agricultural University*, 22: 235-242.

Singh, J. & Kansal. B.D. (1983). Accumulation of heavy metals in soils receiving municipal effluents and effect on soil properties on their availability. In: *Heavy metals in the Environment, 409-412*, CEC Consultants Ltd. Edinburgh, UK.

Tiwana, N.S., Panesar, R.S. and Kansal, B.D. (1987). *Proceedings of National Seminar on Impact of Environmental Protection for Future Development of India, pp 119-126*, Nanital, India.

Totawat, K.L. (1993). Ground water pollution adjoining a smelter's effluent stream. *Journal of Indian Society of Soil Science*, 41:804-806.

Tripathi, D.M.; Tripathi, S. and Tripathi, B.D. (2011). Implications of secondary treated distillery effluent irrigation on soil cellulase and urease activities. *Journal of Environment Protection*, 2: 655-661.

United States Salinity Laboratory Staff (USSL) (1954). *Diagnosis and improvement of saline and alkali soils*, Agriculture Handbook No. 60, U.S. department of Agriculture of agriculture, Washington, D.C.

Verma, K.B. & Kumar, P. (2004). Effect of agro based industrial effluents on the growth and development of wheat (*Triticum aestivum* L. var. RR21), maize (*Zea mays* L.) and mustard (*Brassica compestris* L.). *Indian Journal of Ecology*, 31: 93-96.

Zalawadia, N.M., Raman, S. & Patil, R.G. (1997). Influence of diluted spent wash of sugar industries on yield and nutrient uptake by sugarcane and changes in soil properties. *Journal of Indian Society of Soil Science,* 45:767-769.

# 4

# Recent Advances in Paper Mill Sludge Management

Marko Likon[1] and Polonca Trebše[2]
[1]*Insol Ltd, Postojna,*
[2]*University of Nova Gorica, Nova Gorica,*
*Slovenia*

## 1. Introduction

The management of wastes, in particular of industrial waste, in an economically and environmentally acceptable manner is one of the most critical issues facing modern industry, mainly due to the increased difficulties in properly locating disposal works and complying with even more stringent environmental quality requirements imposed by legislation. In addition, in recent years the need to achieve sustainable strategies has become of greater concern, also because some traditional disposal options, such as landfill, are progressively restricted, and in some cases banned, by legislation. Therefore, the development of innovative systems to maximize recovery of useful materials and/or energy in a sustainable way has become necessary. Industrial wastes are generated through different industrial processes or energy production utilities as additional materials. Industrial symbiosis theory defines non-deliberately produced material as by-products or valuable raw materials which can be exploited in other industrial avenues. Paper industry is a strategic industry in many countries but in the same time, the production of paper consumes high quantities of energy, chemicals and wood pulp. Consequently, the paper production industry produces high environmental emission levels mainly as $CO_2$ due to energy consumption, or solid waste streams which include wastewater treatment sludges, lime mud, lime slaker grits, green liquor dregs, boiler and furnace ash, scrubber sludges and wood processing residuals. In terms of volume, most solids or liquids are those from the treatment of effluents, although waste from wood is also produced in large quantities.

In this chapter the different processes and technologies for conversion of paper mill sludge (PMS) into valuable products with the special emphasis on the technology for conversion of PMS into absorbent for oil spills sanitation, are discussed. The environmental impact of the latest is evaluated with Life Cycle Assesment (LCA).

## 2. Trends in paper mill sludge generation

The pulp and paper industry produce over 304 million tons of paper per year. In 2005, soley in Europe, 99.3 million tons of paper was produced which generated 11 million tons of waste and represented 11 % residue in relationship to the total paper production. The production of recycled paper, during the same period, was 47.3 million tons, generating 7.7 million tons of solid waste which represented 16 % of the total production from this raw material (Monte et al., 2009). Global production in the pulp and paper industry is expected to increase by 77 % by 2020 and over 66 % of paper will be recycled at the same time (Lacour, 2005). On average, the majority of waste generated from paper production and recycling is PMS, which is a by-product of up to 23.4 % per a unit of produced paper, the quantity depending on paper production process (Miner, 1991). The countries joined in the CEPI organization itself produce more than 4.7 million tons of PMS per year and global production of PMS was predicted to rise over the next 50 years by between 48 and 86 % above the current level (Mabee and Roy, 2003). This represents an enormous environmental burden due as more than 69 % of the generated PMS is landfill disposed (Progress in Paper Recycling Staff, 1993; Mabee and Roy, 2003). Flowchart of PMS generation is shown on Figure 1.

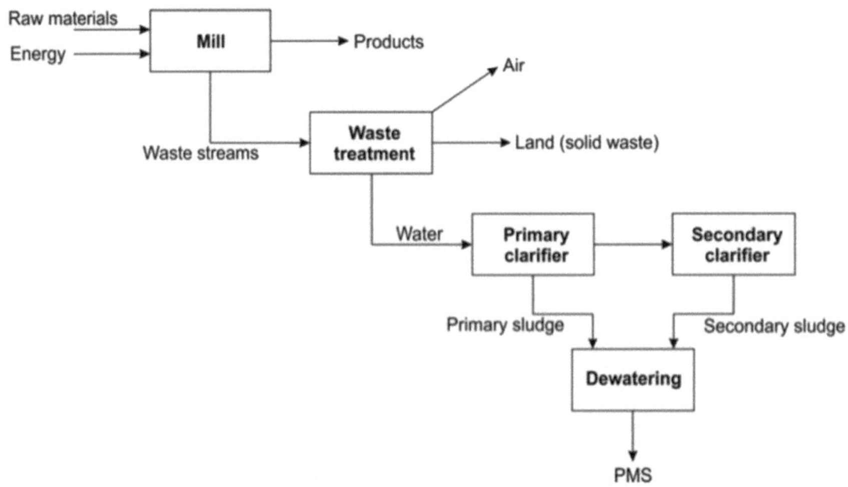


Fig. 1. Scheme of PMS generation

| | PMS origin | Organic content wt % | Ash content wt % | Heat value MJ/kg | CEC cmol/kg |
|---|---|---|---|---|---|
| Low-ash sludges | Primary sludge from pulp mill producing pulp from virgin wood | 94.31 | 5,69 | 20,1 | 30,20 |
| | Primary sludge from paper mill producing paper from virgin wood | 93,79 | 6,21 | 19,8 | 32,41 |
| High-ash sludges | Secondary sludge from paper mill producing paper fom recycled cellulose without deinking process | 67,23 | 32,77 | 16,5 | 17,30 |
| | Primary sludge from paper mill producing paper from recycled cellulose without deinking process | 64,72 | 35,28 | 14,2 | 33,58 |
| | Deinking sludge from papermill that recycled paper not recycled previously | 60,36 | 39,64 | 12 | 18,12 |
| | Deinking sludge from papermill producing newspaper | 59,30 | 40,70 | 12,2 | 19,06 |

Table 1. Composition, heating values and cation exchange characteristics (CEC) for different PMS origins (Méndez et al.; 2009).

Recyclers are producing two to four times more sludge as Virgin pulp mills but the characteristics of PMS vary, depending on pulp and paper mill processes and from raw materials entering into those processes. Produced sludge can be considered to fall into two main types: high-ash sludge (> 30% dry weight) and low-ash sludge (< 30% dry weight). High-ash sludges are chemical flocculation sludges generated by pulp mills, primary sludges generated by production of paper from recycled fibers and deinking sludges generated by paper mills, alternatively, low-ash sludge represents primary, secondary or biological sludges generated by pulp or paper mills (Table 1).

Primary and deinking PMS constitutes a mixture of short cellulosic fibers and inorganic fillers, such as calcium carbonate, china clay, and residual chemicals dissolved in the water (Figure 2).

Fig. 2. SEM image show complex structure of PMS (cellulose fibers and china clay).

Currently, PMS are considered to be a waste material. Increased recycling over the past two decades has consequently increased the amount of material that requires disposal. The paper industry implements several methods to dispose of the sludge that pulp and paper production generates (Figure 3).

## 3. Major concerns with paper mill sludge management

### 3.1 Landfilling

PMS with high organic content (Table 1) in landfill is subjected to aerobic and anaerobic decay. According to Buswell and Mueller (1952), 1 ton of low-ash PMS in landfill theoretically releases into environment approximately 2.69 tons of $CO_2$ and 0.24 ton of $CH_4$ (Likon et al., 2009). Currently, most PMS is dried, spread or deposited onto the landfill (Mabee, 2001). The landfills can be industrial, in that are constructed and operated by the mills, or they can be independently owned, requiring the mills to pay a "tipping fee" for sludge disposal. The European Landfill Directive (1999/31/EC) and upcoming bio-waste directive aims to prevent or reduce, as far as possible, the negative effects of waste landfill on the environment, by introducing stringent technical requirements for wastes and landfill. By increasing landfill fees up to 3.5 €/ton annually until the final landfill use fee will reach 40 €/ton. That fact and adaptation of paper mills to Integrated Pollution Prevention and Control Directive 1996/61/CE forced the paper mill industry to look for processes for their waste minimization as well new technologies for waste reuse or environmental friendly disposal.

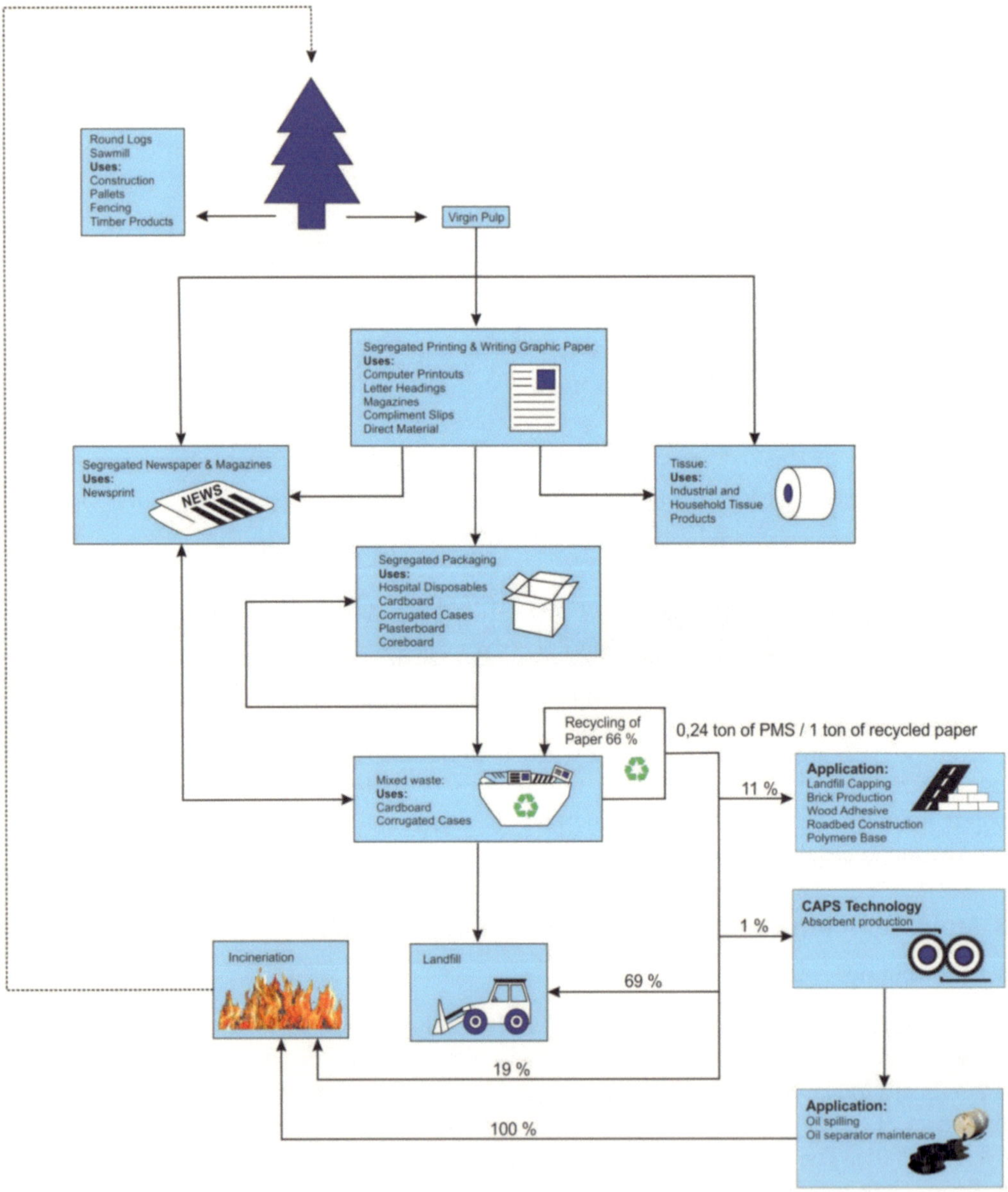


Fig. 3. PMS disposal pathways.

### 3.2 Landspreading

Land surface spreading is one of the possible methods for the application of sludge to forest or agricultural lands for soil enhancement (Christmas, 2002; Ribiero et al., 2010). PMS from pulp mills and secondary sludges are not hazardous due to heavy metals. On the other hand, sludge from deinking and waste paper mills may be relatively high in heavy metals due to the formulations used in ink removal. Compounds that can be found in mill sludges at concentrations above 10 mg/kg of dry sludge include naphthalene, phthalates, chloroform, PCBs, wood extractives or derivates and chlorinated lignin derivates (Thacker, 1985). Sludge as land spreading material must be used very carefully due to an unfavorable C:N ratio, high ion exchange capacity and possible Cr toxicity (Norris and Titshall, 2011). With maximum application rates of 12 tons per 405 ha, even a moderately sized mill will need a large area to spread its sludge (Shields et al., 1985; Legendre et al., 2004). The use of land spreading material adjacent to residential areas is questionable due to odor.

### 3.3 Composting

Composting is one of stabilization techniques used for prevention of uncontrolled PMS decay. The use of compost is questionable due to the same reasons mentioned in land spreading. The C:N ratio in the sludge is up to 930:1 and it is not appropriate for plant growth (Thacker, 1985; Scott and Smith, 1995; Monte et al., 2009). Composting of PMS requires a large land area and additional costs for storage and processing (Gea et al., 2005).

### 3.4 PMS for energy

In practice approx. 19 % of PMS is incinerated on the sites due to energy recovery, but the economics of incineration is questionable because PMS contains 30 to 50 % water and only 30 to 50 % of cellulosic fibers calculated on dry solids. For each additional 1 % of moisture content in PMS, the temperature of combustion must be 10°C higher due to process efficiency (Kraft, 1993). The European Waste Incineration Directive 2000/76/EC intends to prevent or reduce, as much as possible, air, water and soil pollution caused by the incineration or co-incineration of waste, reducing at the same time the risk to human health that incineration processes entail. However, under controlled conditions PMS can become a sustainable fuel for co-generation (Figure 4).

Many technologies to utilize different PMS for energy exploitation have been tested and used. That processes include efficient incineration in fluidized bed and circulating fluid bed combustion chambers (Nickull, 1991; Kraft, 1993; Busbin, 1995; Fitzpatrick and Seiler, 1995; Porteous, 2005; Oral et al., 2005) as well as more sophisticated processes for energy recovery as pyrolysis (Frederik et al., 1996; Kay, 2002; Fytili and Zabaniotou, 2008; Lou et al., 2011, Jiang and Ma, 2011), indirect steam gasification (Durai-Swamy et al., 1991; Demirbas, 2007), wet oxidation (Johnson, 1996; Kay, 2002; Maugans and Ellis, 2004; Fytiliand and Zabaniotou, 2008), super critical water oxidation (Modell et al., 1992; Dahlin, 2002; Kay, 2002) and gasification (CANMET, 2005). Advances and drawbacks of energy recovery from PMS are well described in peer reviewed literature (Monte et al., 2009).

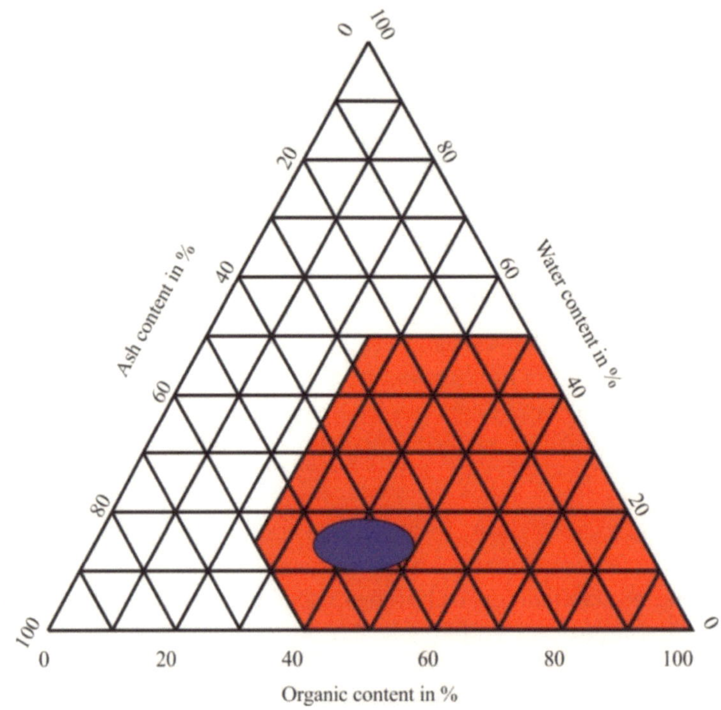


Fig. 4. Fuel triangle for waste from paper industry (blue spot – deinking PMS).

### 3.5 Utilization in brick, light aggregates and cement production

PMS containing a high inorganic fraction (Table 1 - high ash sludges) can be utilized in the production of building materials (Černec et al., 2005). Due to its combustion matrix, it can be used in the brick production industry. The addition of 5-15 % of PMS as raw materials improves both the final product and the processes. First, since its fiber content increases the porosity of the matrix, it enables the manufacture of lighter bricks; second, it saves fuel in the oven, decreasing firing time and makes the product more resistant against cracking during the drying and firing stages (Monte et al., 2009; Furlani et al., 2011). The same advantages can be used in the production of light aggregates for the building industry. (Ducman and Mirtic, 2011). A similar exploitation has been noticed in cement industry. PMS with high organic content (Table 1 - low ash sludges) has an energy level that makes it an efficient alternative fuel in the production of Portland cement. Currently it is classified as Class 2 (liquid alternative fuels) in the Cembureau classification of alternative fuels (Dunster, 2009). Including up to 20 % of deinking sludge into mortar improves the mortar mechanical properties (Yan et al., 2011).

### 3.6 Landfill capping material

Significant progress has been made in the use of paper mill sludge as a material for land fill cover by replacing the clays or geo-composites. PMS behaves similar to a highly organic soil and has good chemical, hydrodynamic and geotechnical properties which make it an efficient impermeable hydrodynamic barrier for the land field cover (Zule et al., 2007).

Combining the paper mill sludge as hydrodynamic impermeable barrier with metal scoria as an oxidizing layer an efficient landfill capping system can be build. More than 21 different studies have been carried out in 2009 on the use of PMS in the landfill body (Rokainen et al. 2009). The characteristics of PMS have been exploited also for road bed construction for light loaded roads and tennis courts (Moo-Young et al., 1996).

## 4. Advanced research and utilisation of PMS

According to modern industrial trends on eco symbiosis economic efficiency is key factor for planning modern technological processes. This includes decreasing the waste streams during production and the use of produced waste as by-products or raw materials with higher added values. As can be seen from Figure 2 PMS is a chemically and physically complex material which can be used in different industrial applications. One of the promising innovations is the use of PMS as heat insulation material with a thermal conductivity factor lower than 0.055 $W/m^2K$. This is comparable to insulation materials available on the market currently. The US company, Minergy, has implemented technology for incineration of PMS under controlled conditions. The product of incineration is a highly efficient water absorbent composed of meta-kaoline granules. Using the PMS, generated by pulp mill, as replacement for virgin fibers in production of fiber-cement sheets is already tested in pilot scale. The cost benefits analysis showed environmental and economic benefits for PMS generator as well as fiber-cement sheets producer (Modolo et al., 2011). Using the PMS for production of construction and insulation boards has been topic of The Waste & Resources Action Programme (Goroyias and Elias; 2004).

The use of the PMS as a paper and wood adhesive is an interesting concept under investigation (Geng et al., 2007). The use of PMS as cat litter is in production stage but without serious market demands. PMS can be dried in mixture with pesticides or fertilizers. Dried mixtures can be used as a pesticide or fertilizer carrier in agriculture which allows more controlled releasing active substances into the soil (Dongieux, 1999). Contamination associated with the sludge would be the ongoing threat in such a disposal route. Conversion of the lignocellulosic portion of PMS into methyltetrahydrofuran, which can be mixed with ethanol and natural gas into cleaner burning fuel, is already developed and has been found economically viable. Bio-conversion of PMS into ethanol is less viable due to toxic impurities in deinking sludges.

## 5. Utilisation PMS for sorbent production

Promising research has been conducted to use PMS as an oil sorbent material and use of the PMS as sorbent material is well documented, but currently the market was non receptive to such sorbent material due to cheap and efficient synthetic absorption material. The results of research studies have shown that PMS can be indirectly used as an active absorbent by converting it into activated carbon (Ben-Reuven, 2007). It can be used as binding material for the removal of heavy metals ions from water (Battaglia et al., 2003; Calace et al., 2003, Hea et al., 2010; Ahmaruzzaman, 2011), removal of phenols (Calace et al., 2002) and as an absorbent for hard surfaces cleaning (Lowe et al., 1988; Eifling and Ebbers, 2006). A variety of the processes and different absorbent products have been developed for commercial purposes. One of the processes which have been developed for the production of a floor absorbent, in the form of a granular product is known as the KAOFIN process and it is described in U.S.

Patent 4343751 (Naresh, 1980). In U.S. Patent 4374794 the sludge is evaporated, extruded into pellets and dried at temperatures ranging from 100 °C to 150 °C, in order to form an oil absorbent material (Kok, 1983). However, modern industry faces frequent and serious oil spills and subsequent sanitation demands high costs for sorbent materials. Offering a cheap and efficient natural material such PMS could become a welcome solution. The CAPS (Conversion of paper mill sludge into absorbent) is an eco-innovation solution in the 'market uptake' phase, therefore prior to expanding industrialization in Slovenia and Finland. The CAPS process uses the surpluses of the thermal energy which paper mills usually waste into the environment for sorbent production. In addition, CAPS uses paper mill waste as a secondary raw material and converts it into a high added value absorbent. The technology is relatively cheap, simple and easily replicable particularly in markets with a developed paper industry. It is based on drying of PMS to the point where it can be efficiently mechanically and/or chemically treated to release cellulosic fibers from its inorganic matrix. The humidity of the deinking and primary paper mill sludge lies between 50 to 70 %, whereas the content of cellulosic fibers is approx. 52 %, the remainder is inorganic. After drying between 70 - 80 % of the solid content, PMS proceeds through special mechanical treatment (unraveling). This stage is crucial for the entire process due to the fact that in this section cellulose fibers are released from the inorganic matrix, which in turn allows material to float on the fluid surface (Figure 5 and 6).

However, the mechanical treatment expanded the surface area, but the sorbency was not linear with regards to the surface area (Table 2).

| | Avg.particle size mm | Active surface area $m^2/g$ | Sorptivity g/g |
|---|---|---|---|
| Untreated PMS | 10 | 4.8096 | 1.23 |
| PMS Grinded | 4.18 | 3.2048 | 2.07 |
| PMS Unraveled | 1.67 | 36.0526 | 4.4 |
| PMS Fluffed | 0.7 | 2.9626 | 7.12 |

Table 2. Particle size, active surface area and sorbency of used mineral oil versus mechanical treatment of PMS.

Fig. 5. Raw PMS (left), unraveled PMS (middle) and fluffed PMS (right).

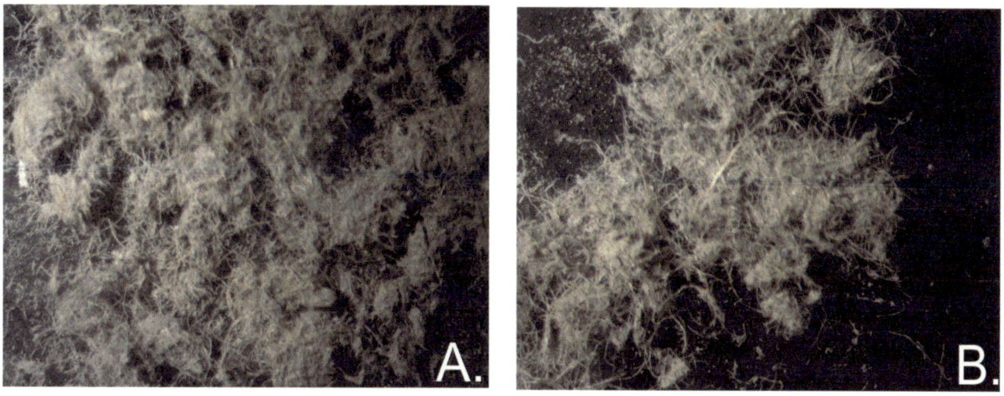


Fig. 6. Microscope pictures of raw PMS (A) and unraveled PMS (B) at 8 x magnification.

As can be seen from figure 7 the same trend was observed in sorbency of different substrates.

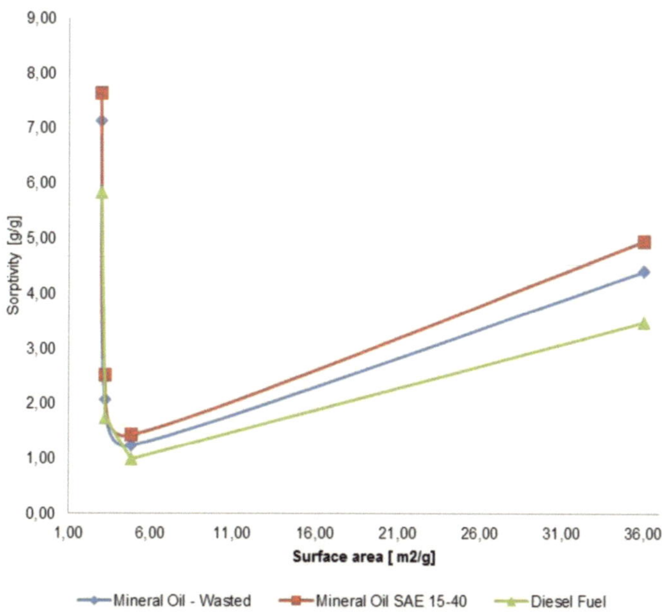


Fig. 7. Active surface area of PMS absorbent samples vs. mechanical treatment.

The mechanical treatment was clearly connected to the breaking of the inorganic matrix into dust. The dust dropped off from the absorbent, which led to the shrinking of the surface area but made the remainder of the surface more accessible to the substrate. More violent mechanical treatment during the production (fluffing) of the absorbent led the isothermal sorption of the absorbent to become similar to the isothermal sorption of the paper standard. The results showed that the appropriate treatment process (especially the mechanical treatment phase) was of the utmost importance for the conversion of the PMS into a sorbent. The possible explanation for this appears to be that the PMS consisted of a fragile net structure composed of cellulosic fibers and an inorganic core, which served as an anchor for linking the cellulosic fibers together. The inorganic core played an important role in the sorption process because it served as absorption points for the hydrophobic substances, while the cellulosic fibers served as a floating skeleton. Chemical treatment (esterification, silanisation) is an option when higher absorption capacity and better buoyancy is required (Likon et al., 2011). Distributed (chemically prepared) paper mill sludge is dried at 130 °C to 150 °C until the humidity oscillates between 1% and 10% in order to obtain a final product with active surface area 36 $m^2/g$, sorption capacity up to 8 g oil/g PMS and capacity to float on the water surface.

Produced sorbent has a calorific value around 3.8 MJ/kg and when it is completely soaked by oil it has a calorific value up to 33.5 MJ/kg and can be used as high quality fuel in the co-generation processes. Incineration under controlled conditions leads to the conversion of the kaolin portion of paper mill sludge into a meta-kaolin substance in the form of vitrified granules. These vitrified granules can be reused as an inert hydrophilic absorbent.

The adsorption kinetic shown (Figure 8), that the mechanism of PMS sorption follows complex combinations of interparticle diffusion at the first stage of the process followed by pseudo-second order adsorption of oil into pores of the inorganic part of PMS.

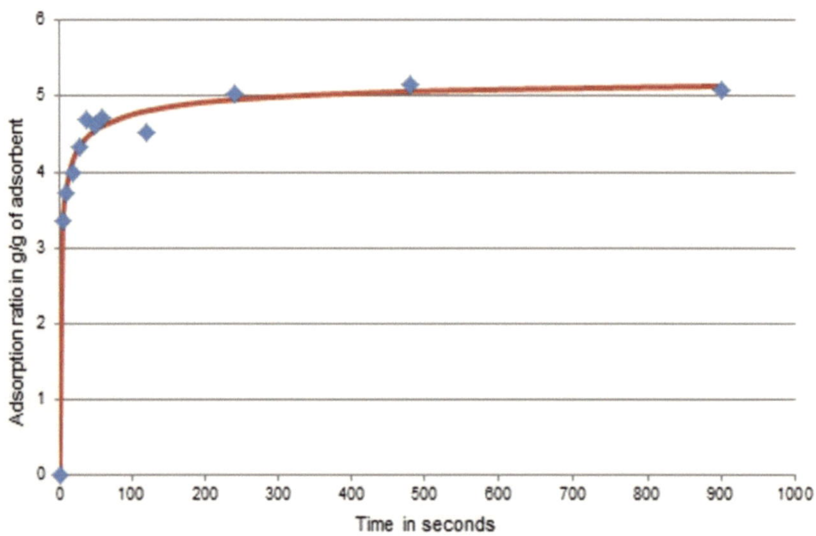


Fig. 8. Sorption kinetics curve for mineral motor oil (SAE 15-40W) sorbing on PMS at 20°C.

In the loose state of adsorbent the adsorption equilibrium is reached after 240 seconds at 20°C and as can we see from the figure 9 PMS absorbs more than 95 % of oil from a water surface within 70 seconds.

## 6. Environmental impact assessment of PMS sorbent versus expanded polypropylene sorbents

LCA model based on figure 10 showed that conversion of PMS from paper production can expand paper life cycle for additional step and helps closing the life cycle in paper production processes. A model presenting the life cycle assessments (LCA) of PMS sorbent

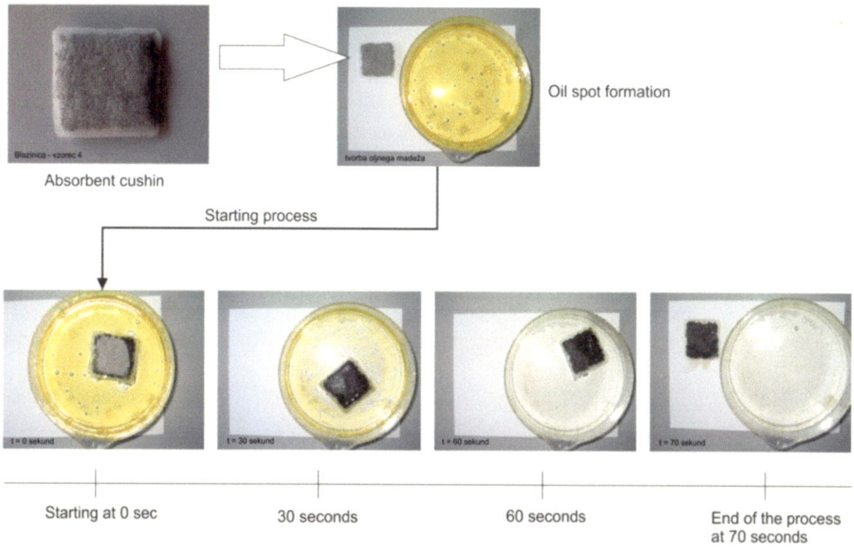


Fig. 9. PMS sorption efficiency in the case of sorption of transformer oil from a water surface (Likon et al. 2011).

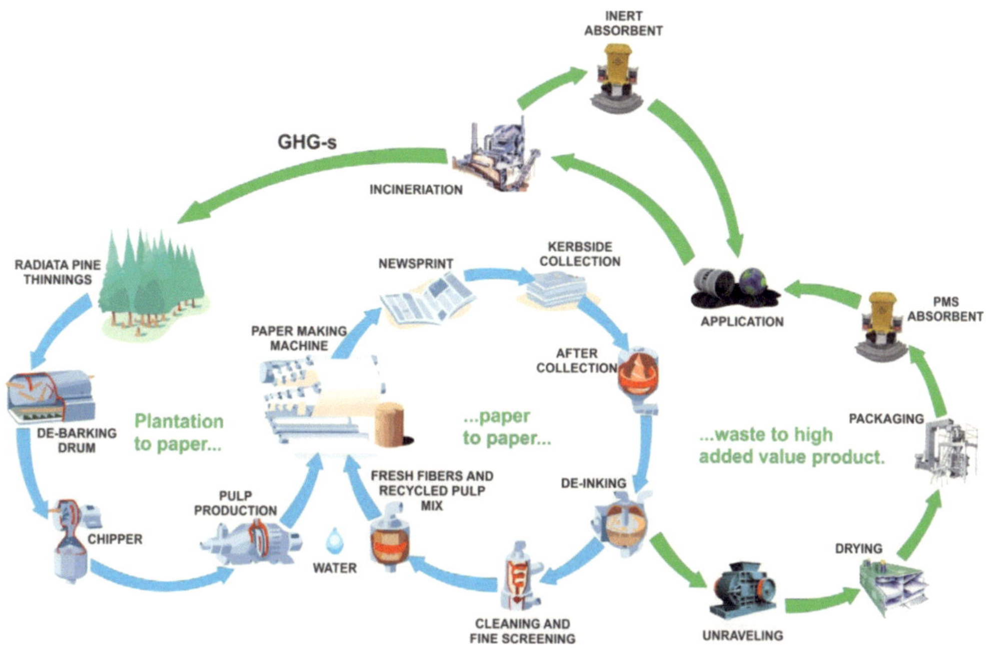


Fig. 10. Life cycle of paper in the case of conversion of PMS into sorbent material.

versus expanded polypropylene sorbents, is a basis for discussion about the sustainability of conversion of industrial waste from the paper industry into value added products. LCA of PMS and expanded polypropylene (EPP) were accomplished by using the BUWAL 250 database included in the software package SimaPro 7.1, and data from peer reviewed literature (Lyondell Basell, 2009; Binder and Woods, 2009; Tabone et al., 2010; Felix et al., 2008).

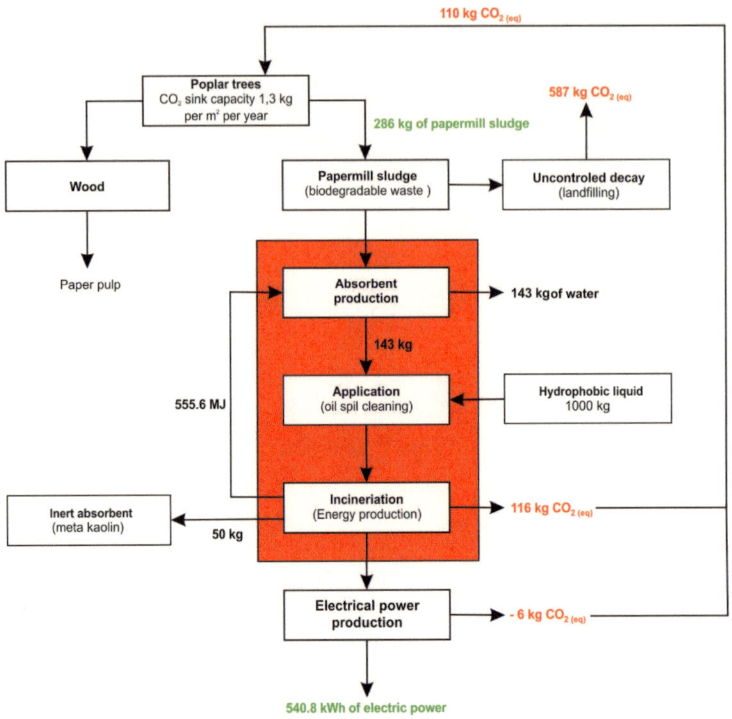


Fig. 11. Life cycle circle for conversion of PMS into absorbent material.

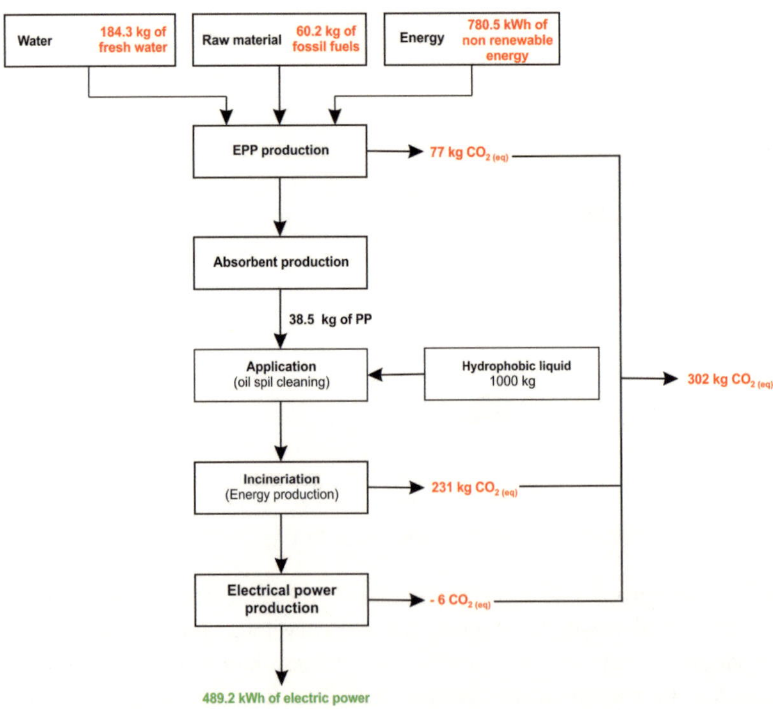


Fig. 12. Life cycle circle production and application of EPP absorbent.

The data for the uncontrolled anaerobic decay were calculated according the equation for anaerobic decay (Buswell and Mueller, 1952) and the data for energy production were calculated according to Martin et al. (Martin et al., 2003). Paper mill sludge was considered as unwanted by-product in the paper and cardboard recycling processes that is without water and energy consumption and without GHG-s generation for its production. EPP was considered as absorbent produced from non-renewable raw materials (propene). The calculations were based on the quantity of absorbent required for the absorption of 1000 kg of oil with the following presumptions: the sorption capacity of absorbent produced from PMS and from EPP were 7 kg of oil/kg sorbent and 26 kg of oil/kg sorbent, respectively. It was considered that used sorbents were incinerated for energy recovery. Absorbed hydrophobic substances were not included in LCA calculations.

As can be seen from Figures 11 and 12, the production and application of PMS as a sorbent for oil spill sanitation reduced carbon footprint by 2.75 times when compared to the production and application of EPP. In addition, considering the production of absorbent instead of landfill disposal the PMS decreased carbon footprint by 5.25 times. Altogether, production of the sorbent material from PMS instead of landfill disposal it and replacing the synthetic EPP absorbent with PMS absorbent for oil spill sanitations reduced the carbon footprint by more than 14 times. The difference in the water balance was 372.3 kg calculated on the quantity of sorbent needed for sanitation of the 1000 kg of oil. The difference in the water balance was due to the consumption of the 184.3 kg of fresh water for EPP production and due to the production of 143 kg of clean technological water in the process of energy recuperation during PMS sorbent production. The energy balance showed the surplus of energy in the production of the PMS absorbent was due to the production of a combustible product from waste and negative energy balance in the case of the production and use of EPP absorbent. The later was due to heavy energy consumption during production of PP from fossil fuels. Additionally, the LCA analysis has shown that conversion of PMS into an absorbent prolongs life cycle of paper products for an additional two cycles and efficiently closes the life cycle circle of paper.

## 7. Conclusion

The modern sustainable management of production processes should be based on the industrial ecology approach, of which an essential element is the eco-symbiosis theory. Pulp and paper industry producing enormous quantities of solid waste what presents huge environmental burden. Appropriate managing with such a waste is most crucial task for modern pulp and paper industry. Many innovative approaches for conversion of the PMS into useful materials have been done in past two decades, but for many of them the markets demands have been too small for successful diverting of PMS from the landfill disposal. The CAPS process for the conversion of PMS into a sorbent for water surface cleaning fulfills all of above mentioned requirements. Moreover, it also includes the surpluses of thermal energy which paper mills usually transmit into the environment. The manufactured natural sorbent may be used by the oil, chemical, logistic and transport industries as well as public bodies such as fire brigades, civil protection and disaster relief institutions. These are institutions which all require an environmental friendly, efficient, cheap and at the same time sustainable product for cleaning of oil spills from water surfaces and/or for oil separators maintenance and on the other hand the pulp and paper industry will expand their product portfolio by using waste as raw material. The main drawback of placing the PMS sorbent on the market is its slow degradation in water. That obstacle can be overcome by chemical treatment, proper application and by mixing PMS sorbent with floating materials but those processes raise the production costs and makes PMS absorbent unattractive to the market. Additional improvement must be made in the future to overcome water degradation of the PMS sorbent and with success the industry can obtain a sustainable, cheap sorbent for oil spill sanitation in unlimited quantities.

## 8. References

Ahmaruzzaman, M. (2011). Industrial wastes as low-cost potential adsorbents for the treatment of wastewater laden with heavy metals. Advances in Colloid and Interface Science, Vol. 166, pp. 36-59.

Battaglia, A., Calace, N., Nardi, E., Maria Petronio, B.M., Pietroletti, M. (2003). Paper mill sludge–soil mixture: kinetic and thermodynamic tests of cadmium and lead sorption capability. Microchemical Journal, Vol. 75, pp. 97–102.

Ben-Reuven, M. (1997). Conversion of Paper-Mill Sludge Into Pelletized, Composite Activated Sorbent; Small Business Innovation Research (SBIR) - Phase I (1997), EPA, Washington, USA.

Binder, M., Woods, L. (2009). Comparative Life Cycle Assessment Ingeo™ biopolymer, PET, and PP Drinking Cups. Final report for Starbucks Coffee Company. Seattle 2009, US, 61 p.

Busbin, S.J. (1995). Fuel specifications-sludge. Environmental Issues and Technology in the pulp and paper industry. A TAPPI press anthology of published papers 1991-1994, pp.349-353.

Buswell, A.M., Mueller, H.F. (1952). Mechanism of Methane Fermentation, Ind. Eng. Chem., Vol. 44 (3), pp. 550-552.

Calace, N., Nardi, E., Petronio, B.M., Pietroletti, M. (2002). Adsorption of phenols by papermill sludges. Environmental Pollution, Vol. 118, pp. 315–319.

Calace, N., Nardi, E., Petronio, B.M., Pietroletti, M., Tosti, G. (2003). Metal ion removal from water by sorption on paper mill sludge, Chemosphere, Vol. 8 (51), pp. 797-803.

Campbell, A. (2007). Literature Review of Worms in Waste Management, Volume 2, Recycled Organics Unit, Vol 2, The University of New South Wales, 55 p.

CANMET Energy Technology Centre, 2005. Pulp and paper sludge to energy – preliminary assessment of technologies. Canada.

Christmas, P. (2002). Building materials from deinking plant residues – a sustainable solution. In: COST Workshop Managing Pulp and Paper Residues, Barcelona, Spain.

Černec, F., Zule, J., Može, A., Ivanuš, A. (2005). Chemical and microbiological stability of waste sludge from paper industry intended for brick production, Waste management & research, Vol. 2 (23), pp. 106-112.

Dahlin, J., 2002. Oxidation on deinking sludge in super critical water in practice. In: COST Workshop on Managing Pulp and Paper Residues, Barcelona, Spain.

Demirbas, A. (2007). Progress and recent trends in biofuels. Prog. Energ.Combust. Sci., Vol. 33 (1), pp.1-18.

Dongieux, P. (1999). Commercially Viable Products that Utilize Pulp and Paper Mill Sludge as a Raw Material Source. International Environmental Conference, TAPPI Proceedings, pp. 745-753.

Ducman, V., Mirtič, B., Lightweight Agreggate Processed from Waste Materials. In: Advances in Materials Science Research. Eds. Wythers, M.C., Vol. 4, pp. 307-323.

Dunster, A.M. (2009). Paper sludge and paper sludge ash in Portland cement manufacture. Characterisation of Mineral Wastes, Resources and Processing technologies - Integrated waste management for the production of construction material, University of Leeds, UK, 8 p.

Durai-Swamy, K., Warren, D.W., Mansour, M.N. (1991). Indirect steam gasification of paper mill sludge waste. TAPPI J., 137-143.

Eifling, R.B., Ebbers, H.J. (2006). Cellulose absorbent, U.S. Patents 7038104 B1.

Felix, E., Tilley, D.R., Felton, G., Flamino, E. (2008). Biomass production of hybrid poplar (Populus sp.) grown on deep-trenched municipal biosolids. Ecological Engineering, Vol. 33, pp. 8 - 14.

Fitzpatrick, J., Seiler, G.S. (1993). Fluid bed incineriation of papermill sludge. Environmental Issues and Technology in the pulp and paper industry. A TAPPI press anthology of published papers 1991-1994, pp. 369-376.

Frederik, W.M.J., Iisa, K., Lundy, J.R., O'Connor, W.K., Reis, K., Scott, A.T., Sinquefield, S.A., Sricharoenchaikul, V., Van Voren, C.A. (1996). Energy and materials recovery from recycled paper sludge. TAPPI J., Vol. 79(6), pp. 123-131.

Furlani, E., Tonello, G., Maschio, S., Aneggi, E., Minichelli D., Bruckner S., Lucchini E. (2011). Sintering and characterisation of ceramics containing paper sludge, glass cullet and different types of clayey materials. Ceramics International, Vol. 37, pp. 1293–1299.

Fytiliand, D., Zabaniotou, A., 2008. Utilization of sewage sludge in EU application of old and new methods - a review. Renew. Sustain.Energy, Vol. 12 (1), 116–140.

Gea, T., Artola, A., Sanchez, A. (2005). Composting of deinking sludge from the recycled paper manufacturing industry. Bioresource Technol., Vol. 96, pp. 1161-1167.

Geng, X., Deng, J., Zhang, S.Y. (2007). Paper mill sludge as a component of wood adhesive formulation. Holzforschung, Vol. 61 (6), 688–692.

Goroyias, G., Elias, R., Fan, M. (2004). Research into using recycled waste paper residues in construction products. WRAP Project PAP009-011. The Waste & Resources Action Programme, 185 p.

Hea, X., Yaoa, L., Lianga Z., Ni, J. (2010). Paper sludge as a feasible soil amendment for the immobilization of $Pb^{2+}$. Journal of Environmental Sciences, Vol. 22 (3), pp. 413-420.

Jiang, J., Ma, X. (2011). Experimental research of microwave pyrolysis about paper mill sludge. Applied Thermal Engineering, doi:10.1016/j.applthermaleng.2011.07.037.

Johnson, W., 1996. Wet oxidation for pulp and paper industry waste. Url: http://www.esemag.com/0796/oxidatio.html1996.

Kay, M. (2002). Development of waste management options for paper sludge. In: 4th Annual Dutch International Paper and Board Technology Event. Pira International.

Kok, J.M. (1983). Process for the preparation of a liquid-absorbing and shock-absorbing material, U.S. Patent 4374794.

Kraft, D.L., Orender, H.C. (1993). Considerations for ussing sludge as a fuel. Tappi, Vol. 76 (3), pp. 175-183.

Lacour P. (2005). Pulp and paper markets inEurope. Prepared for the United Nations Economic Commission for Europe and the Food and Agriculture Organization of the United Nations. September 26th. Available on-line at: http://www.unece.org/trade /timber/docs/tcsessions/tc-63/Presentations_MktDiscussions/Presentations_PDF/ 18_Lacour.pdf.

Legendre, B.L., Bischoff, K.P., Gravois, K.A. Hendrick, R.D., Arceneaux, A.E. (2004). Sugarcane yields and soil analysis following application of paper-mill sludge. Journal American Society Sugar Cane Technologists, Vol. 24, pp. 60-69.

Likon, M., Černec, F., Saarela, J., Zimmie, T.F., Zule, J. (2009). Use of paper mill sludge for absorption of hydrophobic substances, 2nd International Conference on New Developments in Soil Mechanics and Geotechnical Engineering, Near East University, Nicosia, North Cyprus, pp 141-152.

Likon, M., Černec, F., Svegl, F., Saarela, J., Zimmie, T.F. (2011). Papermill industrial waste as a sustainable source for high efficiency absorbent production. Waste Management, Vol. 6 (31), pp. 1350-1356.

Lou, R., Wu, S., Lv, G., Yang, Q. (2011). Energy and resource utilization of deinking sludge pyrolysis. Appl. Energy, doi:10.1016/j.apenergy.2010.12.025.

Lowe, H.E., Yoder, L.R., Clayton, C.N. (1988). Nonclay oil and grease absorbent, US. Patent 4734393.

LyondellBasell (2009). Polypropylene, Environmental Information Document Relevant to Australia. South Yarra Vic 2009, Australia, 6 pages.

Mabee, W. (2001). Study of woody fibre in papermill sludge. Doctor thesys.University of Toronto, Canada, 202 p.

Mabee, W., Roy, D.N. (2003). Modeling the role of papermill sludge in the organic carbon cycle of paper products, Environmental Reviews, Vol. 1 (11), pp. 1-16.

Maugans, C.B., Ellis, C., 2004. Age old solution for today's SO2 and NOx. Pollut. Eng..

Martin, F.M., Roberdo, F.G., Osado,I.I., Ortiz, S.V. (2003). CO2 fixation in poplar-214 plantations aimed at energy production. In: Council on Forest Engineering (COFE) Conference Proceedings:"Forest Operations Among Competing Forest Uses, Bar Harbor.

Miner, R. (1991). Environmental Considerations and Information Needs Associated With an Increased Reliance on Recycled Fiber, In Focus 95+ Proceedings, TAPPI PRESS, Atlatnta, pp. 343-362. Mill waste materials. Journal of analytical and applied pyrolysis, Vol. 86, pp. 66-73.

Méndez, A., Fidalgo, J.M., Guerrero, F., Gascó, G. (2009). Characterisation and pyrolysis behavoiur of different paper

Modell, M., Larson, J., Sobczynski, F. (1992). Supercritical water oxidation of pulp mill sludges. TAPPI J., 195–202.

Modolo, R., Ferreira, V.M., Machado, L.M., Rodrigues, M., Coelho, I. (2011). Construction materials as a waste management solution for cellulose sludge. Waste Management, Vol. 31, pp. 370–377.

Monte, M.C., Fuente, E., Blanco, A., Negro, C. (2009). Waste management from pulp and paper production in the European Union. Waste Management, Vol. 29, pp. 293-308.

Moo-Young, H.K., Zimmie, T.F., (1996). Geotechnical Properties of Paper Mill Sludges for Use in Landfill Covers, J. of Geotechnical Eng., Vol. 9 (122), pp. 768-755.

Monte, M.C., Fuente, E., Blanco, A., Negro, C. (2009). Waste management from pulp and paper production in the European Union. Waste management, Vol. 29, pp. 293-308.

Naresh, K. (1980). Clay aggloomeration process. U.S. Patent 4343751.

Norris, M. Titshall, L.W. (2011). The Potential for Direct Application of Papermill Sludge to Land: A greenhouse study. Int. J. Environ. Res., Vol. 5(3), pp. 673-680.

Nickull, O., Lehtonen, O., Mullen, J. (1991). Burning mill sludge in a fludized-bed incineriator and waste-heat-recovery system. Tappi, Vol. 76 (3), pp. 119-122.

Oral, J., Sikula, J., Puchyr, R., Hajny, Z., Stehlik, P., Bebar, L. (2005). Processing of waste from pulp and paper plant. J. Cleaner Production, Vol. 13, pp. 509-515.

Porteous, A. (2005). Why energy from waste incineriation is an essential component of environmentally responsible waste management. Waste Management, Vol. 25, pp. 451-459.

Progress in Paper Recycling Staff (1993). Utilisation of Mill Residue (Sludge). Mill Survey: Progress in Paper Recycling, Vol. 3(1), pp. 64-70.

Ribeiroa, P., Albuquerquea, A., Quinta-Novab, L., Cavaleiroa, V. (2010). Recycling pulp mill sludge to improve soil fertility using GIS tools. Resources, Conservation and Recycling, Vol. 54 (12), pp. 1303-1311.

Rokainen, N., Kujala, K., Saarela, J. (2009). Use of industrial by-products in landfill cover, Paper in Proceedings Sardinia 2009, Twelth International Waste and Landfill Symposium, S. Margarita di Pula, Caligari, Italy, pp. 5-9.

Scott, G.M., Smith, A. (1995). Sludge Characteristics and Disposal Alternatives for the Pulp and Paper Industry. In: Procedings of the 1995 International Environmental Conference, TAPPI PRESS, pp.269-279.

Shields, W.J., Huddy, M.D., Somers, S.G. (1985). Pulp Mill Sludge Application to a Cottonwood Plantation, Cole, D.W., Henery, C.L. and Nutter, W.L., eds., The Forest Alternative for Treatment and Utilization of Municipal and Industrial Waste, University of Wasinghton Press, Seattle, WA, pp.533-548.

Tabone, M.D., Cregg, J.J, Beckma, E.J., Landis, A.E. (2010). Sustainability Metrics: Life Cycle Assessment and Green Design in Polymers. Environ. Sci. Technol., Vol. 44 (21), pp. 8264–8269.

Thacker, W.E. (1985). Silvicultural Land Application of Waste-water and Sludge From the Pulp and Paper Industry. In: Cole, D.W., Henery, C.L. and Nutter, W.L., eds., The Forest Alternative for Treatment and Utilization of Municipal and Industrial Waste, University of Wasinghton Press, Seattle, WA, pp.41-54.

Zule, J., Černec, F., Likon, M. (2007). Chemical properties and biodegradability of waste paper mill sludges to be used for landfill covering. Waste management & research, Vol. 6 (25), pp. 538-546.

Yan, S., Sagoe-Crentsila, K., Shapiro, G. (2011). Reuse of de-inking sludge from wastepaper recycling in cement mortar products. Journal of Environmental Management, Vol. 92 (8), pp. 2085-2090.

# 5

# Finite-Dimensional Methods for Optimal Control of Autothermal Thermophilic Aerobic Digestion

Ellina Grigorieva[1], Natalia Bondarenko[2],
Evgenii Khailov[2] and Andrei Korobeinikov[3]
[1]*Texas Woman's University*
[2]*Lomonosov Moscow State University*
[3]*University of Limerick*
[1]*USA*
[2]*Russia*
[3]*Ireland*

## 1. Introduction

An activated sludge process (ASP) is a biochemical process employed for treating sewage and industrial waste-water, which uses microorganisms and air (or oxygen) to biologically oxidize organic pollutants, producing a waste sludge (or floc) containing the oxidized material. The optimal operation of the biological treatment processes is a challenging task because of the stringent effluent requirements, the complexity of processes as an object of control and the need to reduce the operation cost. US law has strict requirements on the effluent quality of the ASP; similar strict requirements were adopted during the last decade in Europe and in South Africa (Maine Department of Environmental Protection, Augusta, ME, 2010; Tzoneva, 2007).

Polluted water contains a wide range of inorganic and organic chemical species and microorganisms, including pathogens. Filtered concentrated sludge is produced at initial stages of the treatment. The objectives of the key subsequent stage are eliminating potential pathogenic micro-organisms and reducing organic chemical content, which might act as a substrate for further microbial growth, to an acceptable level. Autothermal thermophilic aerobic digestion (ATAD) is a process that is widely used to achieve these goals. ATAD makes use of bacterial growth within the sludge both to reduce organic chemical content and to kill pathogenic bacteria; aeration of the sludge promotes the growth of aerobic bacteria, which feed on and reduce the organic substrates in the sludge. A review of the ATAD origin, design and operation can be found in Bojarski et al. (2010); Capon-Garcia et al. (2010); Graells et al. (2010); Layden (2007); Rojas & Zhelev (2009).

### 1.1 Waste-water treatment plant operation

This subsection is extracted from Rojas et al. (2010).

ATAD is operated as a batch or semi-batch process. A large reactor containing sludge receives an additional volume of untreated sludge at the start of a batch via a feed inlet (see Figures 1 and 2). During the batch reaction air is pumped continuously into the reactor providing both the oxygenation required for aerobic bacterial growth and the mechanical mixing of the sludge. Digestion of organic substrates proceeds with bacterial growth, predominantly of thermophiles due to elevated temperature. At the end of the batch period, a fraction of the treated volume is removed and is immediately replaced by the next batch of intake sludge. Thus the outflow sludge has the same composition as the sludge in the reactor at the end of the batch reaction time $t$. Batch outflow and inflow cannot be opened at the same time to prevent the untreated inflow sludge might from be drawn off with and taint the outflow sludge. The time between successive batch intakes is typically fixed at 24 hours for staffing

Fig. 1. External view of a reactor.

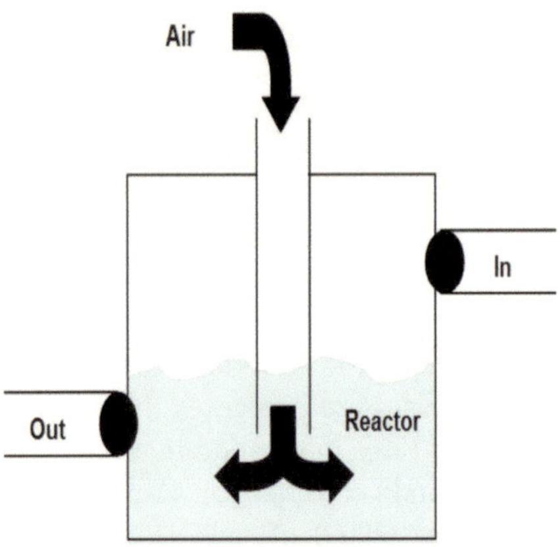


Fig. 2. Scheme of the reactor for waste-water treatment.

reasons. In order to achieve the desired treatment outcomes, treatment plant designs may include a single ATAD reactor stage (as is shown in Figure 1), or two reactors in series with outflow from the first reactor being inflow into the second.

### 1.2 Models of ASP and ATAD

In order to improve efficiency of ATAD process, in recent decades various control strategies for the ASP have been developed by engineers. Simple strategies are typically limited to the maintenance of some desired values for easily determinable process parameters, such as food-microorganism ratio, sludge recycle flow rate, or oxygen concentration in the aeration basin (Holmberg, 1982). More complex models take into consideration that the process also depends on a number of working conditions, such as air pump power to regulate the mean oxygen concentration (Rinaldi et al., 1979). Optimal working conditions and control strategies are usually established with a support of mathematical models (Brune, 1985; Kobouris & Georgakakos, 1991; Moreno, 1999; Moreno et al., 2006; Tzoneva, 2007); relevant studies and comparison of control strategies have been done by Debuscher et al. (1999); Fikar et al. (2005); Lindberg (1998); Lukasse et al. (1998); Potter et al. (1996); Qin et al. (1997).

Obviously, a particular solution depends on the model that was employed; over the last 50 years a variety of models of different level of complexity were proposed for ASP. It should be taken into consideration, however, that there is considerable uncertainty in inferring both the functional description of processes and the accuracy of parameterizations when applying laboratory-derived models to a real-life system. In particular, many existing models of the waste-water treatment are motivated by microbiological and engineering interest and have been constructed with an intention to incorporate as much as possible of the apparent understanding of underlying processes. As a result, such models tend to include too many variables, processes and parameters, which are known with a substantial degree of uncertainty.

Laboratory scale experiments provided understanding (to varying degrees) of many of the microbiological and chemical processes involved. By encapsulating much of this understanding in mathematical models, engineers provided tools for computer simulations that can be used in the design process.

The "Activated Sludge Model 1" (ASM1) (Henze et al., 2000) has achieved a broad level of acceptance in the waste-water treatment community. Based on the dominance of the ASM1 model in the field, its extension to the ATAD process as set out by Gomez et al. (2007) appears to create a de facto standard for modeling this process. The drawbacks of the model are also obvious (Rojas et al., 2010). The full-scale ASM1 model incorporates large numbers of variables and functional responses, and consequently requires many parameters. The motivation for including these particularities in the model is predominantly microbiological, and seems to arise from the urge to include all possible components about which there is some knowledge. However, details for many of these functional responses are unknown, and the accuracy of parameter values is uncertain: they are dependent on the chemical and microbiological make-up of the particular sludge, and therefore typically used as fitting parameters. With such a highly parameterized model, sets of parameter values can be found to fit most data, but it is unclear whether the model remains reliable in such situations. Applying a parameterized model under different (but physically reasonable) operating conditions may give spurious modeling artifacts. This may also cause numerical difficulties in attempting any optimization. Moreover, a mathematical model is not very useful, if it cannot be completely investigated using existing optimization methods. Mathematical optimization can be conducted with the use of optimal control theory.

### 1.3 Optimal control of nonlinear models

Optimal control problem for a system of ordinary differential equations is the problem of finding an extremum of a functional (objective function) under differential restrictions in the infinite dimensional functional space (for example, the space of all bounded piecewise continuous functions). This is a challenging mathematical problem. There are a number of approaches that allow to reduce the problem to the analysis of a finite dimensional problem; one such effective approach is the Pontryagin Maximum Principle (PMP) (Pontryagin et al., 1962), which is a necessary condition for optimality in the original infinitely dimensional problem. It reduces the optimal control problem to the analysis of two point boundary value problem of the Maximum Principle. This boundary value problem is finite in the sense that its solutions, phase vector and adjoint variable, depend either on the initial or terminal conditions, which belong to the elements of a finite dimensional Euclidean space. The optimal control is uniquely determined by the so-called maximum condition. Therefore, a desirable control depends on the variables selected as the initial or terminal conditions of the phase vector and the adjoint variable. Thus, the considered optimal control problem can be reduced to a finite problem, and the well-known methods can be applied to solve it (Fedorenko, 1978). Of course, these arguments exclude from considerations possibilities such as singular arcs in the corresponding optimal control (Bonnard & Chyba, 2003), or situations when the control has an infinite number of switchings on a finite time interval (Bressan & Piccoli, 2007).

Under the assumptions above, look at the boundary value problem of the Maximum Principle on the other side. For a system of ordinary differential equations which is linear in controls, the maximum condition almost everywhere uniquely determines the optimal control in the form of a piecewise constant function with a finite number of switchings. If we take the moments of switching as the corresponding variables, then the phase vector and adjoint variable become dependent on these switchings, and the considered optimal control problem becomes finite in the sense that the required control belongs to the class of piecewise constant functions with an estimated finite number of switchings determined from the analysis of the maximum condition. Grigorieva & Khailov (2010a;b), which are associated with the problem of optimizing the waste-water treatment process, support these arguments. The considered model, which is due to Brune (1985), is sufficiently simple to be investigated analytically, while it adequately describes the principal features of the ASP and water cleaning control process Brune (1985) formulated an optimal control problem for the minimization of the waste concentration in the ASP and offered the PMP for its solution. However, the analysis of the corresponding boundary value problem for the Maximum Principle was not completed; instead the author offered a numerical solution to the problem for different piecewise constant controls. Grigorieva & Khailov (2010a;b) deal with the complete analysis of this model for a different objective functions. Grigorieva & Khailov (2010a) formulated the optimal control problem of minimizing the pollution concentrations at the terminal time and found optimal solutions for this. Authors also investigated how these optimal solutions depend on initial conditions and conducted numerical simulation of the ASP for different parameter values. Grigorieva & Khailov (2010b) stated an optimal control problem of minimizing the water pollution concentrations on a given time interval. The optimal solutions for this problem is found in two stages: firstly, the authors investigated the optimal control problem using the PMP. Secondly, an approach based on Green's Theorem (Hájek, 1991), was applied. As a result of this study the authors obtained possible types of optimal solutions. This study reviled a possibility of the existence of singular arcs at corresponding optimal trajectories (Bonnard & Chyba, 2003); this fact makes an analytical study of the problem extremely difficult. However, the approach based on Green's Theorem enabled the authors to prove the absence of singular arcs at optimal trajectories. In the same time, this approach overestimates the number of switchings for optimal controls. Ultimately, a combination of the results obtained applying these different approaches allowed to solve the considered optimal control problem analytically.

The model studied in these papers is not intended to be the finest or the most precise ASP model; on the contrary, it is robust and rather basic. However, it includes all the essential features of the process. Furthermore, the model is nonlinear and assumes a bounded control; these features make the model very interesting from the mathematical point of view.

An alternative approach, which also allows a reduction of the original optimal control problem to a finite dimensional problem, is a constructing of a finite dimensional parametrization by the moments of switching for some piecewise constant controls of the corresponding attainable set, which fully characterizes the behavior of the original control system. This approach can be applied to solve a range of problems, including:

• estimating of control opportunity, that is investigating where trajectories of the system can lead under different admissible controls;

• solving the controllability problem;

• solving optimal control problems with a terminal functional, such as the problem of minimizing pollution at a terminal time;

• solving a time optimal control problem (that is, how to move a system from an initial state to the terminal state for minimal time).

Methods for solving these problems, except the last one, with particular application to the optimal control for ATAD will be discussed in this Chapter. A time optimal control problem and optimal control problems with an integral functional, such as the problem of minimizing pollution for a specified time interval and the problem of minimizing energy consumption for aeration on a given time interval, are subjects of future research.

### 1.4 Rational and basic assumptions for modeling

The ATAD process is efficient but it can be costly, as it requires continuous aeration that generally is energy-consuming. Optimizing the aeration can significantly reduce the cost of operation. Existing ATAD models are far too large and complex, and this complexity prevents the application of the usual optimization techniques. Our aim is, therefore, to construct a model that retains the fundamental properties of the process (as well as these of the large scale existing models) while being sufficiently simple to allow the use of optimization procedures. With this intention in mind, we formulate and investigate a simplified model of the ATAD reaction based on the existing ASM1 model at thermophilic temperatures.

In this Chapter we formulate a model that (in contrast to a rather complex ASP1 model) we are capable of investigating analytically and optimizing numerically. This model includes the essential mechanisms of the ATAD process while ignoring the issues, which, though microbiologically known, are not material to the process within the likely range of operating conditions. It is our expectation that this model will describe the process with a sufficient degree of accuracy while providing a suitable basis for optimization. A number of simplifying assumptions permit us to reduce the complexity of the model while retaining the essential aspects of the processes. These assumptions are summarized as follows:

• We assume that the reactor in well-stirred (as it is in practice), and hence concentrations of the reagents are homogeneous throughout the volume.

• All the biological activity takes place only in the reactor.

• The batch outflow and inflow stages are of sufficiently short duration, and the biological activity during these stages is negligible.

• Aeration is sufficient, and anaerobic metabolic activity is negligible.

• We consider this process as a reaction with three reagents, namely: oxygen with concentration $x(t)$, organic matter of concentration $y(t)$, and bacteria of concentration $z(t)$.

• The reaction is governed by the low of mass action.

• The aeration rate $u(t)$ is the only control. This control function is bounded.

## 2. Mathematical model

In order to describe the process of aerobic biotreatment, we consider a simple mathematical model, which represent the process as a chemical reaction with three reagents, namely the concentration of oxygen $x(t)$, organic matter with concentration $y(t)$, and the thermophilic aerobic bacteria with concentration $z(t)$. It is assumed that the mass in the reactor is well stirred, and hence the reactant concentrations are homogeneous in the volume. The changes of concentrations of the reagents are described by a three-dimensional nonlinear control system

$$\begin{cases} \dot{x}(t) = -x(t)y(t)z(t) + u(t)(m - x(t)),\ t \in [0,T], \\ \dot{y}(t) = -x(t)y(t)z(t), \\ \dot{z}(t) = x(t)y(t)z(t) - bz(t), \\ x(0) = x_0,\ y(0) = y_0,\ z(0) = z_0,\ x_0 \in (0,m),\ y_0 > 0,\ z_0 > 0. \end{cases} \tag{1}$$

Its nonlinearity is justified by the law of mass action (Krasnov et al., 1995) describing the dependence of the rate of chemical reaction on the concentrations of initial substances. The first equation of system (1) represents the evolution of oxygen concentration: the first term, $-x(t)y(t)z(t)$, describes the process of its absorption in the reaction, whereas the second term describes influx of oxygen (by pumping) into the reactor from outside. Here, $u(t)$ is the rate of aeration, which at the same time is the control function. The second equation describes a decrease of the organic matter in the reaction. The third equation of system (1) shows an evolution of the active biomass concentration; the bacteria mass grows at the rate $x(t)y(t)z(t)$ and decays at a rate $b$. The original system also includes positive initial conditions and a restriction on the rate of pumping air. We consider all possible Lebesgue measurable functions $u(t)$, which for almost all $t \in [0,T]$ satisfy the inequality

$$0 \le u(t) \le u_{\max}, \tag{2}$$

as the admissible controls $D(T)$.

The phase variables $x, y, z$ of system (1) satisfy the following properties.

**Lemma 1.** *Let $u(\cdot) \in D(T)$ be an arbitrary control. Then the corresponding solution $w(t) = (x(t), y(t), z(t))^\top$ of the system (1) is defined on the interval $[0, T]$, the components $x(t)$, $y(t)$, $z(t)$ of which satisfy the following inequalities:*

$$0 < x(t) < x_{max},\ 0 < y(t) < y_{max},\ 0 < z(t) < z_{max},\ t \in [0, T], \tag{3}$$

*where $x_{max}$, $y_{max}$, $z_{max}$ - some positive constants depending on initial conditions $x_0$, $y_0$, $z_0$ and parameters $m$, $b$, $u_{\max}$, $T$ of the original system.*

Here and below, symbol $^\top$ means transpose.

*Proof.* Let $x(t)$, $y(t)$, $z(t)$ be solutions of system (1), which regarding the Existence and Uniqueness Theorem for the system of differential equations (Hartman, 1964), are defined on the biggest semi-interval $\Delta \subseteq [0, T]$.

We will write the solution of the second equation of system (1) as

$$y(t) = y_0 e^{-\int_0^t x(\xi) z(\xi) d\xi}. \tag{4}$$

From this formula it follows that $y(t) > 0$ for all $t \in \Delta$.

Next, by analogy, we obtain the solution of the third equation of the considered system

$$z(t) = z_0 e^{\int_0^t (x(\xi) y(\xi) - b) d\xi}, \tag{5}$$

from which we have that $z(t) > 0$ for all $t \in \Delta$.

Finally, we can write the solution of the first equation of system (1) as

$$x(t) = e^{-\int_0^t (y(\xi) z(\xi) + u(\xi)) d\xi} \left( x_0 + m \int_0^t e^{\int_0^s (y(\xi) z(\xi) + u(\xi)) d\xi} u(s) ds \right). \tag{6}$$

From this formula regarding inequality (2) we find that $x(t) > 0$ for all $t \in \Delta$.

Therefore, we obtain the inequalities:

$$x(t) > 0,\ y(t) > 0,\ z(t) > 0,$$

that are valid for all $t \in \Delta$.

Further, considering positiveness of solutions $x(t)$ and $z(t)$, from formula (4) we find the inequality $y(t) < y_0$ for all $t \in \Delta$.

From formula (6), positiveness of solutions $y(t)$, $z(t)$, and inequality (2) we obtain the relationship

$$x(t) = x_0 e^{-\int_0^t (y(\xi) z(\xi) + u(\xi)) d\xi} + m \int_0^t e^{-\int_s^t (y(\xi) z(\xi) + u(\xi)) d\xi} u(s) ds < x_0 + m u_{max} T.$$

Then we find the inequality $x(t) < x_0 + mu_{max}T$ for all $t \in \Delta$.

Using restrictions on solutions $x(t)$ and $y(t)$ obtained above, from formula (5) we have the relationship

$$z(t) < z_0 e^{\int_0^t x(\xi)y(\xi)d\xi} < z_0 e^{y_0 T(x_0 + mu_{max}T)}.$$

Therefore, we find the inequality $z(t) < z_0 e^{y_0 T(x_0 + mu_{max}T)}$ for all $t \in \Delta$.

Thus, solutions $x(t), y(t), z(t)$ of system (1) on the semi-interval $\Delta \subseteq [0, T]$ are bounded, and so they cannot go to infinity. Regarding Corollary from Lemma (§14) (Demidovich, 1967), solutions $x(t), y(t), z(t)$ of considered system are defined on entire interval $[0, T]$ and satisfy on it inequalities (3). The proof is completed.

Restrictions on the function $x(t)$ are specified as follows.

**Lemma 2.** *For an arbitrary control $u(\cdot) \in D(T)$ the corresponding solution $x(t)$ of the system* (1) *is subject to the inequality*

$$0 < x(t) < m, \; t \in [0, T].$$

*Proof.* Let $v(t) = m - x(t), t \in [0, T]$. Then, using the first equation of system (1), for function $v(t)$ we have Cauchy problem

$$\begin{cases} \dot{v(t)} = -(y(t)z(t) + u(t))v(t) + my(t)z(t), \\ v(0) = v_0 = m - x_0 > 0. \end{cases}$$

Using the variation of a parameter method (Hartman, 1964) we find the solution of this problem as

$$v(t) = e^{-\int_0^t (y(\xi)z(\xi)+u(\xi))d\xi} \left( v_0 + m \int_0^t e^{\int_0^s (y(\xi)z(\xi)+u(\xi))d\xi} y(s)z(s)ds \right).$$

By Lemma 1 from this expression we have relationship $v(t) > 0$ for all $t \in [0, T]$, from which the required inequality follows.

Lemmas 1 and 2 imply that for any control $u(\cdot) \in D(T)$ solutions $x(t), y(t), z(t)$ of the system (1) retain their physical meanings for all $t \in [0, T]$.

## 3. Attainable set and its properties

Let $X(T) \subset R^3$ be the attainable set for system (1) from an initial point $w_0 = (x_0, y_0, z_0)^\top$ at the moment of time $T$; that is, $X(T)$ is the set of all ends $w(T) = (x(T), y(T), z(T))^\top$ of trajectories $w(t) = (x(t), y(t), z(t))^\top$ of system (1) under all possible controls $u(\cdot) \in D(T)$. By Lee & Markus (1967) and Lemma 1, it follows that set $X(T)$ is a compact set in $R^3$ located into the region

$$\left\{ w = (x, y, z)^\top \in R^3 : x > 0, y > 0, z > 0 \right\}.$$

We denote by $\partial Q$ and $\text{int}Q$ the boundary and the interior of the compact set, $Q \subset R^3$, respectively.

Next, we study the boundary of the attainable set $X(T)$. Consider a point $w = (x, y, z)^\top$, such that $w \in \partial X(T)$. It corresponds to control $u(\cdot) \in D(T)$ and a trajectory $w(t) = (x(t), y(t), z(t))^\top$, $t \in [0, T]$, of system (1), such that $w = w(T)$. Then it follows from Lee & Markus (1967) that there exists a non-trivial solution $\psi(t) = (\psi_1(t), \psi_2(t), \psi_3)^\top$, $t \in [0, T]$ of the adjoint system

$$\begin{cases} \dot{\psi}_1(t) = u(t)\psi_1(t) + y(t)z(t)(\psi_1(t) + \psi_2(t) - \psi_3(t)), \\ \dot{\psi}_2(t) = x(t)z(t)(\psi_1(t) + \psi_2(t) - \psi_3(t)), \\ \dot{\psi}_3(t) = x(t)y(t)(\psi_1(t) + \psi_2(t) - \psi_3(t)) + b\psi_3(t), \end{cases} \tag{7}$$

for which, by Lemma 2, the following relationship is valid:

$$u(t) = \begin{cases} 0, & \text{if } \; L(t) < 0, \\ \forall u \in [0, u_{\max}], & \text{if } \; L(t) = 0, \\ u_{\max}, & \text{if } \; L(t) > 0. \end{cases} \tag{8}$$

Here $L(t) = \psi_1(t)$. Function $L(t)$ is the so-called switching function and its behavior determines the type of control $u(t)$.

For convenience in subsequent arguments, we introduce the following auxiliary functions:

$$G(t) = \psi_1(t) + \psi_2(t) - \psi_3(t),\; P(t) = -\psi_3(t),\; d(t) = y(t)z(t) + z(t)x(t) - x(t)y(t),\; t \in [0, T].$$

Using the adjoint system (7), we write for functions $L(t)$, $G(t)$, $P(t)$ the following system of linear differential equations

$$\begin{cases} \dot{L}(t) = u(t)L(t) + y(t)z(t)G(t),\; t \in [0, T], \\ \dot{G}(t) = u(t)L(t) + d(t)G(t) + bP(t), \\ \dot{P}(t) = -x(t)y(t)G(t) + bP(t). \end{cases} \tag{9}$$

The validity of the following Lemma immediately follows from the non-triviality of solution $\psi(t) = (\psi_1(t), \psi_2(t), \psi_3(t))^\top$ of the adjoint system (7).

**Lemma 3.** *The switching function $L(t)$ and the auxiliary functions $G(t)$ and $P(t)$ are nonzero solutions of system* (9).

Lemma 3 allows us to rewrite the relationship (8) in the form

$$u(t) = \begin{cases} 0, & \text{if } \; L(t) < 0, \\ u_{\max}, & \text{if } \; L(t) > 0. \end{cases} \tag{10}$$

At points of discontinuity we will define function $u(t)$ by its limit from the left. Consequently, the control $u(t)$, $t \in [0, T]$, corresponding to a point $w \in \partial X(T)$, is a piecewise constant function, taking values $\{0, u_{\max}\}$. Such type of control is usually called a bang-bang control.

Next, we estimate the number of switchings of control function $u(t), t \in [0, T]$. It follows from (10) that it is sufficient to estimate the number of zeros of the function $L(t)$ on the interval $(0, T)$. The following important statement is valid.

**Lemma 4.** *The switching function $L(t)$ has at most two zeros on the interval $[0, T]$.*

*Proof.* Let us introduce for system (9) new variables:

$$r(t) = L(t),\ v(t) = G(t),\ \mu(t) = P(t) + q_1(t)L(t) + q_2(t)G(t),$$

where functions $q_1(t), q_2(t)$ must be determined. Using new variables $r(t), v(t), \mu(t)$ system (9) has the following form

$$\begin{cases} \dot{r}(t) = u(t)r(t) + y(t)z(t)v(t), \\ \dot{v}(t) = (u(t) - bq_1(t))r(t) + (d(t) - bq_2(t))v(t) + b\mu(t), \\ \dot{\mu}(t) = \Big[\dot{q}_1(t) + (u(t) - b)q_1(t) + u(t)q_2(t) - bq_1(t)q_2(t)\Big] r(t) + \\ \quad + \Big[\dot{q}_2(t) + y(t)z(t)q_1(t) + (d(t) - b)q_2(t) - bq_2^2(t) - x(t)y(t)\Big] v(t) + \\ \quad + b(1 + q_2(t))\mu(t). \end{cases} \tag{11}$$

We will select such functions $q_1(t)$ and $q_2(t)$ that in system (11) the expressions inside brackets by $r(t)$ and $v(t)$ are zero. Now, we have for functions $q_1(t), q_2(t)$ the system of differential equations

$$\begin{cases} \dot{q}_1(t) + (u(t) - b)q_1(t) + u(t)q_2(t) - bq_1(t)q_2(t) = 0, \\ \dot{q}_2(t) + y(t)z(t)q_1(t) + (d(t) - b)q_2(t) - bq_2^2(t) - x(t)y(t) = 0. \end{cases} \tag{12}$$

Then system (11) will be rewritten as

$$\begin{cases} \dot{r}(t) = u(t)r(t) + y(t)z(t)v(t), \\ \dot{v}(t) = (u(t) - bq_1(t))r(t) + (d(t) - bq_2(t))v(t) + b\mu(t), \\ \dot{\mu}(t) = b(1 + q_2(t))\mu(t). \end{cases} \tag{13}$$

In system (13) we will make the following substitutions:

$$\widetilde{r}(t) = r(t),\ \widetilde{v}(t) = v(t) + q_3(t)r(t),\ \widetilde{\mu}(t) = \mu(t),$$

where $q_3(t)$ is the function that must be determined. Then system (13) will be rewritten as

$$\begin{cases} \dot{\widetilde{r}}(t) = (u(t) - y(t)z(t)q_3(t))\widetilde{r}(t) + y(t)z(t)\widetilde{v}(t), \\ \dot{\widetilde{v}}(t) = \Big[\dot{q}_3(t) - bq_1(t) + (u(t) - d(t))q_3(t) + \\ \quad + bq_2(t)q_3(t) - y(t)z(t)q_3^2(t) + u(t)\Big] \widetilde{r}(t) + \\ \quad + (d(t) + y(t)z(t)q_3(t) - bq_2(t))\widetilde{v}(t) + b\widetilde{\mu}(t), \\ \dot{\widetilde{\mu}}(t) = b(1 + q_2(t))\widetilde{\mu}(t). \end{cases} \tag{14}$$

We will choose function $q_3(t)$ so that in system (14) the expression inside brackets by $\widetilde{r}(t)$ will become zero. Now, we have for function $q_3(t)$ the equation

$$\dot{q_3}(t) - bq_1(t) + (u(t) - d(t))q_3(t) + bq_2(t)q_3(t) - y(t)z(t)q_3^2(t) + u(t) = 0. \tag{15}$$

Then system of equations (14) will be rewritten as

$$\begin{cases} \dot{\widetilde{r}}(t) = (u(t) - y(t)z(t)q_3(t))\widetilde{r}(t) + y(t)z(t)\widetilde{v}(t), \\ \dot{\widetilde{v}}(t) = (d(t) + y(t)z(t)q_3(t) - bq_2(t))\widetilde{v}(t) + b\widetilde{\mu}(t), \\ \dot{\widetilde{\mu}}(t) = b(1 + q_2(t))\widetilde{\mu}(t), \end{cases} \tag{16}$$

and the system of differential equations for functions $q_1(t)$, $q_2(t)$, $q_3(t)$, regarding relationships (12) and (15), is following

$$\begin{cases} \dot{q_1}(t) = -(u(t) - b)q_1(t) - u(t)q_2(t) + bq_1(t)q_2(t), \\ \dot{q_2}(t) = -y(t)z(t)q_1(t) - (d(t) - b)q_2(t) + bq_2^2(t) + x(t)y(t), \\ \dot{q_3}(t) = bq_1(t) - (u(t) - d(t))q_3(t) - bq_2(t)q_3(t) + y(t)z(t)q_3^2(t) - u(t). \end{cases} \tag{17}$$

Next, we will rewrite system (17) in a matrix form. At first, we define symmetric matrices $A_1(t)$, $A_2(t)$, $A_3(t)$ as follows:

$$A_1(t) = \begin{pmatrix} 0 & \frac{b}{2} & 0 \\ \frac{b}{2} & 0 & 0 \\ 0 & 0 & 0 \end{pmatrix}, \; A_2(t) = \begin{pmatrix} 0 & 0 & 0 \\ 0 & b & 0 \\ 0 & 0 & 0 \end{pmatrix}, \; A_3(t) = \begin{pmatrix} 0 & 0 & 0 \\ 0 & 0 & -\frac{b}{2} \\ 0 & -\frac{b}{2} & y(t)z(t) \end{pmatrix}.$$

Then we introduce vectors $b_1(t)$, $b_2(t)$, $b_3(t)$ as:

$$b_1(t) = \begin{pmatrix} u(t) - b \\ u(t) \\ 0 \end{pmatrix}, \; b_2(t) = \begin{pmatrix} -y(t)z(t) \\ b - d(t) \\ 0 \end{pmatrix}, \; b_3(t) = \begin{pmatrix} b \\ 0 \\ u(t) - d(t) \end{pmatrix}.$$

At last, we define functions $c_1(t)$, $c_2(t)$, $c_3(t)$ by relationships:

$$c_1(t) = 0, \; c_2(t) = x(t)y(t), \; c_3(t) = -u(t).$$

Then we obtain the following system of equations

$$\begin{cases} \dot{q_1}(t) = q^\top(t)A_1(t)q(t) + b_1^\top(t)q(t) + c_1(t), \\ \dot{q_2}(t) = q^\top(t)A_2(t)q(t) + b_2^\top(t)q(t) + c_2(t), \\ \dot{q_3}(t) = q^\top(t)A_3(t)q(t) + b_3^\top(t)q(t) + c_3(t), \end{cases} \tag{18}$$

where $q(t) = (q_1(t), q_2(t), q_3(t))^\top$.

Now, we will show that for system (18) there exists a solution defined on the entire interval $[0, T]$. Assume the contradiction, i. e. that an arbitrary solution $q(t)$ of system (18) is defined on the interval $[0, t_1)$, $t_1 \in (0, T]$, which is the biggest interval of the existence of this solution. Then, from Lemma (Chapter 4, §14) (Demidovich, 1967) for solution $q(t)$ it follows the relationship

$$\lim_{t \to t_1 - 0} \|q(t)\| = +\infty. \tag{19}$$

Its validity leads to the existence of values $\rho > 0$ and $t_0 \in [0, t_1)$, for which the inclusion $q(t) \in \Omega$ holds for all $t \in [t_0, t_1)$. Here $\Omega = \{q \in \mathrm{R}^3 : \|q\| \geq \rho\}$.

Let us evaluate on the interval $[t_0, t_1)$ the derivative of function $\|q(t)\|$ regarding system (18). We have the equality

$$\begin{aligned} \frac{d}{dt}\Big(\|q(t)\|\Big) = \|q(t)\|^{-1} \cdot \Big(\Big[q_1(t)q^\top(t)A_1(t)q(t) + \\ + q_2(t)q^\top(t)A_2(t)q(t) + q_3(t)q^\top(t)A_3(t)q(t)\Big] + \\ + \Big[q_1(t)b_1^\top(t)q(t) + q_2(t)b_2^\top(t)q(t) + q_3(t)b_3^\top(t)q(t)\Big] + \\ + \Big[c_1(t)q_1(t) + c_2(t)q_2(t) + c_3(t)q_3(t)\Big]\Big). \end{aligned} \tag{20}$$

Next, we will estimate the upper boundary of the expressions inside brackets using inequalities (2), (3).

First, we have the inequality for the terms inside the third brackets

$$c_1(t)q_1(t) + c_2(t)q_2(t) + c_3(t)q_3(t) \leq C \cdot \|q(t)\|,$$

where

$$C = \sqrt{x_{max}^2 y_{max}^2 + u_{max}^2}.$$

Next, we will estimate the terms inside the second brackets. We obtain the relationship

$$q_1(t)b_1^\top(t)q(t) + q_2(t)b_2^\top(t)q(t) + q_3(t)b_3^\top(t)q(t) \leq B \cdot \|q(t)\|^2,$$

where

$$B = \sqrt{7b^2 + 7u_{max}^2 + 9y_{max}^2 z_{max}^2 + 8z_{max}^2 x_{max}^2 + 8x_{max}^2 y_{max}^2}.$$

At last, for the terms inside the first brackets we have the inequality

$$\begin{aligned} q_1(t)q^\top(t)A_1(t)q(t) + q_2(t)q^\top(t)A_2(t)q(t) + q_3(t)q^\top(t)A_3(t)q(t) \leq \\ \leq |q_1(t)| \cdot \|A_1(t)q(t)\| \cdot \|q(t)\| + |q_2(t)| \cdot \|A_2(t)q(t)\| \cdot \|q(t)\| + \\ + |q_3(t)| \cdot |q^\top(t)A_3(t)q(t)|. \end{aligned} \tag{21}$$

Separately, for the first two terms in (21) we obtain the inequalities:

$$\|A_1(t)q(t)\| \leq \frac{b}{2}\|q(t)\|, \ \|A_2(t)q(t)\| \leq b\|q(t)\|.$$

Next, we will find the eigenvalues of matrix $A_3(t)$. We have the formulas:

$$\lambda_1(t) = 0, \ \lambda_2(t) = \frac{y(t)z(t) - \sqrt{y^2(t)z^2(t) + b^2}}{2}, \ \lambda_3(t) = \frac{y(t)z(t) + \sqrt{y^2(t)z^2(t) + b^2}}{2}.$$

Therefore, for the last term from (21) we find the relationship

$$|q^\top(t)A_3(t)q(t)| \le \lambda_3(t)\|q(t)\|^2 \le \frac{y_{max}z_{max} + \sqrt{y_{max}^2 z_{max}^2 + b^2}}{2} \cdot \|q(t)\|^2.$$

Finally, we obtain for expression (21) the following inequality

$$\begin{aligned}|q_1(t)| \cdot \|A_1(t)q(t)\| \cdot \|q(t)\| + |q_2(t)| \cdot \|A_2(t)q(t)\| \cdot \|q(t)\| + \\ + |q_3(t)| \cdot |q^\top(t)A_3(t)q(t)| \le A \cdot \|q(t)\|^3,\end{aligned}$$

where

$$A = \sqrt{\frac{7b^2}{4} + y_{max}^2 z_{max}^2}.$$

Substituting these inequalities into formula (20) we finally find a differential inequality

$$\frac{d}{dt}\Big(\|q(t)\|\Big) \le A\|q(t)\|^2 + B\|q(t)\| + C,\ t \in [t_0, t_1). \tag{22}$$

Now, we will consider the quadratic equation

$$AK^2 - BK + C = 0. \tag{23}$$

Let define the sign of its discriminant. We have a chain of equalities:

$$\begin{aligned} D = B^2 - 4AC = \\ = 7b^2 + 7u_{max}^2 + 9y_{max}^2 z_{max}^2 + 8z_{max}^2 x_{max}^2 + 8x_{max}^2 y_{max}^2 - \\ - 4\sqrt{\frac{7b^2}{4} + y_{max}^2 z_{max}^2} \cdot \sqrt{x_{max}^2 y_{max}^2 + u_{max}^2} = \\ = \Big(\sqrt{7b^2 + 4y_{max}^2 z_{max}^2}\Big)^2 + \Big(\sqrt{4x_{max}^2 y_{max}^2 + 4u_{max}^2}\Big)^2 - \\ - \sqrt{7b^2 + 4y_{max}^2 z_{max}^2} \cdot \sqrt{4x_{max}^2 y_{max}^2 + 4u_{max}^2} + \\ + \Big(3u_{max}^2 + 5y_{max}^2 z_{max}^2 + 8z_{max}^2 x_{max}^2 + 4x_{max}^2 y_{max}^2\Big). \end{aligned}$$

It is easy to see that discriminant $D$ is positive.

Now, we will introduce function $V(q) = \|q\| + K_0$, $q \in \Omega$, where value $K_0$ is defined as a biggest root of equation (23). Then, we have the formula

$$K_0 = \frac{B + \sqrt{B^2 - 4AC}}{2A}.$$

Let us rewrite differential inequality (22) for function $V(q)$. We have the inequality

$$\frac{d}{dt}\Big(V(q(t))\Big) \le A\Big(V(q(t)) - K_0\Big)^2 + B\Big(V(q(t)) - K_0\Big) + C.$$

After necessary transformations regarding relationship $AK_0^2 - BK_0 + C = 0$, we find the differential inequality

$$\frac{d}{dt}\Big(V(q(t))\Big) \le AV^2(q(t)) - \Big(2AK_0 - B\Big)V(q(t)),\ t \in [t_0, t_1). \tag{24}$$

At last, let us consider the auxiliary Cauchy problem

$$\begin{cases} \dot{\chi}(t) = A\chi^2(t) - \Big(2AK_0 - B\Big)\chi(t),\ t \in [t_0, T], \\ \chi(t_0) = \chi_0,\ \chi_0 \ge K_0 + \rho. \end{cases} \tag{25}$$

It is easy to see that for value $\chi_0$ the inequality

$$\chi_0 > 2K_0 - \frac{B}{A} \tag{26}$$

holds.

Let us find a solution of Cauchy problem (25). Solving the corresponding Bernoulli equation and satisfying to the initial condition we obtain the following formula

$$\chi(t) = \Big(\frac{A}{2AK_0 - B} + \Big[\frac{1}{\chi_0} - \frac{A}{2AK_0 - B}\Big]e^{(2AK_0 - B)(t - t_0)}\Big)^{-1},\ t \in [t_0, T]. \tag{27}$$

From (26) we have negativeness of the terms inside brackets in formula (27). Therefore, function $\chi(t)$ is positive and increasing on the interval $[t_0, T]$. Then we find the inequality

$$\chi(t) < \chi(T),\ t \in [t_0, T).$$

Thus, from differential inequality (24), Cauchy problem (25) and the Chaplygin's Theorem (Tikhonov et al., 1985) regarding the condition

$$\chi_0 = V(q(t_0)) = K_0 + \|q(t_0)\|,$$

we have the inequalities:

$$\|q(t)\| < \chi(t) - K_0 < \chi(T) - K_0,\ t \in (t_0, t_1),$$

that contradict to (19). Our assumption was wrong. Therefore, there exists solution $q(t)$ of system (17) on entire interval $[0, T]$.

Then, the solution of system (16) is also defined on the interval $[0, T]$. Using in system (16) the generalization of the Rolle's Theorem (Dmitruk, 1992) we conclude that function $L(t) = \tilde{r}(t)$ has at most two zeroes on the interval $[0, T]$. The proof is completed.

Based on the obtained results, we formulate the following statement.

**Theorem 1.** *Let point $w = (x, y, z)^\top$ belong to the boundary of the attainable set $X(T)$. Then the control $u(t)$, $t \in [0, T]$, which is associated with this point $w$, is a piecewise constant function taking values $\{0, u_{\max}\}$ and having at most two switchings on the interval $(0, T)$.*

## 4. Auxiliary set and its properties

Using Theorem 1, we now can proceed to constructing a parametrization for the attainable set $X(T)$ with moments of switching of piecewise constant controls. For this task, we consider the set

$$\Lambda(T) = \left\{\theta = (\theta_1, \theta_2, \theta_3)^\top \in \mathrm{R}^3 : 0 \leq \theta_1 \leq \theta_2 \leq \theta_3 \leq T\right\}.$$

For every point $\theta \in \Lambda(T)$ we form the control $u_\theta(\cdot) \in D(T)$ by formula

$$u_\theta(t) = \begin{cases} u_{\max}, & \text{if } 0 \leq t \leq \theta_1, \\ 0, & \text{if } \theta_1 < t \leq \theta_2, \\ u_{\max}, & \text{if } \theta_2 < t \leq \theta_3, \\ 0, & \text{if } \theta_3 < t \leq T. \end{cases} \tag{28}$$

We denote by $w_\theta(t)$, $t \in [0, T]$ the solution of system (1) corresponding to the control $u_\theta(t)$. Finally, we define the mapping $F(\cdot, T) : \Lambda(T) \to \mathrm{R}^3$ as

$$F(\theta, T) = w_\theta(T),\ \theta \in \Lambda(T).$$

For this mapping we have the following proposition.

**Lemma 5.** *The mapping $F(\cdot, T)$ is continuous on the set $\Lambda(T)$.*

*Proof.* Let consider arbitrary values $\theta, \tau \in \Lambda(T)$. We extend controls $u_\theta(t)$, $u_\tau(t)$, defined by (28), to the interval $(T, T+\delta)$ for some $\delta > 0$ by value of zero. Corresponding trajectories $w_\theta(t)$, $w_\tau(t)$ of system (1) are also extended to this interval. Then we transform the Cauchy problems (1) for trajectories $w_\theta(t)$, $w_\tau(t)$ to the corresponding integral equations for all $t \in (T, T+\delta)$. Then, by Lemma 1, we evaluate the difference $\|w_\theta(t) - w_\tau(t)\|$, $t \in (T, T+\delta)$. Applying the Gronwall's inequality (Robinson, 2004), we obtain as a result the relationship

$$\|w_\theta(t) - w_\tau(t)\| \leq L_w \|\theta - \tau\|,\ t \in [T, T+\delta),$$

where $L_w$ is a positive constant. Assuming that $t = T$ in the inequality, we find that mapping $F(\cdot, T)$ satisfies the Lipschitz condition (Robinson, 2004) on the set $\Lambda(T)$. The required continuity of the mapping $F(\cdot, T)$ on the set $\Lambda(T)$ immediately follows from this fact. The proof is completed.

Using the mapping $F(\cdot, T)$, we introduce the auxiliary set $Z(T) = F(\Lambda(T), T)$, which consists of all ends $w_\theta(T)$ of trajectories $w_\theta(t)$ of system (1) under all possible controls $u_\theta(t)$, $t \in [0, T]$, defined by formula (28). Every element of set $Z(T)$ is a result of a bang-bang control $u_\theta(t)$, $t \in [0, T]$, with at most three switchings on the interval $(0, T)$.

Now, we have to discuss some properties of auxiliary set $Z(T)$. Considering a point $\theta \in \mathrm{int}\Lambda(T)$, its corresponding control $u_\theta(t)$ defined by (28) and a trajectory $w_\theta(t)$, $t \in [0, T]$, we can reformulate the Cauchy problem (1) in the form

$$\begin{cases} \dot{w}_\theta(t) = A w_\theta(t) + \varphi(w_\theta(t)) c + u_\theta(t) g(w_\theta(t)),\ t \in [0, T], \\ w_\theta(0) = w_0 = (x_0, y_0, z_0)^\top, \end{cases} \tag{29}$$

where $A$ is a $3 \times 3$ matrix, $c \in R^3$, and functions $g(w)$ and $\varphi(w)$ are a vector and a scalar functions, respectively, such that

$$A = \begin{pmatrix} 0 & 0 & 0 \\ 0 & 0 & 0 \\ 0 & 0 & -b \end{pmatrix}, \quad c = \begin{pmatrix} -1 \\ -1 \\ 1 \end{pmatrix}, \quad g(w) = \begin{pmatrix} m - x \\ 0 \\ 0 \end{pmatrix}, \quad \varphi(w) = xyz.$$

For the system (29), we define a function $\Phi_\theta(t)$, $t \in [0, T]$, as a solution of the matrix Cauchy problem

$$\begin{cases} \dot{\Phi}_\theta(t) = \left( A + c \left( \frac{\partial \varphi}{\partial w}(w_\theta(t)) \right)^\top + u_\theta(t) \frac{\partial g}{\partial w}(w_\theta(t)) \right) \Phi_\theta(t), \; t \in [0, T], \\ \Phi_\theta(T) = E, \end{cases} \tag{30}$$

where $E$ is the identity matrix. Let us evaluate the derivatives $\frac{\partial w_\theta}{\partial \theta_i}(T)$, $i = \overline{1,3}$. Using an approach which is due to Hájek (1991), one can find that the derivatives satisfy the following equalities:

$$\frac{\partial w_\theta}{\partial \theta_i}(T) = (-1)^{i-1} u_{\max} \Phi_\theta^{-1}(\theta_i) g(w_\theta(\theta_i)), \; i = \overline{1,3}. \tag{31}$$

Now we are in a position to state the following theorem.

**Theorem 2.** *The following equalities hold:*

$$F(\mathrm{int}\Lambda(T), T) = \mathrm{int}Z(T), \; F(\partial\Lambda(T), T) = \partial Z(T), \tag{32}$$

*and the restriction of mapping $F(\cdot, T)$ onto the interior of set $\Lambda(T)$ is one-to-one.*

*Proof.* Firstly, we consider the set $\mathrm{int}\Lambda(T)$. The mapping $F(\cdot, T)$ is continuously differentiable on the set $\mathrm{int}\Lambda(T)$, and for every point $\theta \in \mathrm{int}\Lambda(T)$, by (31), the following equalities hold:

$$\frac{\partial F}{\partial \theta_i}(\theta, T) = (-1)^{i-1} u_{\max} \Phi_\theta^{-1}(\theta_i) g(w_\theta(\theta_i)), \; i = \overline{1,3}. \tag{33}$$

The continuity of these derivatives on the set $\mathrm{int}\Lambda(T)$ is determined by a continuous dependence of the trajectory $w_\theta(t)$ and solution $\Phi_\theta(t)$ of the matrix Cauchy problem (30) in variables $\theta_i$, $i = \overline{1,3}$. It is established by arguments, which are similar to the arguments presented in Lemma 5.

We have to show that the Jacobi matrix of the restriction of mapping $F(\cdot, T)$ onto $\mathrm{int}\Lambda(T)$ is nonsingular. Suppose the opposite. Then there is a point $\bar{\theta} \in \mathrm{int}\Lambda(T)$ for which vectors $\frac{\partial F}{\partial \theta_i}(\bar{\theta}, T)$, $i = \overline{1,3}$, are linearly dependent. With respect to (33), it means the existence of a nonzero vector $q \in R^3$ such that the equalities:

$$(g(w_\theta(\bar{\theta}_i)), \eta(\bar{\theta}_i)) = 0, \; i = \overline{1,3}, \tag{34}$$

hold. Here $\eta(t) = (\Phi_\theta^{-1}(t))^\top q$. By (30), we can see that function $\eta(t)$, $t \in [0,T]$, satisfies the adjoint system (7), which is written as

$$\dot{\psi}(t) = -\left(A + c\left(\frac{\partial\varphi}{\partial w}(w_\theta(t))\right)^\top + u_\theta(t)\frac{\partial g}{\partial w}(w_\theta(t))\right)^\top \psi(t),$$

where $\psi(t) = (\psi_1(t), \psi_2(t), \psi_3(t))^\top$. Then, applying Lemmas 2 and 4 to the function $r(t) = (g(w_\theta(t)), \eta(t))$, we find that function $r(t)$ has two zeros on interval $(0,T)$ at most. This fact contradicts the equalities:

$$r(\bar{\theta}_i) = 0,\ i = \overline{1,3},$$

resulting from (34). Therefore, the assumption is wrong, and hence the proposition is true. By this and by the Theorem on the invariance of interior points (Partasarathy, 1983), the first equality of (32) follows.

Furthermore, set $\mathrm{int}\Lambda(T)$ is a convex set, and the set $\mathrm{int}Z(T)$ is path connected. Indeed, the mapping $F(\cdot,T)$ transforms any segment of $\mathrm{int}\Lambda(T)$ into a curve located completely inside $\mathrm{int}Z(T)$. For every point of $\mathrm{int}\Lambda(T)$ the Local Theorem on an implicit function (Partasarathy, 1983) holds. Then the last statement of the proposition follows from the Global Theorem 3 on an implicit function (Shigeo, 1985). Hence the validity of the second equality of (32) follows. The proof is completed.

**Remark.** *We extend by continuity the derivatives* $\frac{\partial F}{\partial \theta_i}(\theta,T)$, $i = \overline{1,3}$ *of the mapping* $F(\cdot,T)$ *onto the boundary of the set* $\Lambda(T)$. *As a result, we have continuous partial derivatives of the mapping* $F(\cdot,T)$ *on the entire set* $\Lambda(T)$.

From the definitions of attainable set $X(T)$ and the auxiliary set $Z(T)$, and Theorems 1 and 2, the following inclusions hold:

$$Z(T) \subseteq X(T),\ \partial X(T) \subseteq \partial Z(T). \tag{35}$$

These explain why the set $Z(T)$ plays such an important role in the study of the attainable set $X(T)$.

Further investigation of the auxiliary set $Z(T)$ involves the study of its supplement $\mathrm{R}^3 \setminus Z(T)$. The following statement is valid.

**Theorem 3.** *The set* $\mathrm{R}^3 \setminus Z(T)$ *is path connected.*

*Proof.* Let $F_i(\theta,T)$, $i = \overline{1,3}$, be the components of mapping $F(\cdot,T)$. We define the following values:

$$F^i_{\min} = \min_{\theta\in\Lambda(T)} F_i(\theta,T),\ F^i_{\max} = \max_{\theta\in\Lambda(T)} F_i(\theta,T),\ i = \overline{1,3}.$$

By Lemma 5 and the Extension Theorem of Brouwer-Urysohn (Hausdorff, 1962), we construct continuous mapping $\Pi(\cdot,T)$ defined on the whole space $\mathrm{R}^3$, which coincides with the mapping $F(\cdot,T)$ for all points of the set $\Lambda(T)$. In addition, for the components $\Pi_i(\theta,T)$, $i = \overline{1,3}$, of this mapping at each point $\theta \in \mathrm{R}^3$ the following inequalities hold:

$$F^i_{\min} \le \Pi_i(\theta,T) \le F^i_{\max},\ i = \overline{1,3}.$$

Such bounded continuous mapping $\Pi(\cdot, T)$ is called an extension of the mapping $F(\cdot, T)$ from the set $\Lambda(T)$ on the whole space $R^3$.

We now define continuous functions $\xi_i(\theta), i = \overline{1,3}$, as

$$\xi_i(\theta) = \begin{cases} \theta_i + 1, & \text{if} \quad \theta_i < 0, \\ 1, & \text{if} \quad 0 \le \theta_i \le T,\ i = \overline{1,3}, \\ \theta_i - T + 1, & \text{if} \quad \theta_i > T. \end{cases}$$

With these functions, we define a continuous mapping $\Psi(\cdot, T)$ of the whole space $R^3$ onto whole space $R^3$ as

$$\Psi_i(\theta, T) = \xi_i(\theta)\Pi_i(\theta, T),\ i = \overline{1,3},$$

where $\Psi_i(\theta, T), i = \overline{1,3}$, are the components of mapping $\Psi(\cdot, T)$.

Furthermore, the open set $R^3 \setminus \Lambda(T)$ is path connected, and hence it is a connected set (Hall & Spencer, 1955). By Theorem 2, the continuous mapping $\Psi(\cdot, T)$ transfers the set $R^3 \setminus \Lambda(T)$ onto the set $R^3 \setminus Z(T)$, which is also open and connected (Hall & Spencer, 1955). Then the set $R^3 \setminus Z(T)$ simultaneously is a path connected set. This completes the proof.

## 5. Parametric description of attainable set

Finally, we now able to establish the validity of the main result of this paper.

**Theorem 4.** *For the attainable set $X(T)$ and the auxiliary set $Z(T)$, the equality $X(T) = Z(T)$ holds.*

*Proof.* It follows from the first inclusion in (35) that in order to prove the hypothesis it is sufficient to show the validity of the inclusion $X(T) \subseteq Z(T)$. Let us assume the opposite, i.e. assume that there exists a point $\widetilde{w}$ such that

$$\widetilde{w} \notin Z(T),\ \widetilde{w} \in X(T)$$

holds. Consider a point $\widehat{w} \notin X(T)$.

The arguments presented in Theorems 2 and 3 show that the boundary of set $Z(T)$ divides $R^3$ into two path connected subsets $\text{int}Z(T)$ and $R^3 \setminus Z(T)$. The path connectedness of the second set ensures the existence of a continuous curve $\sigma(s)$, $s \in [0, 1]$, as well as $\widetilde{w} = \sigma(0)$, $\widehat{w} = \sigma(1)$, and $\sigma(s) \notin Z(T)$ for all $s \in (0, 1)$. By Theorem 36 on "transition through customs" in (Schwartz, 1967), there is a value $s_\star \in (0, 1)$ such that $\sigma(s_\star) \in \partial X(T)$. Therefore, there is a defined point $\bar{w} = \sigma(s_\star)$, such that the relationships:

$$\bar{w} \in \partial X(T),\ \bar{w} \notin \partial Z(T),$$

simultaneously hold. This contradicts to the second inclusion in (35). Hence the assumption is incorrect, and the required inclusion holds. The proof is completed.

We have obtained analytically the properties of the attainable set $X(T)$. The moments of switching of controls $u_\theta(t)$ from (28), which form the set $\Lambda(T)$, together with the mapping $F(\cdot, T)$ play the role of parametrization for the set $X(T)$ (its interior and boundary). This implies that each point on the boundary of the attainable set $X(T)$ can be reached by a bang-bang control $u_\theta(t)$ with at most two switchings, and every point of the interior of the set $X(T)$ is the result of such control with precisely three switchings.

**Remark.** *To establish these results we utilize an approach that was developed for another class of control systems by Grigorieva & Khailov (2001; 2005).*

## 6. Numerical simulations of attainable set

Figures 3 to 9 show examples of attainable sets $X(T)$, constructed with MATLAB using Theorem 4.

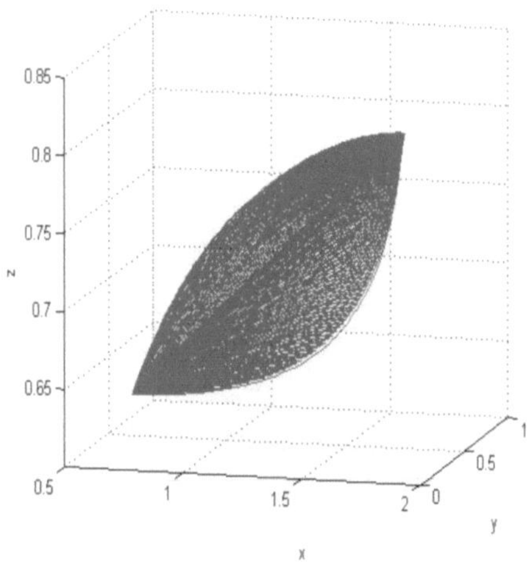


Fig. 3. Attainable set $X(T)$ for Example 1.

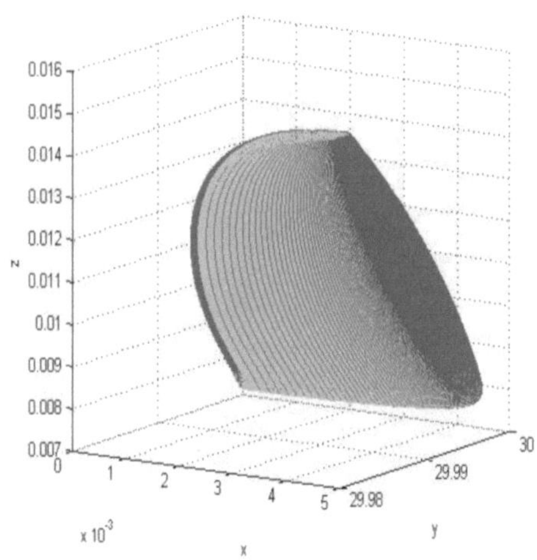


Fig. 4. Attainable set $X(T)$ for Example 2.

**Example 1.** *(Figure 3) The initial conditions and parameters of the system* (1) *are:*

$$x_0 = 1.00,\ y_0 = 1.00,\ z_0 = 1.00,\ m = 2.00,\ b = 1.00,\ u_{\max} = 4.00,\ T = 1.00.$$

**Example 2.** *(Figure 4) The initial conditions and parameters of the system* (1) *are:*

$$x_0 = 0.0002,\ y_0 = 30.00,\ z_0 = 0.03,\ m = 0.005,\ b = 0.24,\ u_{\max} = 4.00,\ T = 6.00.$$

**Example 3.** *(Figure 5) The initial conditions and parameters of the system* (1) *are:*

$$x_0 = 0.0019,\ y_0 = 2.498,\ z_0 = 0.0874,\ m = 0.048,\ b = 0.24,\ u_{\max} = 4.00,\ T = 6.00.$$

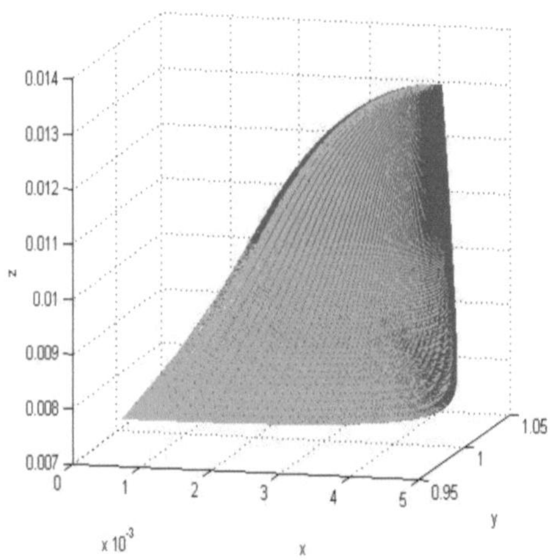


Fig. 5. Attainable set $X(T)$ for Example 3.

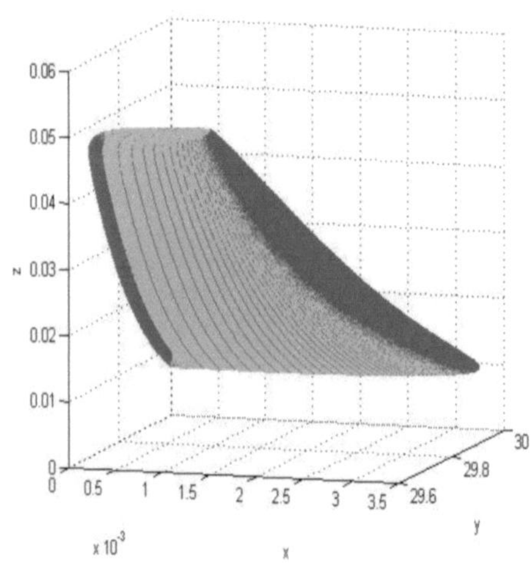


Fig. 6. Attainable set $X(T)$ for Example 4.

**Example 4.** *(Figure 6) The initial conditions and parameters of the system* (1) *are:*

$$x_0 = 0.0011,\ y_0 = 38.3406,\ z_0 = 0.1643,\ m = 0.0274,\ b = 0.24,\ u_{\max} = 4.00,\ T = 6.00.$$

**Example 5.** *(Figure 7) The initial conditions and parameters of the system* (1) *are:*

$$x_0 = 0.0192,\ y_0 = 74.94,\ z_0 = 0.0874,\ m = 0.048,\ b = 0.24,\ u_{\max} = 4.00,\ T = 6.00.$$

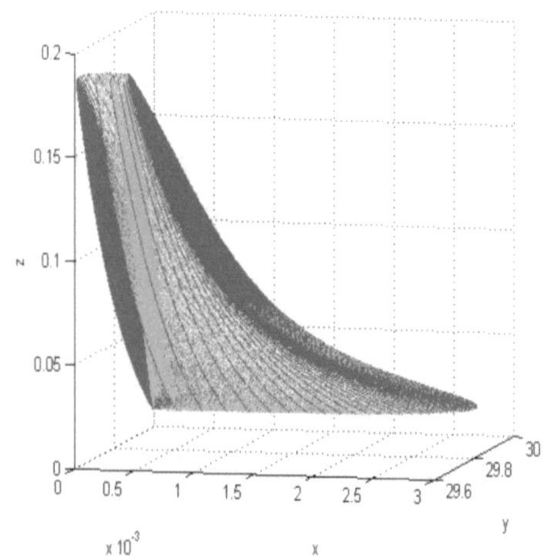


Fig. 7. Attainable set $X(T)$ for Example 5.

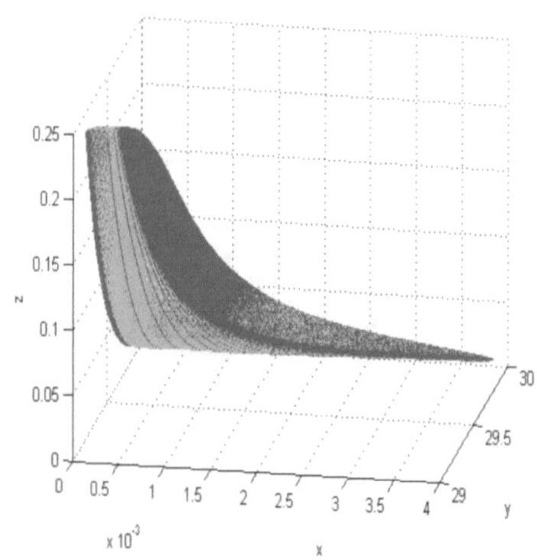


Fig. 8. Attainable set $X(T)$ for Example 6.

**Example 6.** *(Figure 8) The initial conditions and parameters of the system* (1) *are:*

$$x_0 = 0.0019,\ y_0 = 74.94,\ z_0 = 0.0874,\ m = 0.048,\ b = 1.00,\ u_{\max} = 4.00,\ T = 12.00.$$

**Example 7.** *(Figure 9) The initial conditions and parameters of the system* (1) *are:*

$$x_0 = 0.001,\ y_0 = 146.9694,\ z_0 = 0.1715,\ m = 0.0245,\ b = 0.50,\ u_{\max} = 4.00,\ T = 20.00.$$

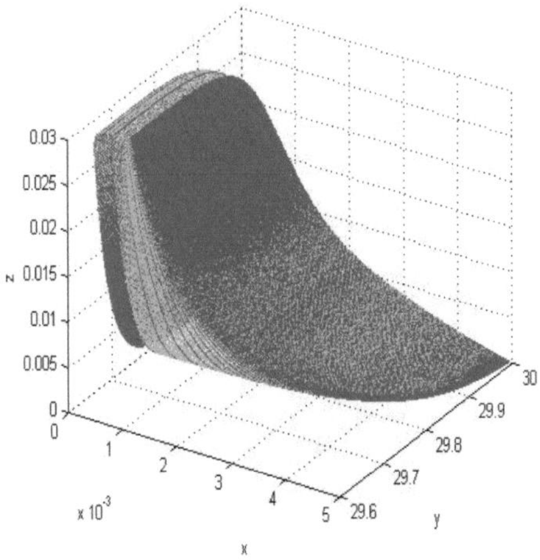


Fig. 9. Attainable set $X(T)$ for Example 7.

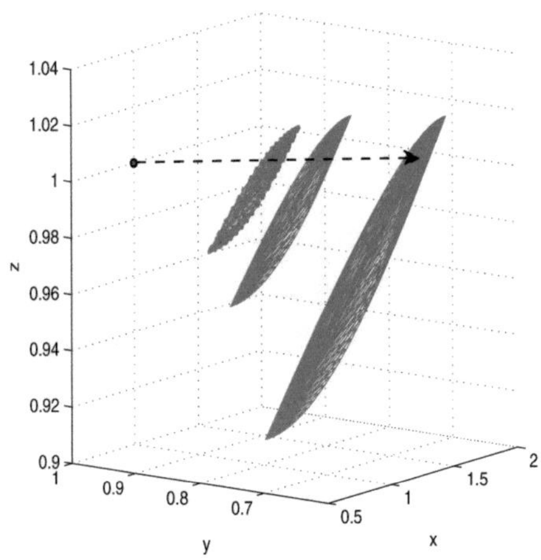


Fig. 10. Dynamics of the attainable set $X(T)$.

## 7. Controllability problem

Let us consider for system (1) the controllability problem. To do this, we add to it given positive terminal conditions:

$$x(T) = x_1,\ y(T) = y_1,\ z(T) = z_1,\ x_1 > 0,\ y_1 > 0,\ z_1 > 0. \tag{36}$$

The controllability problem consists of finding the terminal time $T$ and the control $u(\cdot) \in D(T)$, which steers system (1) from initial point $w_0$ to the terminal point $w_1 = (x_1, y_1, z_1)^\top$.

Using terminology of the attainable set $X(T)$, the considered problem is reformulated as follows. It is necessary to find such terminal time $T$, for which the inclusion $w_1 \in X(T)$ holds. Using the parametrization of the set $X(T)$ constructed above, the controllability problem is rewritten as the problem of finite dimensional minimization of the auxiliary function

$$G(\theta, T) = 0.5 \cdot \|F(\theta, T) - w_1\|^2 \tag{37}$$

in variables $\theta = (\theta_1, \theta_2, \theta_3)^\top \in \Lambda(T)$ and $T > 0$.

An important property of this problem is the validity of the fact following from the arguments above.

**Lemma 6.** *Let the controllability problem for system* (1) *with condition* (36) *has a solution. Then the minimum of the function* $G(\theta, T)$ *on the set* $\Lambda(T) \times (0, +\infty)$ *equals zero.*

The problem of the minimization of the function $G(\theta, T)$ is solved numerically. In order to find the minimum of this function, the iterative gradient projection method (Vasil'ev, 2002) is used. Corresponding numerical algorithm is written in $C++$.

The rule of stop of the iterative process following from Lemma 6 is an execution on some $k$-iterative step of the following condition $G(\theta^k, T_k) \leq \varepsilon$, where $\theta^k$, $T_k$ are values of variables of the minimization on this step. Positive value $\varepsilon$ is an accuracy of calculations of the minimum of the function $G(\theta, T)$.

The convergence of the using method at the considered problem depends on a validity of two following conditions:

• continuous differentiability of the function $G(\theta, T)$ in variables $\theta = (\theta_1, \theta_2, \theta_3)^\top, T$ on the set $\Lambda(T) \times [T_{\min}, T_{\max}]$;

• satisfaction for the gradient of the function $G(\theta, T)$ the Lipschitz condition

$$\left\| \left(\frac{\partial G}{\partial \theta}(\theta^2, T_2), \frac{\partial G}{\partial T}(\theta^2, T_2)\right) - \left(\frac{\partial G}{\partial \theta}(\theta^1, T_1), \frac{\partial G}{\partial T}(\theta^1, T_1)\right) \right\| \leq L_G(\|\theta^2 - \theta^1\| + |T_2 - T_1|), \tag{38}$$

where $L_G$ is a some positive constant. Here $T_{\min}$, $T_{\max}$ are given positive parameters. From the analysis of formula (37) it follows that the conditions formulated above hold if the function $F(\theta, T)$ is continuous differentiable on the set $\Lambda(T) \times [T_{\min}, T_{\max}]$ and the Lipschitz condition is valid for this function and its partial derivatives $\frac{\partial F}{\partial \theta_i}(\theta, T), i = \overline{1,3}$, $\frac{\partial F}{\partial T}(\theta, T)$:

$$\|F(\theta^2, T_2) - F(\theta^1, T_1)\| \leq L_F(\|\theta^2 - \theta^1\| + |T_2 - T_1|), \tag{39}$$

$$\left\| \frac{\partial F}{\partial \theta_i}(\theta^2, T_2) - \frac{\partial F}{\partial \theta_i}(\theta^1, T_1) \right\| \leq K_\theta(\|\theta^2 - \theta^1\| + |T_2 - T_1|), \ i = \overline{1,3}, \tag{40}$$

$$\left\| \frac{\partial F}{\partial T}(\theta^2, T_2) - \frac{\partial F}{\partial T}(\theta^1, T_1) \right\| \leq K_T(\|\theta^2 - \theta^1\| + |T_2 - T_1|), \tag{41}$$

where $L_F$, $K_\theta$, $K_T$ are also some positive constants. In the inequalities (38)-(41) the values $\theta^1$, $\theta^2$, $T_1$, $T_2$ are arbitrary and satisfy the inclusions: $\theta^1$, $\theta^2 \in \Lambda(T)$; $T_1$, $T_2 \in [T_{\min}, T_{\max}]$. Relationships (39)-(41) are proved by arguments, which are similar to the arguments presented in Lemma 5 and Theorem 2.

| | $x_1$ | $y_1$ | $z_1$ | $\theta_1$ | $\theta_2$ | $\theta_3$ | $T$ | $\varepsilon$ |
|---|---|---|---|---|---|---|---|---|
| **Example 1** | 3.0625 | 4.3694 | 0.1528 | 0.8403 | 1.2201 | 2.2958 | 2.5340 | 0.0001 |
| | 5.5556 | 4.5139 | 6.1111 | 0.2722 | 1.9042 | 1.9736 | 3.4368 | 0.0001 |
| | 6.7361 | 4.8611 | 6.4583 | 0.2076 | 1.5896 | 1.9208 | 2.6875 | 0.0001 |
| **Example 2** | 0.0044 | 29.9900 | 0.0152 | 4.5649 | 4.5749 | 4.5849 | 4.5949 | 0.0001 |
| | 0.0010 | 29.9800 | 0.0100 | 1.8849 | 1.8949 | 3.1448 | 4.9021 | 0.0001 |
| | 0.0015 | 29.9850 | 0.0200 | 0.5756 | 0.5856 | 1.0639 | 3.9393 | 0.0001 |
| **Example 3** | 0.0010 | 2.4960 | 0.0500 | 0.0100 | 1.5909 | 1.6009 | 1.6109 | 0.0001 |
| | 0.0016 | 2.4930 | 0.0600 | 0.0100 | 1.0051 | 1.0151 | 1.0337 | 0.0001 |
| | 0.0020 | 2.4900 | 0.0070 | 4.4211 | 4.5145 | 4.7195 | 7.9649 | 0.0001 |
| **Example 4** | 0.0047 | 38.2000 | 2.2080 | 1.9424 | 1.9524 | 1.9624 | 1.9753 | 0.0001 |
| | 0.0008 | 38.2500 | 0.1955 | 1.2110 | 1.2210 | 1.2310 | 1.3060 | 0.0001 |
| | 0.0046 | 38.1764 | 0.2091 | 2.0709 | 2.0809 | 2.2736 | 2.3402 | 0.0001 |
| **Example 5** | 0.0012 | 74.6300 | 0.2780 | 1.9992 | 2.1024 | 2.1267 | 2.3000 | 0.0001 |
| | 0.0001 | 74.7676 | 0.1338 | 1.0001 | 2.0999 | 2.2002 | 3.3000 | 0.0001 |
| | 0.0001 | 74.8003 | 1.1499 | 0.2184 | 0.2284 | 0.8765 | 1.9992 | 0.0001 |
| **Example 6** | 0.0166 | 74.7600 | 0.0321 | 1.1003 | 3.1998 | 3.7001 | 4.0000 | 0.0001 |
| | 0.0009 | 74.6093 | 0.1030 | 0.0997 | 0.2003 | 2.6997 | 3.0000 | 0.0001 |
| | 0.0016 | 74.0408 | 0.1212 | 0.1000 | 0.2000 | 6.7438 | 7.0000 | 0.0001 |
| **Example 7** | 0.0019 | 146.8420 | 0.1684 | 1.4071 | 1.4171 | 1.4271 | 1.4371 | 0.0001 |
| | 0.0019 | 146.9000 | 0.1700 | 0.7174 | 0.7274 | 0.7374 | 0.7474 | 0.0001 |
| | 0.0023 | 146.5565 | 0.1667 | 0.1027 | 0.2974 | 5.0406 | 5.0506 | 0.0001 |

Table 1. Results of solving of controllability problem for Examples 1-7.

In Table 1 the results of corresponding numerical calculations are demonstrated for Examples 1-7 considered above. Here $(x_1, y_1, z_1)^\top$ are coordinates of the terminal point $w_1$; $(\theta_1, \theta_2, \theta_3)^\top$ are the required moments of switching of the piecewise constant control $u_\theta(t)$, defined by formula (28), which steers system (1) from initial point $w_0$ to the terminal point $w_1$; $T$ is the required terminal time such that on the interval $[0, T]$ this transfer occurs; $\varepsilon$ is the accuracy of calculations of the minimum of the function $G(\theta, T)$.

For Example 1 the dynamics of the attainable set $X(T)$ for solving the controllability problem is shown in Figure 10.

## 8. Minimizing of pollution concentration at terminal time

Let us consider for system (1) the problem of minimizing of pollution concentration at the terminal time $T$ as

$$J(u) = y(T) \rightarrow \min_{u(\cdot) \in D(T)}. \tag{42}$$

The existence of the optimal control $u_*(t)$ and its corresponding optimal trajectory $w_*(t) = (x_*(t), y_*(t), z_*(t))^\top$, $t \in [0, T]$ for the problem (1), (42) follows from Lemma 1 and (Bressan & Piccoli, 2007). We denote by $J_*$ the minimum of the functional $J(u)$, i.e. $J_* = J(u_*)$.

Then the optimal control problem (1), (42) we reformulate as an equivalent problem of finding minimum on the attainable set $X(T)$ of the type

$$J(w) = (p_0, w(T)) \to \min_{w \in X(T)}, \tag{43}$$

where $p_0 = (0,1,0)^\top \in R^3$.

Using the parametrization of the set $X(T)$ constructed above, the problem (43) is rewritten as the problem of finite dimensional minimization

$$J(\theta) = (p_0, w_\theta(T)) = (p_0, F(\theta, T)) \to \min_{\theta \in \Lambda(T)}. \tag{44}$$

Obviously, the components $\theta_i^*$, $i = \overline{1,3}$ of the vector $\theta^* \in \Lambda(T)$ that minimize the function $J(\theta)$ are the moments of switching of the optimal control $u_*(t)$.

For numerical solution of the problem (44) again the iterative gradient projection method (Vasil'ev, 2002) is used. Corresponding numerical algorithm is written in $C++$. The basing of the convergence of this method is presented in the previous section.

In Table 2 the results of the corresponding numerical solution of the optimal control problem (1), (42) are demonstrated for Examples 1-7 considered above. Here $(\theta_1^*, \theta_2^*, \theta_3^*)^\top$ are the required moments of switching of the piecewise constant optimal control $u_\theta^*(t) = u_*(t)$, defined by formula (28); $J_*$ is the minimum of the function $J(\theta)$; $\varepsilon$ is the accuracy of calculations of this minimum.

| | $\theta_1^*$ | $\theta_2^*$ | $\theta_3^*$ | $J_*$ | $\varepsilon$ |
|---|---|---|---|---|---|
| **Example 1** | 1.0000 | 1.0000 | 1.0000 | 0.219408 | 0.0001 |
| **Example 2** | 6.0000 | 6.0000 | 6.0000 | 29.984309 | 0.0001 |
| **Example 3** | 6.0000 | 6.0000 | 6.0000 | 2.451992 | 0.0001 |
| **Example 4** | 6.0000 | 6.0000 | 6.0000 | 37.896490 | 0.0001 |
| **Example 5** | 6.0000 | 6.0000 | 6.0000 | 73.949615 | 0.0001 |
| **Example 6** | 12.0000 | 12.0000 | 12.0000 | 73.319410 | 0.0001 |
| **Example 7** | 20.0000 | 20.0000 | 20.0000 | 146.285096 | 0.0001 |

Table 2. Results of solving of problem of minimizing of pollution concentration at terminal time for Examples 1-7.

In Figures 11-14 the graphs of the optimal control $u_*(t)$ for Examples 1-7 are presented. The graphs imply that for the considered parameters of system (1) the minimization of pollution concentration at terminal time $T$ can be achieved only at the maximal rate of aeration during the entire time interval $[0, T]$. Numerical calculations were made for other parameters of the system. It was found that at the given accuracy of calculations $\varepsilon$ the corresponding optimal control $u_*(t)$ was a piecewise constant function with one and two switchings.

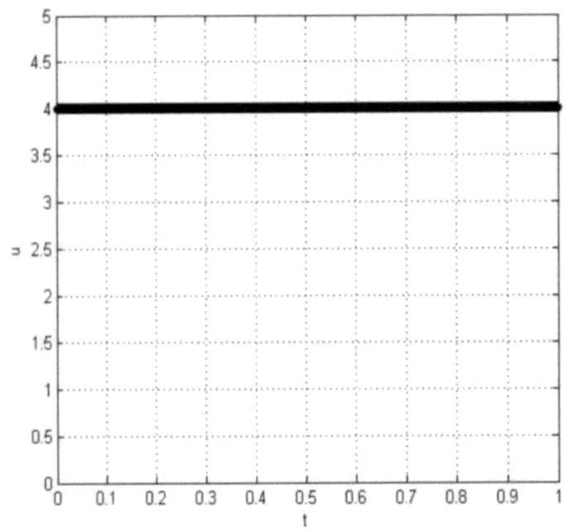


Fig. 11. Optimal control $u_*(t)$ for Example 1.

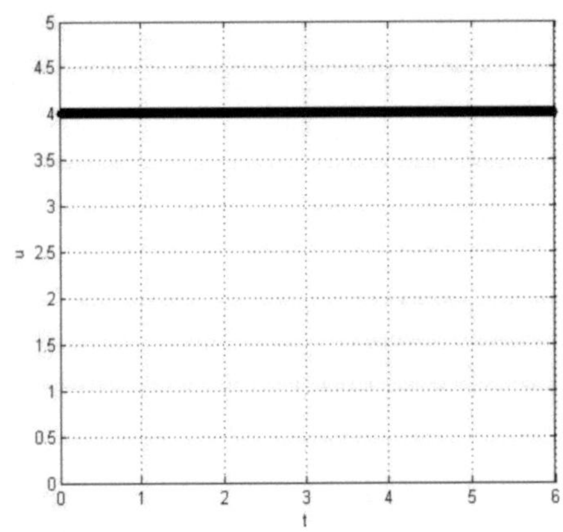


Fig. 12. Optimal control $u_*(t)$ for Examples 2-5.

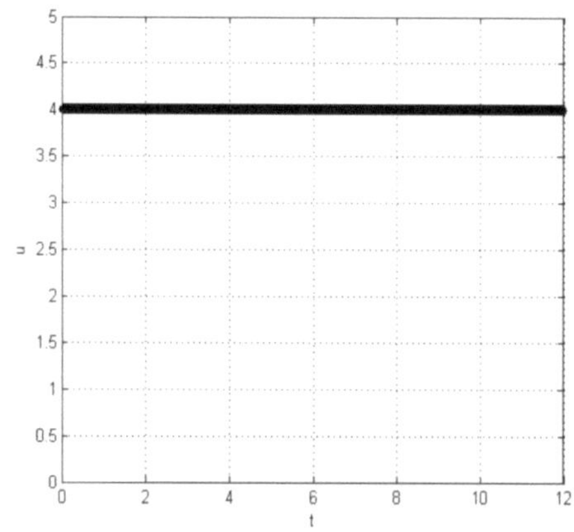


Fig. 13. Optimal control $u_*(t)$ for Example 6.

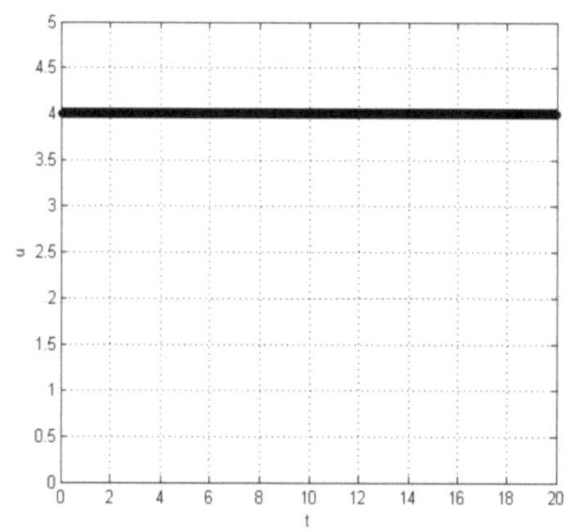


Fig. 14. Optimal control $u_*(t)$ for Example 7.

## 9. Conclusions

The ultimate aim of this Chapter is to reduce the operational cost of the aerobic biotreatment process via an increasing of its energy efficiency. In order to achieve this objective, we attempted constructing an optimal control for this process. Results of this Chapter can be immediately applied to practical ATAD reaction design. Having in mind the goal of developing the optimal control, in this Chapter we analytically obtained the detailed structure of an attainable set $X(T)$ for the model of the process. Furthermore, we succeeded in proving that the optimal control for this particular process is a bang-bang process, with at most two switchings. The moments of switching of the controls $u_\theta(t)$ in (28), which form the set $\Lambda(T)$, together with the mapping $F(\cdot, T)$, play the role of parametrization for the set $X(T)$ (for both its interior and boundary). We also proved that each point on the boundary of set $X(T)$ can be reached by a control from the above mentioned class (that is, a bang-bang control with at most two switchings), and that every point of the interior of set $X(T)$ is the result of a bang-bang control with precisely three switchings. This results serve as a basis for a computer code, which allows to construct attainable sets for a variety of initial conditions and the system parameters.

All these results are of instantaneous practical importance, as they, firstly, immensely narrow the class of functions, which should be considered as candidates for optimal control. This result is highly nontrivial, taking into consideration the nonlinearity of the model and it three dimensions, and it enables us to use a computer-assisted design for constructing an optimal control for a real-life situation.

In this Chapter we consider a particularly simple mathematical model of the process, which is composed of three variables. We postulate that the reaction is governed by the mass action law. Our objective is to present a rigorous mathematical analysis rather than a straightforward numerical simulation of the process so that more complicated models which incorporate specific features of ATAD are not addressed here. Nonetheless, we obtain a number of results that are of direct relevance to common practice. It is shown that the bang-bang control is applicable for this nonlinear model (it is a non-trivial result), and that at most two switchings are needed. However, it can be expected that the same results will be valid for a more complex system, such as Michaelis-Menten kinetics (functional responses with saturation) or models with larger numbers of variables.

Applied to more complicated models, these results may significantly reduce a choice for possible optimal controls, and thus considerably decrease the amount of computations which are needed to find these optimal controls numerically. It is also noteworthy that the general approach, which in this Chapter we applied to a particular problem of the optimal control for the aerobic biotreatment process, can be successfully used for a considerably wider range of real-life problems. Thus, this Chapter provides a general notion of how modern optimal control theory can be used in everyday practice.

## 10. Acknowledgements

A. Korobeinikov is supported by the Mathematics Applications Consortium for Science and Industry (www.macsi.ul.ie) funded by the Science Foundation Ireland Mathematics Initiative Grant 06/MI/005.

## 11. References

Bojarski, A. D., Rojas, J. & Zhelev, T. (2010). Modelling and sensitivity analysis of ATAD, *Computers & Chemical Engineering* 34(5): 802–811.

Bonnard, B. & Chyba, M. (2003). *Singular Trajectories and their Role in Control Theory*, Vol. 40 of *Mathematics & Applications*, Springer-Verlag, Berlin-Heidelberg-New York.

Bressan, A. & Piccoli, B. (2007). *Introduction to the Mathematical Theory of Control*, Vol. 2 of *Series on Applied Mathematics*, AIMS, USA.

Brune, D. (1985). Optimal control of the complete-mix activated sludge process, *Environmental Technology Letters* 6(11): 467–476.

Capon-Garcia, E., Rojas, J., Zhelev, T. & Graells, M. (2010). Operation scheduling of batch autothermal thermophilic aerobic digestion process, *Computer Aided Chemical Engineering* 28: 1117–1182.

Debuscher, D., Vanhooren, H. & Vanrolleghem, P. A. (1999). Benchmarking two biomass loading control strategies for activated sludge WWTPs, *Proceedings of 13th Forum on Applied Biotechnology*, University of Gent, 64/5a, Gent, Belgium, pp. 127–132.

Demidovich, B. P. (1967). *Lectures on Stability Theory*, Nauka, Moscow (in Russian).

Dmitruk, A. V. (1992). A generalized estimate on the number of zeroes for solutions of a class of linear differential equations, *SIAM Journal of Control and Optimization* 30(5): 1087–1091.

Fedorenko, R. P. (1978). *Approximate Solution of Optimal Control Problems*, Nauka, Moscow (in Russian).

Fikar, C. F., Chachuat, B. & Latifi, M. A. (2005). Optimal operation of alternating activated sludge processes, *Control Engineering Practice* 13(7): 853–861.

Gomez, J., de Gracia, M., Ayesa, E. & Garsia-Heras, J. L. (2007). Mathematical modeling of autothermal thermophilic aerobic digesters, *Water Research* 41(5): 959–968.

Graells, M., Rojas, J. & Zhelev, T. (2010). Energy efficiency optimization of wastewater treatment. Study of ATAD, *Computer Aided Chemical Engineering* 28: 967–972.

Grigorieva, E. V. & Khailov, E. N. (2001). On the attainability set for a nonlinear system in the plane, *Moscow University, Computational Mathematics and Cybernetics* (4): 27–32.

Grigorieva, E. V. & Khailov, E. N. (2005). Discription of the attainability set of a nonlinear controlled system in the plane, *Moscow University, Computational Mathematics and Cybernetics* (3): 23–28.

Grigorieva, E. V. & Khailov, E. N. (2010a). Minimization of pollution concentration on a given time interval for the waste water cleaning plant, *Journal of Control Sciences and Engineering* 2010: Article ID 712794, 10 pages.

Grigorieva, E. V. & Khailov, E. N. (2010b). Optimal control of a waste water cleaning plant, *Proceedings of 8th Mississippi State - UAB Conference on Differential Equations and Computational Simulations*, Electronic Journal of Differential Equations, Conference 19, pp. 161–175.
URL: *http://ejde.math.txstate.edu or http://ejde.math.unt.edu*

Hájek, O. (1991). *Control Theory in the Plane*, Vol. 153 of *Lecture Notes on Control and Information Sciences*, Springer-Verlag, Berlin-Heidelberg-New York.

Hall, D. W. & Spencer, G. L. (1955). *Elementary Topology*, Jorn Wiley & Sons, New York.

Hartman, P. (1964). *Ordinary Differential Equations*, Jorn Wiley & Sons, New York.

Hausdorff, F. (1962). *Set Theory*, AMS Chelsea Publishing, New York.

Henze, M., Gujer, W., Mino, T. & van Loosdrecht, M. (2000). *Activated Sludge Models ASM1, ASM2, ASM2d and ASM3*, IWA Publishing, London.

Holmberg, A. (1982). Modeling of the activated sludge process for microprocessor-based state estimation and control, *Water Research* 16(7): 1233–1246.

Kobouris, J. C. & Georgakakos, A. P. (1991). Optimal real-time activated sludge regulation, *Proceedings of 1991 Georgia Water Resources Conference*, Georgia Institute of Technology, Athens, Georgia, USA.
URL: *http://hdl.handle.net/1853/32021*

Krasnov, K. S., Vorob'ev, N. K., Godnev, I. N. & et al. (1995). *Physical Chemistry 2*, Vysshaya Shkola, Moscow (in Russian).

Layden, N. M. (2007). An evaluation of autothermal thermophilic aerobic digestion (ATAD) of municipal sludge in Ireland, *Journal of Environmental Engineering and Science* 6(1): 19–29.

Lee, E. B. & Markus, L. (1967). *Foundations of Optimal Control Theory*, Jorn Wiley & Sons, New York.

Lindberg, C.-F. (1998). Multivariable modeling and control of an activated sludge process, *Water Science and Technology* 37(12): 149–156.

Lukasse, L. J. S., Keesman, K. J., Klapwijk, A. & van Straten, G. (1998). Optimal control of N-removal in ASPs, *Water Science and Technology* 38(3): 255–262.

Maine Department of Environmental Protection, Augusta, ME (2010). *Aerated Lagoons - Wastewater Treatment*, Maine Lagoon Systems Task Force, Accessed 2010-07-11.

Moreno, J. (1999). Optimal time control of bioreactors for the wastewater treatment, *Optimal Control Applications & Methods* 20(3): 145–164.

Moreno, J. A., Betankur, M. J., Buitron, G. & Moreno-Andrade, I. (2006). Event-driven time optimal control for a class of discontinuous biorectors, *Biotechnology & Bioengineering* 94(4): 803–814.

Partasarathy, T. (1983). *On Global Univalence Theorems*, Vol. 977 of *Lecture Notes in Mathematics*, Springer-Verlag, Berlin-Heidelberg-New York.

Pontryagin, L. S., Boltyanskii, V. G., Gamkrelidze, R. V. & Mishchenko, E. F. (1962). *Mathematical Theory of Optimal Processes*, Jorn Wiley & Sons, New York.

Potter, T. G., Koopman, B. & Svoronos, S. A. (1996). Optimization of a perodic bilogical process for nitrogen removal from wastewater, *Water Research* 30(1): 142–152.

Qin, S. J., Martinez, V. M. & Foss, B. A. (1997). An interpolating model predictive control strategy with application to a waste treatment plant, *Computers & Chemical Engineering* 21(supplement): 881–886.

Rinaldi, S., Soncini-Sessa, R., Stehfest, H. & Tamura, H. (1979). *Modeling and Control of River Quality*, McGraw-Hill Inc., London.

Robinson, R. C. (2004). *An Introduction to Dynamical Systems: Continuous and Discrete*, Pearson Prentice Hall, New Jersey.

Rojas, J., Burke, M., Chapwanya, M., Doherty, K., Hewitt, I., Korobeinikov, A., Meere, M., McCarthy, S., O'Brien, M., Tuoi, V. T. N., Winstenley, H. & Zhelev, T. (2010). Modeling of autotermal thermophilic aerobic digestion, *Mathematics-in-Industry Case Studies Journal* 2: 34–63.

Rojas, J. & Zhelev, T. (2009). Maximizing the capacity of wastewater treatment process - study of autothermal thermophilic aerobic digestion, *Chemical Engineering Transactions* 18: 881–886.

Schwartz, L. (1967). *Analyse Mathematique 1*, Hermann, Paris.

Shigeo, I. (1985). A note on global implicit function theorems, *IEEE Transactions on Curcuits and Systems* 32(5): 503–505.

Tikhonov, A. N., Vasil'eva, A. B. & Sveshnikov, A. G. (1985). *Differential Equations*, Springer-Verlag, Berlin-Heidelberg-New York.

Tzoneva, R. (2007). Method for real time optimal controlof the activated sludge process, *Proceedings of the 15th Mediterranean Conference on Control and Automation*, Athens, Greece, T28-008, 6 pages.

Vasil'ev, F. P. (2002). *Optimization Methods*, Factorial Press, Moscow (in Russian).

# 6

# Use of Agro-Industrial Wastes in Solid-State Fermentation Processes

Solange I. Mussatto, Lina F. Ballesteros, Silvia Martins and José A. Teixeira
*IBB – Institute for Biotechnology and Bioengineering, Centre of Biological Engineering, University of Minho, Portugal*

## 1. Introduction

Large amount of wastes is generated every year from the industrial processing of agricultural raw materials. Most of these wastes are used as animal feed or burned as alternative for elimination. However, such wastes usually have a composition rich in sugars, minerals and proteins, and therefore, they should not be considered "wastes" but raw materials for other industrial processes. The presence of carbon sources, nutrients and moisture in these wastes provides conditions suitable for the development of microorganisms, and this open up great possibilities for their reuse in solid-state fermentation (SSF) processes, for example. Agro-industrial wastes can be used as solid support, carbon and/or nutrient source in SSF processes for the production of a variety of value-added compounds.

SSF is a technology lesser explored than submerged fermentation systems, but that has been proved to be able to give higher product yields and productivities, which is of great interest for industrial activities. In addition, costs are much lower due to the efficient utilization and value-addition of wastes. The reuse of agro-industrial wastes in SSF processes is of particular interest due to their availability and low cost, besides being an environment friendly alternative for their disposal. The present chapter deals with the use of agro-industrial wastes in SSF processes. Initially, an overview about the generation of these wastes and their chemical composition is presented. In the sequence, the characteristics of SSF systems and variables that affect the product formation by this process are reviewed and discussed. Finally, potential applications of agro-industrial wastes in SSF processes for the obtainment of value-added compounds are described.

## 2. Agro-industrial wastes: generation and chemical composition

Agro-industrial wastes are generated during the industrial processing of agricultural or animal products. Those derived from agricultural activities include materials such as straw, stem, stalk, leaves, husk, shell, peel, lint, seed/stones, pulp or stubble from fruits, legumes or cereals (rice, wheat, corn, sorghum, barley…), bagasses generated from sugarcane or sweet sorghum milling, spent coffee grounds, brewer's spent grains, and many others. These wastes are generated in large amounts throughout the year, and are the most abundant renewable resources on earth. They are mainly composed by sugars, fibres, proteins, and minerals, which are compounds of industrial interest. Due to the large availability and composition rich in compounds that could be used in other processes, there is a great interest on the reuse of these wastes, both from economical and environmental view points. The economical aspect is based on the fact that such wastes may be used as low-cost raw materials for the production of other value-added compounds, with the expectancy of reducing the production costs. The environmental concern is because most of the agro-industrial wastes contain phenolic compounds and/or other compounds of toxic potential; which may cause deterioration of the environment when the waste is discharged to the nature.

Large amount of the agro-industrial wastes are mainly composed by cellulose, hemicellulose and lignin, being called "lignocellulosic materials". In the lignocellulosic materials, these three fractions are closely associated with each other constituting the cellular complex of the vegetal biomass, and forming a complex structure that act as a protective barrier to cell destruction by bacteria and fungi. Basically, cellulose forms a skeleton which is surrounded by hemicellulose and lignin (Fig. 1).

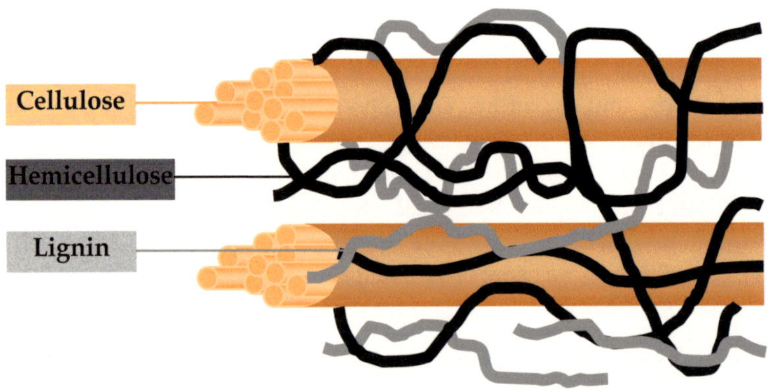


Fig. 1. Schematic representation of the cellulose, hemicelluloses and lignin fractions in the lignocellulosic materials

The cellulose structure is composed only by glucose units, i.e., it is a homopolymer where units of cellobiose (two anhydrous glucose rings joined via a β-1,4 glycosidic linkage) are sequentially repeated (Fig. 2) (Klemm et al., 1998). The long-chain of cellulose polymers, which may have until 10,000 glucose units, are linked together by hydrogen and van der Walls bonds, which cause the cellulose to be packed into microfibrils (Ha et al., 1998). By forming these hydrogen bounds, the chains tend to arrange in parallel and form a crystalline structure (as represented in Fig. 1). Cellulose microfibrils have both highly crystalline regions (around 2/3 of the total cellulose) and less-ordered amorphous regions. More ordered or crystalline cellulose is less soluble and less degradable, being strongly resistant to chemicals (Taherzadeh & Karimi, 2008).

On the contrary of the cellulose, hemicellulose is a heterogeneous polymer usually composed by five different sugars (L-arabinose, D-galactose, D-glucose, D-mannose, and D-xylose) and some organic acids (acetic and glucuronic acids, among others). The structure of the hemicellulose is linear and branched. The backbone of the hemicellulose chain can be formed by repeated units of the same sugar (homopolymer) or by a mixture of different

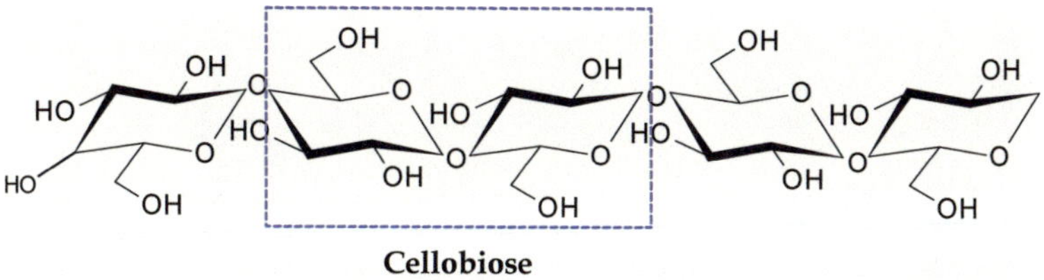


Fig. 2. Representation of the cellulose structure

sugars (heteropolymer). According to the main sugar in the backbone, hemicellulose has different classifications e.g., xylans, glucans, mannans, arabinans, xyloglucans, arabinoxylans, glucuronoxylans, glucomannans, galactomannans, galactoglucomannans and β-glucans. Fig. 3 shows an example of hemicellulose structure formed by a xylan backbone (repeated units of xylose sugar). Besides the differences in the chemical composition, hemicellulose also differs from cellulose structure in other aspects, including: 1) the size of the chain, which is much smaller (it contains approximately 50-300 sugar units); 2) the presence of branching in the main chain molecules, and 3) to be amorphous, being less resistant to chemicals (Fengel & Wegener, 1989).

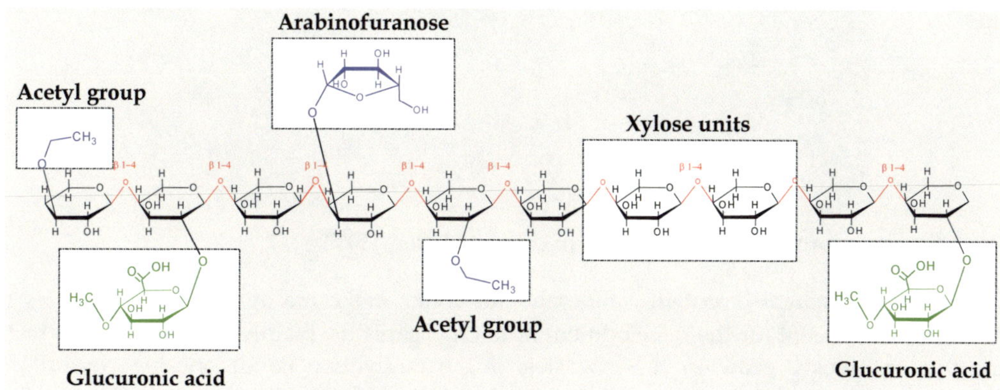


Fig. 3. Representation of a hemicellulose structure formed by a xylan backbone

The lignin structure is not formed by sugar units, but by phenylpropane units linked in a large and very complex three-dimensional structure. Three phenyl propionic alcohols are usually found as monomers of lignin, which include the alcohols p-coumaryl, coniferyl, and sinapyl. Lignin is closely bound to cellulose and hemicellulose and its function is to provide rigidity and cohesion to the material cell wall, to confer water impermeability to xylem vessels, and to form a physic-chemical barrier against microbial attack (Fengel & Wegener, 1989). An example of structure proposed for the lignin is given in Fig. 4. Due to its molecular configuration, lignin is extremely resistant to chemical and enzymatic degradation.

The percentage of cellulose, hemicellulose and lignin is different to each waste since it varies from one plant species to another, and also according to the process that the agricultural material was submitted. In addition, the ratios between various constituents in a single plant may also vary with age, stage of growth, and other conditions. Usually, cellulose is the dominant fraction in the plant cell wall (35–50%), followed by hemicellulose (20–35%) and lignin (10–25%). Average values of the main components in some lignocellulosic wastes are shown in Table 1.

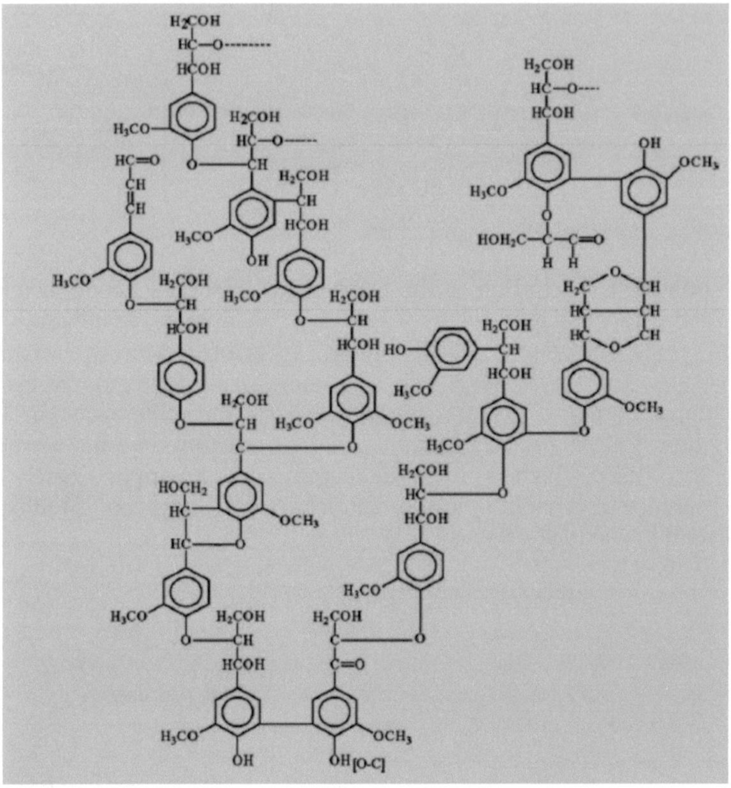

Fig. 4. Structure proposed for the lignin of conifer (Adler, 1977).

The presence of sugars, proteins, minerals and water make the agro-industrial wastes a suitable environment for the development of microorganisms, mainly fungal strains, which are able to quickly grow in these wastes. If the cultivation conditions are controlled, different products of industrial interest may be produced, avoiding the loss of potential energy sources.

| Lignocellulosic waste | Cellulose (wt %) | Hemicellulose (wt %) | Lignin (wt %) |
|---|---|---|---|
| Barley straw | 33.8 | 21.9 | 13.8 |
| Corn cobs | 33.7 | 31.9 | 6.1 |
| Corn stalks | 35.0 | 16.8 | 7.0 |
| Cotton stalks | 58.5 | 14.4 | 21.5 |
| Oat straw | 39.4 | 27.1 | 17.5 |
| Rice straw | 36.2 | 19.0 | 9.9 |
| Rye straw | 37.6 | 30.5 | 19.0 |
| Soya stalks | 34.5 | 24.8 | 19.8 |
| Sugarcane bagasse | 40.0 | 27.0 | 10.0 |
| Sunflower stalks | 42.1 | 29.7 | 13.4 |
| Wheat straw | 32.9 | 24.0 | 8.9 |

Table 1. Main components of some lignocellulosic wastes (Nigam et al., 2009)

## 3. Solid-State Fermentation

Solid-state fermentation (SSF) consists of the microbial growth and product formation on solid particles in the absence (or near absence) of water; however, the substrate contains sufficient moisture to allow the microorganism growth and metabolism (Pandey, 2003). This bioprocess has been subject of several studies and it has been proved that SSF has the important advantage of leading to higher yields and productivities or better product characteristics than submerged fermentation (SmF), which is characterized by the cultivation of the microorganisms in a liquid medium. Another great advantage of SSF compared to SmF is the lower capital and operating costs due to the utilization of low cost agricultural and agro-industrial wastes as substrates. The low water volume used in SSF has also a large impact on the economy of the process mainly because of the smaller fermenter-size, the reduced downstream processing, the reduced stirring and lower sterilization costs (Hölker & Lenz, 2005; Nigam, 2009). Several advantages and disadvantages of SSF over SmF are summarized in Table 2.

### 3.1 Important technical aspects

Despite the numerous processing and biological advantages that SSF has over SmF, the scaling-up of this bioprocess is not well developed and has a drawback associated to control of operations (heat transfer and culture homogeneity problems) and fermentation variables (mainly pH and temperature). Three types of bioreactors are commonly used in SSF processes: packed-bed, horizontal drum and fluidized bed. However, these bioreactors have their own advantages and disadvantages, and there is a necessity to develop novel bioreactors with better design. In order to overcome all these difficulties, research attention has been directed towards the development and implementation of effective control strategies of bioreactors operating in SSF conditions in large-scale and their design (Durand, 2003; von Meien et al., 2004; Mitchell et al., 2000). Also an economical evaluation of the overall process is needed in order to determine its feasibility to a specific purpose.

Separation of biomass represents another challenge in SSF, being essential for the kinetic studies. Several indirect methods have been employed including glucosamine estimation, ergosterol estimation, protein (Kjeldahl) estimation, DNA estimation, dry weight changes and $CO_2$ evolution; however, all of these methodologies have their own weaknesses. The estimation of oxygen intake and carbon dioxide evolution rate are considered to be most accurate for the determination of the microorganism growth (Pandey et al., 2007). Digital image processing has also been developed for measuring biomass in SSF (Couri et al., 2006).

There are various important factors to be considered for the development of a successful bioprocess under SSF conditions. Some of the most important include the selection of a suitable microorganism and solid support to be used. A diversity of microorganisms, including fungi, yeasts and bacteria can be used in SSF processes. Due to the low moisture content in the SSF media, fungi are the most commonly used microorganisms because of their ability to growth in these environments. Table 3 summarizes some of the most recent studies in SSF, the microorganisms and solid supports employed. Note that a large variety of solid supports have been used in these processes, and fungi are effectively the most used microorganisms. The selection of the microorganism to be used in a SSF process also depends on the desired end product. For example, filamentous fungi have great potential to produce bioactive compounds and thermostable enzymes of high scientific and commercial value by SSF, and therefore they are the most used microorganisms for this purpose.

| Advantages | Disadvantages |
|---|---|
| Similar or higher yields than those obtained in the correspondent submerged cultures | Biomass determination is very difficult |
| The low availability of water reduces the possibilities of contamination by bacteria and yeast. This allows working in aseptic conditions in some cases | Usually the substrates require pre-treatment (reduction of size by grinding, rasping or chopping, homogenization, physical, chemical or enzymatic hydrolysis, steam treatment) |
| The environmental conditions are similar to those of the natural habitats of fungi, which constitute the main group of microorganisms used in SSF | Only microorganisms able to grow at low moisture levels can be used |
| Higher levels of aeration, especially adequate in processes that demand an intensive oxidative metabolism | The solid nature of the substrate causes problems in the monitoring of the process parameters (pH, moisture content, oxygen and biomass concentration) |
| The inoculation with spores (in those processes that involve fungi) facilitates their uniform dispersion through the medium | Agitation may be very difficult. For this reason static conditions are preferred |
| Culture media are often quite simple. The substrate usually provides all the nutrients necessary for growth | Frequent need of high inoculum volumes |
| Simple design reactors with few spatial requirements can be used due to the concentrated nature of the substrates | Many important basic scientific and engineering aspects are yet poor characterized. Information about the design and operation of reactors on a large scale is scarce |
| Low energetic requirements (in some cases autoclaving or steam treatment, mechanical agitation and aeration are not necessary) | Possibility of contamination by undesirable fungi |
| Small volumes of polluting effluents are generated. Fewer requirements of solvents are necessary for the product extraction due to its high concentration | The removal of metabolic heat generated during growth may be very difficult |
| The low moisture availability may favour the production of specific compounds that may not be produced or may be poorly produced in SmF | Extracts containing products obtained by leaching of fermented solids are often of viscous nature |
| In some cases, the products obtained have slightly different properties (e.g. more thermotolerance) when produced in SSF in comparison to SmF | Mass transfer limited to diffusion<br>In some SSF, aeration can be difficult due to the high solid concentration |
| Due to the concentrated nature of the substrate, smaller reactors in SSF with respect to SmF can be used to hold the same amounts of substrate | Spores have longer lag times due to the need for germination<br>Cultivation times are longer than in SmF |

Table 2. Advantages and disadvantages of SSF compared to SmF (Pérez-Guerra et al., 2003)

| Microorganism | Solid support | Reference |
|---|---|---|
| **Fungi** | | |
| *Aspergillus niger* | Lemon peel, orange peel, apple pomace, pistachio shell, wheat bran, coconut husk, pecan nutshell, creosote bush leaves, bean residues | Orzua et al. (2009) |
| *Aspergillus niveus* | Sugar cane bagasse | Guimarães et al. (2009) |
| *Aspergillus oryzae* | Red gram plant waste | Shankar & Mulimani (2007) |
| *Aspergillus sojae* | Crushed maize, maize meal, corncob | Ustok et al. (2007) |
| *Bjerkandera adusta* *Ganoderma applanatum* *Phlebia rufa* *Trametes versicolor* | Wheat straw | Dinis et al. (2009) |
| *Phanerochaete chrysosporium* | Rice straw | Yu et al. (2009) |
| *Penicillium sp.* | Soybean bran | Wolski et al. (2009) |
| *Sporotrichum thermophile* | Sesame oil cake | Singh & Satyanarayana (2008) |
| *Trichosporon fermentans* | Rice straw | Huang et al. (2009) |
| **Yeast** | | |
| *Baker yeast AF37X* | Sweet sorghum | Yu et al. (2008) |
| *Saccharomyces cerevisiae* | Mahula flowers Corn stover | Mohanty et al. (2009) Zhao & Xia (2009) |
| **Bacteria** | | |
| *Nocardia lactamdurans* | Wheat bran, rice, soybean oil cake, soybean flour | Kagliwal et al. (2009) |
| *Bacillus sphaericus* | Wheat bran | El-Bendary et al. (2008) |
| *Pseudomonas aeruginosa* | Jatropha curcas seed cake | Mahanta et al. (2008) |
| *Streptomyces* | Coffee pulp | Orozco et al. (2008) |

Table 3. Recent studies on SSF (Martins et al., 2011)

The process variables such as pre-treatment and particle-size of substrates, medium ingredients, supplementation of growth medium, sterilization of SSF medium, moisture content, water activity (aw), inoculum density, temperature, pH, agitation and aeration, have also to be considered for an efficient SSF process (Nigam & Pandey, 2009). Generally the production yields can be improved with a suitable choice of the substrate or mixture of substrates with appropriate nutrients. As a whole, the support material must present characteristic favourable for the microorganism development, be of low cost, and present a good availability, which might involve the screening of several agro-industrial wastes. The moisture content and aw play also an important role in SSF. In general, the substrates have a water content oscillating between 30 and 85%. Low values might induce the sporulation of the microorganism, while more elevated levels may reduce the porosity of the system, which can produce oxygen transfer limitation. In fact, the water requirements of the microorganism have been considered to be better defined in terms of aw rather than water content in the solid substrate (Raimbault, 1998), since the aw is the water available or accessible for the growth of the microorganism, and affects the biomass development, metabolic reactions, and the mass transfer processes.

Finally, the selection of the most appropriate downstream process for the final product obtained is also crucial when SSF processes are performed. The product obtained by SSF may be recovered from the solid fermented mass by extraction with solvents (aqueous or other solvents mixtures). The type of solvent and its concentration, as well as the ratio of solvent to the solid and pH are important variables that influence the product extraction. In addition, since the metabolites diffuse throughout the solid mass during the culturing, long extraction-times may be required for complete product recovery. The cost of purification depends on the quality of the obtained extract. For example, the presence and concentration of inert compounds in the extract increase the cost of purification and therefore the cost of recovery is increased. Particularly those secondary metabolites which are used in bulk in the pharmaceutical and health industry and whose purity is governed by stringent regulations need to go through specific purification strategy (Nigam, 2009).

### 3.2 Modelling in SSF

Modelling is an important tool to improve the chances of successfully transforms an SSF process from laboratory to a scaled-up commercial level. Basically, the emergence of modelling in SSF underlie on the need of searching mathematical expressions that represent the system under consideration, focusing the following main problems: (i) the representation of the microbial activity (kinetic patterns and thermodynamic concerns), (ii) studies of the problems of heat accumulation and heterogeneous distribution in complex gas-liquid-solid multiphase bioreactor (or fermenter) systems, (iii) the connection between the two above systems and (iv) the selection of the best type of the fermenter (Koncsag & Kirwan, 2010; Singhania et al., 2009). The ability to predict the behaviour of a SSF process can ascertain the validity of the system, while establishing different parameters that characterize the process, and appropriate mechanisms for its development and control. The modelling of bioreactors employed in SSF processes can play a crucial role in the analysis, design and development of bioprocesses for the treatment of agro-industrial wastes. The development of models can also significantly reduce the number of wet experiments, and in consequence will have a strong impact on time saving and costs of the process.

Several research studies have been performed in order to develop mathematical models capable of characterizing SSF bioreactors. For example, a mathematical model has been developed for a packed bed SSF bioreactor employing the N-tanks in series approach to analyze the production of protease by *Aspergillus niger* (Sahir et al., 2007). Another simple mathematical model was studied in order to quantify the performance of continuous solid-state bioreactors having two different solid substrate flow patterns, plug flow and completely mixed flow, where plug flow shown to have superior performance when high product concentration was needed (Khanahmadia et al., 2006). The modelling and optimization of simultaneous influence of temperature and moisture on microbial growth through a quantitative description has been also investigated, as well as the effect of microbial biomass on the isotherm of the fermenting solids in SSF. More recent studies evaluated the modelling of different phases of the bacterial growth curve and the production of α-amylase by *Bacillus* sp. KR-8104 in a solid-state fermentation process based on variation in dry weight using wheat bran (Hashemi et al., 2011). Another recent subject of research was the determination of the kinetics of microbial growth related to pectinase and xylanase synthesis during the growth of *Aspergillus niger* F3 in a horizontal SSF drum bioreactor charged with 2 kg of dried citrus peel (Rodríguez-Fernández et al., 2011).

It is clear that all the results obtained by these studies are of great importance in the scale-up and optimization of SSF bioreactors configuration for the production of a desired end product using agro-industrial wastes as solid support. Nevertheless, several issues still need to be analyzed and solved for improved modelling of SSF. Information regarding modelling in SSF has been limited because of the unavailability of suitable methods for direct measurement of the microorganism growth. This is due to the difficulty to separate the microorganism from the substrate, and problems related to the determination of the substrate utilization rate. An alternative to overcome these problems is the determination of the microorganism cell growth by measuring the change in gaseous compositions inside the bioreactor, since it is well known that the fermentation kinetics is very sensitive to the variation in ambient and internal gas compositions (Singhania et al., 2009).

## 4. Potential applications of agro-industrial wastes in SSF processes for the obtainment of value-added compounds

SSF has two potential areas of application. One of them is for environmental control such as for the production of compost, ensiling and animal feed from solid wastes, bioremediation and biodegradation of hazardous compounds, and biological detoxification of agro-industrial wastes. On the other hand, SSF may be utilized to obtain value-added compounds such as enzymes, mushrooms, amino acids, biopesticides, biofuels, biosurfactants, organic acids, flavours, colorants, aromatic compounds, biologically active secondary metabolites, and other substances of interest to the food industry. The following sections describe some of the most important applications of agro-industrial wastes in SSF processes, the microorganisms used and products obtained.

### 4.1 Organic acids

Organic acids are widely used in food and beverages industries because of their excellent characteristic to prevent deterioration and extending the shelf life of food. SSF has been employed for many years to produce citric acid and lactic acid to large-scale production being a successful process. The production of oxalic acid, gluconic acid and gallic acid by SSF has also been reported. Table 4 shows some examples of organic acids produced by SSF, the microorganisms and solid supports utilized.

| Product | Microorganism | Solid support |
|---|---|---|
| Citric acid | *Aspergillus niger* | corncob, sugarcane bagasse, coffee husk, kiwi fruit peel, wheat bran, rice bran, pineapple waste, mixed fruit waste, apple pomace, sawdust with rice hulls, cassava fibrous residue, potato starch residue |
| | *Aspergillus foetidus* | pineapple waste |
| Lactic acid | *Lactobacillus delbrueckii* | cassava bagasse and sugarcane bagasse |
| | *Rhizopus oryzae* | sugarcane bagasse |
| | *Lactobacillus paracasei* | sweet sorghum |
| | *Rhizopus oryzae* | carrot-processing waste |
| Oxalic acid | *Aspergillus oryzae* | wheat kernels |
| Gluconic acid | *Aspergillus niger* | tea waste with sugarcane molasses |
| Gallic Acid | *Rhizopus oryzae* | gallo seeds cover |

Table 4. Organic acids production by SSF

### 4.1.1 Citric acid

Citric acid is the most important organic acid produced at industrial level, and is extensively used in foods, beverages, cosmetics, pharmaceuticals and chemical products. In the food industry, citric acid has been applied as an additive, being utilized as preservative, flavour enhancer, antifoam, or antioxidant. In the chemical industry, it is exploited as plasticizer, softener, and for the treatment of textiles. This acid has also been widely used in the detergent industry for replacement of polyphosphates, decreasing the production costs.

SmF has been substantially studied to produce citric acid. *Aspergillus niger* is one of the microorganisms commercially used to produce citric acid under SSF conditions, being cultivated on agro-industrial wastes such as corncob, sugarcane bagasse, coffee husk, kiwi fruit peel, wheat bran, rice bran, pineapple waste, mixed fruit waste, maosmi waste sugar beet molasses, sawdust with rice hulls, cassava fibrous residue, apple pomace, and potato starch residue. In a study on the production of citric acid by *A. niger* LPB 21 cultivated on three cellulosic supports (cassava bagasse, sugarcane bagasse, and vegetable sponge), cassava bagasse was the best substrate, giving 27 g citric acid per 100 g dry substrate under optimum fermentation conditions, which corresponded to 70% yield (based on sugars consumed). When *A. niger* NRRL 2001 was cultivated on sugarcane bagasse, coffee husk and cassava bagasse, cassava bagasse was also the best support for SSF, giving the highest yield of citric acid (88 g/kg dry matter) (Pandey et al., 2000a).

It merits emphasizing that the acid production by SSF is directly influenced by the nitrogen source used in the fermentation, and ammonium salts such as urea, ammonium chloride, and ammonium sulphate are the most used to obtain high yields. Potassium dihydrogen phosphate has been found to be the most suitable phosphorus source and low levels of phosphorus were found to favour citric acid production. It has been generally found that addition of methanol, ethanol, isopropanol or methyl acetate, copper and zinc enhance the production of citric acid, while magnesium result essential for growth as well as for citric acid production (Krishna, 2005).

### 4.1.2 Lactic acid

Lactic acid plays an important role in various biochemical processes. It is used, for example, as acidulant and preservative of many food products such as cheese, meat, beer and jellies. Besides the applications in the food industry, lactic acid has also wide uses in pharmaceutical, leather and textile industries. One of the most important applications of this acid is in the synthesis of biodegradable plastics and coatings, but it is also used in the manufacture of cellophane, resins, some herbicides and pesticides.

Lactic acid production by SSF has been carried out using fungal as well as bacterial strains. *Rhizopus* sp. and *Lactobacillus* sp. strains are the most utilized microorganisms to produce this acid; and sugarcane bagasse, sugarcane press mud and carrot-processing wastes have been used as substrates in these processes. The production of L(+)-lactic acid by *R. oryzae* in SSF conditions using sugarcane bagasse as support material was demonstrated to promote slightly higher productivity than the production in SmF conditions (Couto & Sanromàn, 2006). *L. Paracasei* and *L. delbrueckii* bacteria have also been reported to efficiently produce this acid by SSF.

### 4.1.3 Other acids

Oxalic acid and gluconic acid are other examples of organic acids produced by SSF. Some studies report the production of oxalic acid by *Aspergillus oryzae* using wheat kernels as support (Biesebeke et al., 2002), and the production of gluconic acid by *Aspergillus niger* ARNU-) using tea waste as support and sugarcane molasses as carbon source (Sharma et al., 2008). Both compounds have great industrial applications. The main application of the oxalic acid, for example, includes cleaning or bleaching, especially for the removal of rust. It is also used in the restoration of old wood and is an important reagent in lanthanide chemistry. Gluconic acid is utilized as a food additive, acting as acidity regulator. It is also used in cleaning products where it dissolves mineral deposits especially in alkaline solution.

## 4.2 Production of flavour and aroma compounds

Aroma compounds can be found in food, wine, spices, perfumes and essential oils, but over a quarter of them are used in the food industry. These compounds play an important role in the production of flavours, which are used to improve food quality and give it value-added. Most of the flavouring compounds are presently produced via chemical synthesis or extraction from natural materials. However, since the consumers prefer food free of chemical substances, the microbial biosynthesis or bioconversion systems result as a promising substitute to produce aroma compounds.

Both fungi and bacteria have ability to synthesise aroma compounds by SSF. Fungi from the genus *Ceratocystis* produce a large range of fruit-like or flower-like aromas such as peach, pineapple, banana, citrus, and rose, depending on the strain and the culture conditions. Another interesting fungal strain for use on the production of aroma compounds is *C. fimbriata*. This strain has an enormous potential for ester synthesis since it grows rapidly, has a great sporulation capacity and produces a wide variety of aromas. Fruity aroma production by *C. fimbriata* in solid cultures have been reported using different agro-industrial wastes such as cassava bagasse (used in combination with soya bean or apple pomace), apple pomace, amaranth and soya bean. Production of a strong pineapple aroma is reported when this fungus is cultivated on SSF using coffee husk as substrate. It has also been demonstrated that the addition of glucose to the solid medium (wheat bran, cassava bagasse and sugarcane bagasse) results in the production of a fruity aroma by *C. fimbriata*, while the addition of leucine or valine causes a strong banana aroma (Christen et al., 1997).

Production of aroma compounds by SSF has also been reported using *Kluyveromyces marxianus* cultivated on different solid substrates such as cassava bagasse, giant palm bran, apple pomace, sugarcane bagasse and sunflower seeds. This strain was able to produce nine and eleven compounds (alcohols, esters and aldehydes) from palm bran and cassava bagasse, respectively; among of which, the esters were responsible for the fruity aroma (Medeiros et al., 2000). In fact, esters are considered the source of the aromas. Pyrazine, for example, is a heterocyclic aromatic organic compound found in a wide variety of food, which possess nutty and roasty flavour.

Other fungal strains reported to produce aroma compounds are the edible fungus *Rhizopus oryzae*, *Neurospora* sp., *Zygosaccharomyces rouxii*, and *Aspergillus* sp. Among the bacterial strains, *Bacillus natto* and *B. subtilis* are reported to produce aroma compounds by SSF.

### 4.3 Production of enzymes

The production of enzymes has evolved rapidly and nowadays, enzymes are the most important products obtained for human needs through microbial sources. Enzymes have application in a variety of areas including food biotechnology, environment, animal feed, pharmaceutical, textile, paper and others technical and chemical industries. Due to the large industrial application and significant cost, there is a necessity to develop processes able to minimize the production costs. In this sense, the production of enzymes by SSF has been greatly explored and it has been proved that this technology is able to promote higher yields than the production by SmF, for the same microorganism strain. Additionally, a variety of agro-industrial wastes may be used as support material, carbon and nitrogen sources for the production of different enzymes by SSF. This is an important aspect since allows the reuse of a variety of low cost wastes for the production of this value-added product, which contribute for the wastes reuse and to decrease the production costs. Some of the most relevant enzymes obtained by SSF, the microorganisms and solid supports used are shown in the Table 5.

### 4.4 Fructooligosaccharides production

Fructooligosaccharides (FOS), also called oligofructose or oligofructan, are oligosaccharides that, when ingested, promote enormous benefits to the human health. They can be used as artificial or alternative sweetener and are considered a small dietary fibre with low caloric value. Additionally, FOS has important functional properties due to their capacity of serving as a substrate for microflora in the large intestine, increasing the overall gastrointestinal tract health. FOS promotes also the calcium and magnesium absorption in animals and human gut, and increases the levels of phospholipids, triglycerides and cholesterol.

| Product | Microorganism | Solid support |
|---|---|---|
| Glucoamylase | *Aspergillus niger* | Wheat bran |
| Cellulases | *Bacillus subtilis* | Banana fruit stalk |
| | *Aspergillus ustus,*<br>*Sporotrichum pulverulentum,*<br>*Trichoderma sp.*<br>*Botrytis sp.* | Wheat bran, rice straw |
| | *Trichoderma aureoviride* | Leached beet pulp |
| | *Penicillium citrinum* | Rice husks |
| | *Trichoderma reesi* | Sweet sorghum silage |
| | *Bacillus subtilis* | Banana fruit stalk |
| | *Trichoderma viride* | Coconut pith |
| Lipases | *Candida rugosa*<br>*Candida sp.*<br>*Monascus fulginosus*<br>*Neurospora sitophila*<br>*A. niger* | Rice bran, wheat bran, peanut<br>Press cake and coconut oil cake |
| Proteases | *Bacillus licheniformis* | Rice straw |
| | *Penicillium sp.* | Defatted soybean cake |
| Xylanase | *Thermoascus aurantiacus* | Corn silage |
| | *Penicillium decumbens* | Corn straw |
| α-galactosidase and β-galactosidase | *Aspergillus niger*<br>*A. oryzae*<br>*Candida sp.*<br>*Neurospora sitophila*<br>*P. candidum*<br>*Mucor sp.*<br>*Kluyveromyces lactis* | Wheat bran, soybean cake waste |
| α -amylase and β -amylase | *Aspergillus sp.*<br>*Rhizopus sp.*<br>*Mucor sp.*<br>*Bacillus sp.*<br>*Saccharomyces sp*<br>*Bacillus subtilis*<br>*Aeromonas caviae* | Rice husk, coconut cake, tea waste, cassava, cassava bagasse, sugarcane bagasse, banana waste, corn flour |
| Ligninase | *Pleurotus sp.*<br>*Phanerochaete chrysosporium* | Wheat straw and bagasse |
| Tannase | *A. niger* | Palm kernel cake |
| Phytase | *A. ficuum, A. carbonarius* | Wheat bran |

Table 5. Enzymes produced by SSF

Currently, FOS is mainly produced on industrial scale by SmF from the disaccharide sucrose by microbial enzymes with transfructosylating activity (β-fructofuranosidases, EC.3.2.1.26, also designed as fructosyltransferases EC.2.4.1.9). However, the interest to produce FOS using different fungus such as *Aspergillus, Aureobasidum,* and *Penicillium,* in which these enzymes have been found, has increased in the last years. *Aspergillus japonicus* has been considered a potential strain for industrial production of FOS by SSF (Chien et al., 2001; Mussatto et al., 2009), and the development of a viable and economic process that permits to obtain high volumetric productivity is one of the main challenges to be overcome.

In a recent study, some agro-industrial wastes including corn cobs, coffee silverskin, and cork oak were used as support and nutrient source during the FOS production by *Aspergillus japonicus* under SSF conditions. Among the wastes, coffee silverskin was the most suitable support for FOS production. Furthermore, the highest enzymatic activity results were also achieved when using coffee silverskin as solid support (Mussatto & Teixeira, 2010). These results were considered of great importance for the development of an efficient strategy to produce FOS on industrial scale with higher yield and productivity than that currently obtained.

### 4.5 Bioactive compounds

Bioactive compounds are extra nutritional constituents used as ingredients in food and cosmetic industries. Most common bioactive compounds include secondary metabolites such as mycotoxins, bacterial endotoxins, alkaloids, plant growth factors, antibiotics, immuno-suppressive drugs, food grade pigments, and phenolic compounds. In the last decades, there has been an increasing trend towards the utilization of the SSF to produce bioactive compounds, since this process has been shown more efficient than SmF. Table 6 shows some examples of bioactive compounds produced with significantly higher yields by SSF than by SmF. Besides the higher yields, SSF has also been reported to produce secondary metabolites in shorter times than SmF, with capital costs significantly lesser.

| Product | Microorganism |
|---|---|
| 6-pentyl-alpha-pyrone | *Trichoderma harzianum* |
| Bafilomycin B1 + C1 | *Streptomyces halstedii K122* |
| Benzoic acid | *Bjerkandera adusta* |
| Benzyl alcohol | *Bjerkandera adusta* |
| Cephamycin C | *Streptomyces clavuligerus* |
| Coconut aroma | *Trichoderma sp.* |
| Ergot alkaloids | *Claviceps fusiformis* |
| Giberellic acid | *Giberella fujikuroi* |
| Iturin | *Bacillus subtilis* |
| Ochratoxin | *Aspergillus ochraceus* |
| Oxytetracycline | *Streptomyces rimossus* |
| Penicillin | *Penecillium chrysogenum* |
| Rifamycin-B | *Amycolatopsis mediterranei* |
| Tetracycline | *Streptomyces viridifaciens* |

Table 6. Examples of secondary metabolites produced with higher yield by SSF than by SmF (Hölker et al., 2004)

Alkaloids are secondary metabolites synthesized usually from amino acids. The production of total ergot alkaloids by *Claviceps fusiformis* in SSF was reported to be 3.9 times higher than that in SmF. The production of antibiotics by SSF has also been reported to occur with higher yields and in shorter times when compared to SmF. Among the antibiotics produced by SSF are penicillin, chlorotetracyclines, cephalosprin, tetracyclines, oxytetracyclines, iturin, surfcatin, actinorhodin, methylenomycin and monorden. The important factors in antibiotic production by SSF include the type of strain used, the fermenter design, the general methodology, and control of parameters (Krishna, 2005; Pandey et al., 2000b).

Phenolic compounds, which present a large diversity of biological effects, including anti-inflammatory, antimicrobial and antioxidant activities, have also been efficiently produced by SSF. Many researchers have been focused on finding natural sources, including fruits, vegetables, plants, and agro-industrial wastes, for the production of these compounds. Among these sources, SSF of soy flour-supplemented guava residue by *Rhizopus oligosporus* was considered a good strategy to obtain phenolic compounds. *R. oligosporus* was also demonstrated to be an important food-grade fungus for use in SSF systems with others fruit wastes (cranberry pomace and pineapple waste) as carbon sources (Correia et al., 2004).

Most of the bioactive compounds are produced by SSF using agro-industrial wastes as substrate, even though some studies report the use of sugarcane bagasse or agar as inert solid supports (Table 7). Iturin, for instance, known as a potent antifungal peptide antibiotic, effective in suppressing phytopathogens, has been produced by bacterial strains from okara (soybean waste) and wheat bran (Balakrishnan & Pandey, 1996), resulting in a six to eight times more efficient process by SSF compared to SmF. Antifungal and antibacterial metabolites from a sclerotium-colonizing isolate of *Mortierella vinacea* grown under SSF conditions has also been reported.

| **Microorganism** | **Solid support** | **Product/Function** |
|---|---|---|
| *Gibberella fujikuroi, Fusarium moniliforme* | Corn cob, sugarcane bagasse[a], Cassava flour | Gibberellic acid/ Plant growth hormone |
| *B.subtilis* | Impregnated loam based compost | Antifungal/Antifungal compounds |
| *Bacillus thuringiensis* | Coconut waste | Bacterial endotoxins/ Insecticide |
| *Penicillium chrysogenum* | Sugarcane bagasse[a] | Penicillin/ Antibiotic |
| *S. rimosus* | Corn cob | Oxytetracycline/Antibiotic |
| *S. viridifaciens* | Sweet potato waste | Tetracycline chlorotetracycline/Antibiotic |
| *Monascus purfureus* | Sugarcane bagasse | Pigments |
| *R. oligosporus* | Pineapple waste, cranberry pomace, guava, soy flour | Phenolic antioxidant compound |
| *Streptomyces* sp | Coffee pulp waste | Polyphenols, tannins, chlorogenic acids |
| *B. subtilis* Antibiotic | Soybean waste Okara | Surfactin/Antibiotic |

Table 7. Bioactive compounds produced by SSF. [a] Inert solid support

### 4.6 Bioinsecticides

Biological pest control agents have received considerable attention as a potential alternative to develop eco-friendly pesticides and provide a sustainable agriculture. The identification of a suitable fungal strain that possess pesticide activity is the most important aspect to take into account when developing pesticides. In the last years, bio-pesticides agents for controlling insects and pests have been produced with *entomopathogenic* and *mycoparasitic* fungi. Besides the microorganism, the understanding of the molecular aspects of fungus-fungus, and fungus-insect interactions, the role of hydrolytic enzymes, especially chitinases in killing processes, and the possible use of chitin synthesis inhibitors are crucial aspects to be taken into consideration while making fungi, either singly or in combination, as an effective biopesticide agent.

Several agro-industrial wastes (refused potatoes, coffee husks and sugarcane bagasse) have been used in SSF to produce spores from *Beauveria bassiana* to obtain biopesticides for biocontrol of pests of banana, sugarcane, soybean, and coffee (Santa et al., 2005). *Colletotrichum truncatum* is another fungus that has been studied under SSF and that possesses characteristics to be used as mycoherbicide against the difficult weed *Sesbania exaltata* (Pandey et al., 2000b).

### 4.7 Bioethanol

Bioethanol production by SSF using agro-industrial wastes have been considered as an excellent alternative for reusing these wastes with additional technological and economic advantages, since this process is of easy operation and save energy. The production of bioethanol by SSF is a field that has not been much explored yet; however, recent studies reveal that this technology merits to be better explored.

Ethanol production by SSF using grape and sugar beet pomaces (Rodríguez et al., 2010), and apple pomace (Hang et al., 2006; Joshi & Devrajan, 2008) as solid substrates, has been recently evaluated. When grape pomace and sugar beet pomace were used for cultivation of the yeast *Saccharomyces cerevisiae*, the obtained ethanol production yields were greater than that obtained by SmF (Rodríguez et al., 2010). Therefore, and considering the importance of the ethanol production in the actual world economy, it is expected to observe an increase in the researches for the development of a suitable process for ethanol production by SSF.

## 5. Conclusion

Agro-industrial wastes are generated in large amounts every year. Finding alternatives for the reuse of these wastes is an objective that has been strongly taken into account by countries around the world, considering environmental and economical aspects. The agro-industrial wastes reuse in SSF processes is of particular interest due to their availability, low cost, and characteristics that allow obtaining different value-added compounds, besides being an environment friendly alternative for their disposal. Additionally, the agro-industrial wastes may be used in these processes as solid support, carbon, nitrogen and/or mineral sources, which would allow obtaining more economical fermentation processes avoiding the use of expensive chemical components in the media formulation. As a consequence, more economical processes could be established for implementation on industrial scale.

## 6. References


Adler, E. (1977). Lignin chemistry - Past, present and future. *Wood Science and Technology*, 11, 169-218

Balakrishnan, K., & Pandey, A. (1996). Production of biologically active secondary metabolites in solid state fermentation. *Journal of Scientific & Industrial Research*, 55, 365-372

Biesebeke, R., Ruijter, G., Rahardjo, Y.S.P., Hoogschagen, M.J., Heerikhuisen, M., Levin, A., van Driel, K.G., Schutyser, M.A., Dijksterhuis, J., Zhu, Y., Weber, F.J., de Vos, W.M., van den Hondel, K.A., Rinzema, A., & Punt, P.J. (2002). *Aspergillus oryzae* in solid state and submerged fermentations. Progress report on a multi-disciplinary project. *FEMS Yeast Research*, 2, 245-248

Chien, C.S., Lee, W.C., & Lin, T.J. (2001). Immobilization of *Aspergillus japonicus* by entrapping cells in gluten for production of fructooligosaccharides. *Enzyme and Microbial Technology*, 29, 252-257

Christen, P., Meza, J., & Revah, S. (1997). Fruity aroma production in solid state fermentation by Ceratocystis fimbriata: influence of the substrate type and the presence of precursors. *Mycological Research*, 101, 911-919

Correia, R.T.P., McCue, P., Magalhães, M.M.A., Macêdo, G.R., & Shetty, K. (2004). Production of phenolic antioxidants by the solid-state bioconversion of pineapple waste mixed with soy flour using *Rhizopus oligosporus*. *Process Biochemistry*, 39, 2167-2172

Couri S., Merces, E.P., Neves, B.C.V., & Senna, L.F. (2006). Digital image processing as a tool to monitor biomass growth in *Aspergillus niger* 3T5B8 solid-state fermentation: preliminary results. *Journal of Microscopy*, 224, 290-297

Couto, S.R., & Sanromàn, M. (2006). Application of solid-state fermentation to food industry--A review. *Journal of Food Engineering*, 76, 291-302

Dinis, M.J., Bezerra, R.M.F., Nunes, F., Dias, A.A., Guedes, C.V., Ferreira, L.M.M., Cone, J.W., Marques, G.S.M., Barros, A.R.N., & Rodrigues, M.A.M. (2009). Modification of wheat straw lignin by solid state fermentation with white-rot fungi. *Bioresource Technology*, 100, 4829-4835

Durand, A. (2003). Bioreactor designs for solid state fermentation. *Biochemical Engineering Journal*, 13, 113–125

El-Bendary, M.A., Moharam, M.E., & Foda, M.S., (2008). Efficient mosquitocidal toxin production by *Bacillus sphaericus* using cheese whey permeate under both submerged and solid state fermentations. *Journal of Invertebrate Pathology*, 98, 46-53

Fengel, D., & Wegener, G. (Eds.) (1989). *Wood: Chemistry, ultrastructure, reactions*, Walter de Gruyter, New York.

Guimarães, L.H.S., Somera, A.F., Terenzi, H.F., Polizeli, M.L.T.M., & Jorge, J.A. (2009). Production of β-fructofuranosidases by *Aspergillus niveus* using agroindustrial residues as carbon sources: Characterization of an intracellular enzyme accumulated in the presence of glucose. *Process Biochemistry*, 44, 237-241

Ha, M-.A., Apperley, D.C., Evans, B.W., Huxham, I.M., Jardine, W.G., Vietor, R.J., Reis, D., Vian, B., & Jarvis, M.C. (1998). Fine structure in cellulose microfibrils: NMR evidence from onion and quince. *The Plant Journal*, 16, 183-190

Hang, Y., Lee, C., & Woodams, E. (2006). A solid state fermentation system for production of ethanol from apple pomace. *Journal of Food Science*, 47, 1851-1852

Hashemi, M., Mousavi, S.M., Razavi, S.H., & Shojaosadati, S.A. (2011). Mathematical modeling of biomass and α-amylase production kinetics by *Bacillus* sp. in solid-state fermentation based on solid dry weight variation. *Biochemical Engineering Journal*, 53, 159-164

Hölker, U., & Lenz, J. (2005). Solid-state fermentation: are there any biotechnological advantages? *Current Opinion in Microbiology*, 8, 301-306

Hölker, U., Höfer, M., & Lenz, J. (2004). Biotechnological advantages of laboratory-scale solid-state fermentation with fungi. *Applied Microbiology and Biotechnology*, 64, 175-186

Huang, C., Zong, M.-H., Wu, H., & Liu, Q.-P. (2009). Microbial oil production from rice straw hydrolysate by *Trichosporon fermentans*. *Bioresource Technology*, 100, 4535-4538

Joshi, V.K. & Devrajan, A. (2008). Ethanol recovery from solid state fermented apple pomace and evaluation of physico-chemical characteristics of rhe residue. *Nature Product Radiance*, 7, 127-132

Kagliwal, L.D., Survase, S.A., & Singhal, R.S. (2009). A novel medium for the production of cephamycin C by *Nocardia lactamdurans* using solid-state fermentation. *Bioresource Technology*, 100, 2600-2606

Khanahmadia, M., Mitchell, D.A., Beheshtic, M., Roostaazadc, R., & Sanchezd, R.L. (2006). Continuous solid-state fermentation as affected by substrate flowpattern. *Chemical Engineering Science*, 61, 2675-2687

Klemm, D., Philipp, B., Heinze, T., Heinze, U., & Wagenknecht, W. (Eds.) (1998). *Comprehensive Cellulose Chemistry*, Wiley VCH, Chichester.

Koncsag, C.I., & Kirwan, K. (2010). Heat and mass transfer study during wheat straw solid substrate fermentation with *P. Ostreatus*. *Chemical Bulletin of "Politehnica"* University of Timisoara, 55, 1-4

Krishna, C. (2005). Solid-state fermentation systems-an overview. *Critical Reviews in Biotechnology*, 25, 1-30

Mahanta, N., Gupta, A., & Khare, S.K. (2008). Production of protease and lipase by solvent tolerant *Pseudomonas aeruginosa* PseA in solid-state fermentation using *Jatropha curcas* seed cake as substrate. *Bioresource Technology*, 99, 1729-1735

Martins, S., Mussatto, S.I., Martínez-Avila, G., Montañez-Saenz, J., Aguilar, C.N., & Teixeira, J.A. (2011). Bioactive phenolic compounds: Production and extraction by solid-state fermentation. A review. *Biotechnology Advances*, 29, 365-373

Medeiros, A.B.P., Pandey, A., Freitas, R.J.S., Christen, P., & Soccol, C.R. (2000). Optimization of the production of aroma compounds by *Kluyveromyces marxianus* in solid-state fermentation using factorial design and response surface methodology. *Biochemical Engineering Journal*, 6, 33-39

Mitchell, D.A., Krieger, N., Stuart, D.M., & Pandey, A. (2000). New developments in solid-state fermentation: II. Rational approaches to the design, operation and scale-up of bioreactors. *Process Biochemistry*, 35, 1211-1225

Mohanty, S.K., Behera, S., Swain, M.R., & Ray, R.C. (2009). Bioethanol production from mahula (*Madhuca latifolia* L.) flowers by solid-state fermentation. *Applied Energy*, 86, 640-644

Mussatto, S.I., & Teixeira, J.A. (2010). Increase in the fructooligosaccharides yield and productivity by solid-state fermentation with *Aspergillus japonicus* using agro-industrial residues as support and nutrient source. *Biochemical Engineering Journal,* 53, 154-157

Mussatto, S.I., Aguilar, C.N., Rodrigues, L.R., & Teixeira, J.A. (2009). Fructooligosaccharides and β-fructofuranosidase production by *Aspergillus japonicus* immobilized on lignocellulosic materials. *Journal of Molecular Catalysis B: Enzymatic,* 59, 76-81

Nigam, P.S. (2009). Production of bioactive secondary metabolites, In: *Biotechnology for Agro-Industrial Residues Utilization,* Nigam, P.S., & Pandey, A. (Eds.) pp. 129-145, Springer, Netherlands.

Nigam, P.S., & Pandey, A. (2009). Solid-state fermentation technology for bioconversion of biomass and agricultural residues, In: *Biotechnology for Agro-Industrial Residues Utilization,* Nigam, P.S., & Pandey, A. (Eds.) pp. 197-221, Springer, Netherlands.

Nigam, P.S., Gupta, N., & Anthwal, A. (2009). Pre-treatment of agro-industrial residues, In: *Biotechnology for Agro-Industrial Residues Utilization,* Nigam, P.S., & Pandey, A. (Eds.) pp. 13-33, Springer, Netherlands.

Orozco, A.L., Pérez, M.I., Guevara, O., Rodríguez, J., Hernández, M., González-Vila, F.J., Polvillo, O., & Arias, M.E. (2008). Biotechnological enhancement of coffee pulp residues by solid-state fermentation with *Streptomyces*. Py–GC/MS analysis. *Journal of Analytical and Applied Pyrolysis,* 8, 247-252

Orzua, M.C., Mussatto, S.I., Contreras-Esquivel, J.C., Rodriguez, R., De la Garza, H., Teixeira, J.A., & Aguilar, C.N. (2009). Exploitation of agro industrial wastes as immobilization carrier for solid-state fermentation. *Industrial Crops and Products,* 30, 24-27

Pandey, A. (2003). Solid state fermentation. *Biochemical Engineering Journal,* 13, 81-84

Pandey, A., Soccol, C.R., & Larroche, C. (2007). Current Developments in Solid-state Fermentation, In: *Current Developments in Solid-state Fermentation,* Pandey, A., Soccol, C.R., & Larroche, C. (Eds.), pp.13-25, Springer Science/Asiatech Publishers, Inc., New York, USA/New Delhi, India.

Pandey, A., Soccol, C.R., & Mitchell, D. (2000b). New developments in solid state fermentation: I-bioprocesses and products. *Process Biochemistry,* 35, 1153-1169

Pandey, A., Soccol, C.R., Nigam, P., Soccol, V.T., Vandenberghe, L.P.S., & Mohan, R. (2000a). Biotechnological potential of agro-industrial residues. II: cassava bagasse. *Bioresource Technology,* 74, 81-87

Pérez-Guerra, N., Torrado-Agrasar, A., López-Macias, C., & Pastrana, L. (2003). Main characteristics and applications of solid substrate fermentation. *Electronic Journal of Environmental, Agricultural and Food Chemistry,* 2, 343-350

Raimbault, M. (1998). General and microbiological aspects of solid substrate fermentation. *Electronic Journal of Biotechnology,* 1, 174-188

Rodríguez, L.A., Toro, M.E., Vazquez, F., Correa-Daneri, M.L., Gouiric, S.C., & Vallejo, M.D. (2010). Bioethanol production from grape and sugar beet pomaces by solid-state fermentation. *International Journal of Hydrogen Energy,* 35, 5914-5917

Rodríguez-Fernández, D.E., Rodríguez-León, J.A., Carvalho, J.C., Sturm, W., & Soccol C.R. (2011). The behaviour of kinetic parameters in production of pectinase and xylanase by solid-state fermentation. *Bioresource Technology,* doi: 10.1016/j.biortech.2011.08.106

Sahir, A.H., Kumar, S., & Kumar, S. (2007). Modelling of a packed bed solid-state fermentation bioreactor using the N-tanks in series approach. *Biochemical Engineering Journal*, 35, 20-28

Santa, H.S.D., Santa, O.R.D., Brand, D., Vandenberghe, L.P.S., & Soccol, C.R. (2005). Spore production of beauveria bassiana from agroindustrial residues. *Brazilian Archives of Biology and Technology*, 48, 51-60

Shankar, S.K., & Mulimani, V.H. (2007). α-Galactosidade production by *Aspergillus oryzae* in solid-state fermentation. *Bioresource Technology*, 98, 958-961

Sharma, A., Vivekanand, V., & Singh, R.P. (2008). Solid-state fermentation for gluconic acid production from sugarcane molasses by *Aspergillus niger* ARNU-4 employing tea waste as the novel solid support. *Bioresource Technology*, 99, 3444-3450

Singh, B., & Satyanarayana, T. (2008). Improved phytase production by a thermophilic mould *Sporotrichum thermophile* in submerged fermentation due to statistical optimization. *Bioresource Technology*, 99, 824-830

Singhania, R.R., Patel, A.K., Soccol, C.R., & Pandey, A. (2009). Recent advances in solid-state fermentation: review. *Biochemical Engineering Journal*, 44, 13-18

Taherzadeh, M.J., & Karimi, K. (2008). Pretreatment of lignocellulosic wastes to improve ethanol and biogas production: a review. *International Journal of Molecular Science*, 9, 1621-1651

Ustok, F.I., Tari, C., & Gogus, N. (2007). Solid-state production of polygalacturonase by *Aspergillus sojae* ATCC 20235. *Journal of Biotechnology*, 127, 322-334

von Meien, O.F., Luz Jr, L.F.L., Mitchell, D.A., Pérez-Correa, J.R., Agosin, E., Fernández-Fernández, M., & Arcas, J.A. (2004). Control strategies for intermittently mixed, forcefully aerated solid-state fermentation bioreactors based on the analysis of a distributed parameter model. *Chemical Engineering Science*, 59: 4493-4504

Wolski, E., Menusi, E., Remonatto, D., Vardanega, R., Arbter, F., Rigo, E., Ninow, J., Mazutti, M.A., Di Luccio, M., Oliveira, D., & Treichel, H. (2009). Partial characterization of lipases produced by a newly isolated *Penicillium sp.* in solid state and submerged fermentation: A comparative study. *Lebensmittel-Wissenschaft und-Technologie*, 42, 1557-1560

Yu, J., Zhang, X., & Tan, T. (2008). Ethanol production by solid state fermentation of sweet sorghum using thermotolerant yeast strain. *Fuel Processing Technology*, 89, 1056-1059

Yu, M., Zeng, G., Chen, Y., Yu, H., Huang, D., & Tang, L. (2009). Influence of *Phanerochaete chrysosporium* on microbial communities and lignocellulose degradation during solid-state fermentation of rice straw. *Process Biochemistry*, 44, 17-22

Zhao, J., & Xia, L. (2009). Simultaneous saccharification and fermentation of alkaline-pretreated corn stover to ethanol using a recombinant yeast strain. *Fuel Processing Technology*, 99, 1193-1197

# 7

# Types of Waste for the Production of Pozzolanic Materials – A Review

A. Seco, F. Ramirez, L. Miqueleiz,
P. Urmeneta, B. García, E. Prieto and V. Oroz
*Dept. of Projects and Rural Engineering,
Public University of Navarre,
Spain*

## 1. Introduction

Currently there are large volumes of materials considered as wastes or by-products produced by industrial activities. In most cases, these wastes have not possibility of reuse, or low economic value for the companies that generate it. This adds to the cost of its management and disposal. Current production models have associated inefficient practices such as:

1. Resource consumption. Materials and energy.
2. Waste generation
3. Economic costs

It is important to realize that the generation of wastes is an inherent part of productive systems. The quantities and characteristics of the waste generated depend on the technologies used by companies. Sometimes companies need to restructure their productive systems in order to treat the waste they generate and many times its elimination is not easy, especially in sectors where production technologies are very matured. In this context, one of the possibilities for the recovery of these materials is their reuse and recycling in the construction sector. The construction sector annually consumes large volumes of materials, which clearly gives this sector potential to absorb and give value to the large quantities of wastes produced in the industry. On the one hand, this situation can achieve the environmental protection demanded by Society, and on the other hand allows to the companies to operate more sustainable productive systems. The reuse of waste may enable the attainment of a "sustainable construction" procedure, which can be defined as a set of constructive actions which take into account technical, economical, environmental and social aspects.

The development of new urban projects and the construction of new buildings need to consume huge quantities of materials, such as paving block, bricks, tiles, cement, aggregates, etc. The manufacture of these materials involves the consumption of natural raw materials and energy. Taking into account these conditions, the development of new sustainable materials is a very effective line of work for the modernization of the construction sector. One of the most acceptable possibilities in this area is the development of pozzolanic materials based on wastes, which can be a real alternative to conventional construction materials. Currently, pozzolanic materials are frequently used in the construction sector. In civil engineering soil stabilization technique uses these materials as binders. In building the use of pozzolanic materials is also very important as structural elements, building enclosure, etc. In both applications the most useful pozzolanic additive is Ordinary Portland Cement (OPC), one of the main current construction materials, with an annual production estimated in 2010 at 3,300,000 Tons (USGS, 2011). This material is an import economic factor, used as an indicator of the development and economic activity of countries (Daugherty, 1973).

OPC is currently under discussion, not only for its cost but also for its environmental effects during manufacture. Thus, the production of 1 Ton of OPC supposes the consumption of 1.5 Tons of quarry material, energy consumption of 5.6 Gj/Ton and an emission of nearly 0.9 Ton of $CO_2$ , representing 5 % of total anthropogenic $CO_2$ emission (Reddy et al., 2006; O'Rourke et al., 2009; Juenger et al., 2011; Billong et al., 2011).

Currently, other Alternative Hydraulic Binders (AHB) based on pozzolanic materials are being developed. The American Concrete Institute in its report ACI 232.1R-00 defines pozzolanic material as "siliceous and aluminous material, which in itself possesses little or no cementitious value but will, in finely divided form and in the presence of moisture, chemically react with calcium hydroxide at ordinary temperatures to form compounds possessing cementitious properties".

Some industrial wastes, usually coming from the combustion process, such as fly ash and bottom ash, have pozzolanic properties. The pozzolanity of these materials is set in the ASTM C 593 and ASTM C 618 standards, and depends mainly on the amount of Ca, Si and Al oxides present, the ratio between them and their reactivity (Pacheco-Torgal et al., 2008; Ravikumar, 2010; Billong et al., 2011). The addition of pozzolans to OPC or their use together with lime allow the partial substitution or in certain applications full substitution of OPC. This supposes a reduction in waste, in energy consumption and in $CO_2$ emission, lower production cost and sometimes improved engineering properties. This converts the AHB into a potentially competitive material with OPC both in developed and in developing countries (Singh and Garg, 1999; Shen et al., 2007; Ercikdi et al., 2009; Billong et al., 2011; Kinuthia and Nidzam, 2011).

## 2. Chemical principles of the pozzolanic reactions

Pozzolanic reactions take place when significant quantities of reactive CaO, $Al_2O_3$ and $SiO_2$ are mixed in presence of water. Usually CaO is added as lime or cement meanwhile $Al_2O_3$ and $SiO_2$ can be present in the material to develop cementation gels to be added as cement or, for example, with a pozzolan. In this process the hydration of the CaO liberates $OH^-$ ions, which causes an increase pH values up to approximate 12.4. Under these conditions pozzolanic reactions occur: the Si and Al combine with the available Ca, resulting in cementitious compounds called Calcium Silicate Hydrates (CSH) and Calcium Aluminate Hydrates (CAH) (Dermatas and Meng, 2003; Nalbantoglu, 2004; Guney et al., 2007; Yong and Ouhadi, 2007; Chen and Lin, 2009). A simplified qualitative representation of these reactions is summarized below:

$$Ca(OH)_2 \rightarrow Ca^{2+} + 2OH^-$$
$$Ca^{2+} + 2OH^- + SiO_2 \rightarrow CSH$$
$$Ca^{2+} + 2OH^- + Al_2O_3 \rightarrow CAH$$

These compounds are responsible for improving the mechanical properties of the mix, due to the increasing development of pozzolanic reactions over time; some authors point up this may take place over years (Wild et al., 1998).

Where these oxides are not available in sufficient quantities in the materials to be cemented, they must be incorporated with the binder. In these cases it is particularly advantageous to use stabilizers which, like OPC, are rich in $SiO_2$ and $Al_2O_3$ as well as in CaO (Wild et al., 1998; Degirmenci et al., 2007) or for example the use of lime and pozzolan mixes. When these oxides are present in the material which needs to be cemented it is not necessary to add them as a binder. This situation is most usual in the stabilization of clay soils where oxides are present in the clay matrix. They are naturally rich in Si and Al oxides which become soluble at high pH conditions and then become available for the development of the pozzolanic reactions. The improved mechanical capacities achieved in each case depend on the amount, reactivity and concentration of the oxides, the size and shape of the particles and also on the curing conditions (Misra et al., 2005; Yarbasi et al., 2007; Göktepe et al., 2008).

One adverse effect related to the pozzolanic reactions is caused by the presence of sulfate ($SO_4^{2-}$). It may often appear both in industrial wastes and in natural soils, provoking the formation of a highly hydrated and expansive mineral called ettringite ($Ca_6Al_2(SO_4)_3(OH)_{12}{\cdot}26H_2O$). The mechanics of the formation of ettringite are not yet well established (Mohamed, 2000), although the conditions for its formation are well known (Ouhadi and Yong, 2003) and are:

1. High pH,
2. presence of soluble Al,
3. presence of soluble Ca,
4. presence of soluble sulfate and
5. availability of water.

It is also known that the rate of ettringite formation is accelerated by high temperatures (Rajasekaran, 2005) and that given the right conditions for rapid formation, appears even in few seconds. Mohamed (2000) determined the total time of formation of ettringite in a natural soil as 48 hours. In their experiments Ouhadi and Yong (2003) and (2008) established the formation of this mineral over one month in one case, and in another experiment 48 hours after the mixture of the soil with lime . From a chemical point of view, the reactions involved in the formation of ettringite are (Ouhadi and Yong, 2008):

$$CaO + H_2O \rightarrow Ca(OH)_2$$

$$Ca(OH)_2 \rightarrow Ca^{2+} + 2(OH)^-$$

$$Al_2Si_4O_{10}(OH)_2 \cdot nH_2O + 2(OH)^- + 10H_2O \rightarrow 2\{2Al(OH)_4^- + 4H_4SiO_4\} + nH_2O$$

$$M_xSO_4 \cdot nH_2O \rightarrow XM^y + SO_4 + nH_2O$$
$$where$$
$$x = 1, y = 2$$
$$or$$
$$x = 2, y = 1$$

$$6Ca^{2+} + 2Al(OH)_4^- + 4OH^- + 3(SO_4^{2-}) + 26H_2O \rightarrow Ca_6Al_2(SO_4)_3(OH)_{12} \cdot 26H_2O$$

As can be seen, ettringite has 26 $H_2O$ molecules in it, which is why it is very expansive during its formation and can cause the destruction of the cemented materials. This is the case, for example, of the well documented cases of premature concrete deterioration or the expansivity of stabilized soils (Collepardi, 2003; Ouhadi and Yong, 2008; Thomas et al., 2008).

## 3. Interesting wastes for the production of pozzolanic materials

As explained previously, the most interesting waste substances for the production of pozzolanic materials are these that contain appreciable amounts of reactive Ca, Si or Al oxides. Although the majority of them are obtained in combustion processes, there are also other possible origins for the pozzolans. The composition and the quality of wastes with potential pozzolanic properties can vary from high reactivity to none, depending on: (1) the original materials and (2) the production processes. Apart from them, other wastes with no pozzolanic activity can be added as a target during the production of pozzolanic-based materials, to reach up to as much as 90% of the total mass of the final product. In the following sections of this chapter the main available pozzolanic wastes useful as binder materials are reviewed.

### 3.1 Fly Ash (FA)

Fly ash is the finely divided residue resulting from the combustion of coal in electric power generating stations. Its pozzolanic properties are well documented (Temimi et al., 1998; Senol et al., 2006; Samaras et al., 2008 among others) and it is usually employed in Portland Cement production. The large available quantities of this waste makes it necessary to dump it, causing environmental effects. Standard ASTM C 615 classifies it as F class or C class based on its reactivity and chemical composition.

### 3.2 Ground Granulate Blast-furnace Slag (GGBS)

GGBS is a by-product obtained during the manufacture of pig iron in the blast furnace and is formed by the combination of iron ore with limestone flux. If the molten slag is cooled and solidified by rapid water quenching to a glassy state, little or no crystallization occurs. This process results in the formation of sand size fragments, usually with some friable clinker-like material known as GGBS. The physical structure and gradation of GGBS depend on the chemical composition of the slag, the temperature at the time of water quenching and the method of production. GGBS, when it is activated with small amounts of lime, has a high pozzolanic activity, and it has been widely used in various engineering applications (Wild et al, 1998; Binici et al., 2007; Oti et al., 2009 among others).

| OXIDES | CLASS C FA | CLASS F FA | GGBS | SILICA-FUME | RHA | PG | CW |
|---|---|---|---|---|---|---|---|
| CaO | 14.8 | 2.39 | 41.99 | 0.8-1.2 | | 32.04 | 9.45 |
| $Al_2O_3$ | 25.5 | 26.67 | 11.59 | 1-3 | | 0.88 | 14.16 |
| $SiO_2$ | 47.4 | 54.45 | 35.35 | 85-95 | 99 | 3.44 | 52.36 |
| MgO | 2.7 | 1.12 | 8.04 | 1-2 | | | 1.97 |
| $SO_3$ | 2.8 | 0.10 | 0.23 | | | 44.67 | |
| $Fe_2O_3$ | | 6.91 | 0.35 | | | 0.32 | 6.22 |
| $Na_2O$ | | 0.21 | | | | 0.13 | 1.02 |
| $K_2O$ | | 2.03 | | | | | 1.85 |

Table 1. Oxide composition of the waste materials

### 3.3 Silica-fume

Silica-fume is an industrial waste produced from the smelting process for silicon metal and ferrosilicon alloy production. It contains high amounts of extremely fine and amorphous $SiO_2$ particles (see table 1). Silica-fume has been used in civil engineering works as a binder in combination with cement for soil stabilization giving great results (Taylor et al, 1996).

### 3.4 Rice Husk Ash (RHA)

Rice husk is a common agricultural by-product around the world. In many countries this husk is used as fuel for power plants, generating from its combustion large quantities of RHA. These ashes are riches in amorphous $SiO_2$ as is shown in table 1, having great pozzolanic properties (Behak and Perez, 2008; Seco et al., 2011a; Seco et al., 2011b among others).

### 3.5. Phosphogypsum (PG)

Phosphogypsum is a byproduct of the chemical reaction between sulfuric acid and phosphate rock to produce phosphoric acid. Nowadays large volumes of PG are dumped in tips, causing economic loss and environmental concerns. Some authors have used it, combined with cement, lime and fly ash, to stabilize soils in spite of its high sulfate content (Shen et al., 2007; Degirmenci et al., 2007).

### 3.6 Ceramic Wastes (CW)

Ceramic wastes include all the waste from bricks, tiles and other fired-clay based materials. These materials when ground up have pozzolanic properties, known since Phoenician times (Baronio and Binda, 1997). This is because they are produced from clay, and the thermal process leaves the Al and Si oxides in an amorphous state.

### 3.7 Sewage Sludge (SS)

Sewage Sludge is the generic name of all kinds of fresh or incinerated wastewater treatment wastes. Its composition and available quantities vary depending on its origin (industrial, water treatment, tannery, etc.). The majority of these materials often have an appreciable content of heavy metals, making their use as construction material more difficult. Although the majority of the experiences of their use in construction is as a target in fired clay materials (Weng et al., 2003; Lin et al., 2005; Chen and Li, 2009) other authors have demonstrated the possibility of using them in the production of pozzolanic materials (Tay and Show, 1997; Cyr et al., 2007; Liu et al., 2011).

## 4. Construction materials based on pozzolanic wastes

This section includes a review of the most interesting construction materials created from pozzolanic wastes, such as soil stabilized layers, unfired bricks and blocks, and masonry mortars.

### 4.1 Soil stabilized layers

Wild et al. (1998) improved the strength properties of a soil with sulfates by stabilizing it with lime and GGBS. Nalbantoglu (2004) used C class fly ash as a binder to stabilize an expansive soil, reducing its plasticity and swelling capacity. Kolias et al. (2005) demonstrated the benefits of soils stabilized with cement and fly ash compared with conventional flexible pavements. Degirmenci et al. (2007) showed the possibility of stabilizing soil with a mix of phosphogypsum, cement and fly ash. They improved the compressive strength as well as the Atterberg limits. Lin et al. (2007) demonstrated the improvement of the mechanical properties of a soil when it was treated with a mix of lime and sewage sludge, increasing the strength resistance and reducing the swelling capacity. Shen et al. (2007) showed the possibility of using a lime-fly ash-phosphogypsum binder for road base construction. Yarbasi et al. (2007) used silica-fume and fly ash to increase the durability of the tested samples of granular soils for road construction. Chen and Lin (2008) used a mix of cement and incinerated sewage sludge to improve the basic properties (strength and swelling) of subgrade soils. Samaras et al. (2008) demonstrated the potential use of a mix of sewage sludge, fly ash and lime as soil stabilizer.

### 4.2 Unfired bricks and blocks

Kumar (2002) studied fly ash-lime-gypsum bricks, observing that the resistance of these materials and durability make them available as low cost housing materials. Poon and Chan (2006) demonstrated the possibility of making paving blocks from concrete and ceramic wastes. Chindraprasirt and Pimraksa (2008) demonstrated that heavy metals were retained in a fly ash-lime unfired brick. Degirmenci (2007) demonstrated the possibility of producing adobes with phosphogypsum as stabilizer, obtaining good strength values, although this material was susceptible to water damages. Oti et al. (2009) demonstrated the increase in the strength and durability of unfired clay bricks with GGBS with lime or cement as binder. Wattanasiriwech et al. (2009) produced paving blocks with a ceramic waste that reached the standard requirements after a curing period of between 7 and 28 days. Oti et al. (2010) demonstrated that unfired clay bricks based on a clay soil stabilized with a mix of GGBS or lime comply with the thermal design requirements for masonry. Liu et al. (2011) analyzed the effect of water sludge waste in unfired brick production, demonstrating that by combining it with cement it was possible to achieve all the performance criteria for unfired bricks.

### 4.3 Masonry mortars

Tay and Show (1997) created a waste-based mortar by incineration of municipal wastewater sludge. This material was useful as masonry mortar, replacing up to 30% of Portland Cement. McCarthy and Dhir (1999) stated that fly ash can be used to improve the properties of Portland Cement. Papayianni and Stefanidou (2006) observed lower strength and good durability of lime-pozzolan mortars compared with cement based ones. Rodriguez de Sensale (2006) demonstrated the increase of strength in cement mortars at 90 days curing age when rice husk ash was added. Cyr et al. (2007) analyzed mortars containing sewage sludge ash, discovering a long-term positive effect because of its slight pozzolanic activity. Zerbino et al. (2011) observed the possibility of replacing cement by rice husk ash by up to 25% in concrete without loss of resistance.

## 5. Conclusions

This review has aimed to improve our knowledge on the application of different industrial wastes in the construction field. Much industrial wastes can be absorbed by civil engineering and building construction, reducing at the same time both the consumption of Portland Cement and the environmental problems provoked by the dumping of wastes. There are industrial wastes with pozzolanic characteristics which can at least partially substitute Portland Cement without loss of resistance, and even increase other properties like durability. All of them have a potential use as a sustainable, green, and low cost binders. Nowadays these sustainable binders are being used in soil stabilization and in the building industry in the development of new sustainable construction materials. These new materials have demonstrated their capacity to reach all the technical requirements. Thus the potential for industrial waste utilization in the field of construction has a potential so far not fully exploited, that may transform current construction practices into new sustainable construction, socially and environmentally more responsible.

## 6. References

[1] American Concrete Institute. Use of raw or processed natural pozzolans in concrete. ACI 232.1R-00. ACI committee 232. December 2000.

[2] Baronio, G., Binda, L. Study of the pozzolanicity of some bricks and clays, Construction and Building Materials, Volume 11, Issue 1, February 1997, Pages 41-46.

[3] Behak, L., Perez, N., Characterization of a material comprised of sandy soil, rice husk ash and potentially useful lime in pavements. Journal Ingeniería de construcción. Volume 23 Issue 1 April 2008 Pages 34-41.

[4] Billong, N., Melo, U.C., Kamseu, E., Kinuthia, J.M., Njopwouo, D. Improving hydraulic properties of lime–rice husk ash (RHA) binders with metakaolin (MK), Construction and Building Materials, Volume 25, Issue 4, April 2011, Pages 2157-2161.

[5] Binici H., Temiz, H., Köse, M. M. The effect of fineness on the properties of the blended cements incorporating ground granulated blast furnace slag and ground basaltic pumice, Construction and Building Materials, Volume 21, Issue 5, May 2007, Pages 1122-1128.

[6] Chen, L., Lin, D-F., Stabilization treatment of soft subgrade soil by sewage sludge ash and cement, Journal of Hazardous Materials, Volume 162, Issue 1, 15 February 2009, Pages 321-327.
[7] Chindaprasirt, P., Pimraksa, K. A study of fly ash–lime granule unfired brick, Powder Technology, Volume 182, Issue 1, 15 February 2008, Pages 33-41.
[8] Collepardi, M. A state-of-the-art review on delayed ettringite attack on concrete, Cement and Concrete Composites, Volume 25, Issues 4-5, May-July 2003, Pages 401-407
[9] Daugherty, K.E. Cement as an indicator of economic development. Cement and Concrete Research. Volume 3, Issue 2, 1973. Pages 163-183.
[10] Cyr, M., Coutand, M., Clastres, P. Technological and environmental behavior of sewage sludge ash (SSA) in cement-based materials, Cement and Concrete Research, Volume 37, Issue 8, August 2007, Pages 1278-1289.
[11] Degirmenci, N., Okucu, A., Turabi, A., Application of phosphogypsum in soil stabilization, Building and Environment, Volume 42, Issue 9, September 2007, Pages 3393-3398.
[12] Dermatas, D. Meng, X. Utilization of fly ash for stabilization/solidification of heavy metal contaminated soils, Engineering Geology, Volume 70, Issues 3-4, November 2003, Pages 377-394,
[13] Ercikdi, B Cihangir, F Kesimal, A Deveci, H Alp, İ. Utilization of industrial waste products as pozzolanic material in cemented paste backfill of high sulphide mill tailings, Journal of Hazardous Materials, Volume 168, Issues 2-3, 15 September 2009, Pages 848-856,
[14] Goktepe, A. B, Sezer, A., Sezer, G.I., Ramyar, K., Classification of time-dependent unconfined strength of fly ash treated clay, Construction and Building Materials, Volume 22, Issue 4, April 2008, Pages 675-683.
[15] Guney Y., Sari D., Cetin M., Tuncan M., Impact of cyclic wetting-drying on swelling behavior of lime-stabilized soil, Building and Environment, Volume 42, Issue 2, February 2007, Pages 681-688.
[16] Higgins D. D., Kinuthia J.M., Wild, S. Soil stabilization using lime activated ground granulated blast furnace slag. Proc. 6th CANMET/ACI International Conference on fly ash, Silica Fume, slag and natural pozzolans in concrete, 1998, Thailand, ACI publication SP178-55.
[17] Juenger, M.C.G., Winnefeld, F., Provis, J.L., Ideker, J.H. Advances in alternative cementitious binders. Cement and concrete research. Volume 41, Issue 12, December 2011, Pages 1232-1243.
[18] Kalinski, M E., Yerra, P. K. Hydraulic conductivity of compacted cement–stabilized fly ash, Fuel, Volume 85, Issue 16, November 2006, Pages 2330-2336,
[19] Kinuthia, J.M., Nidzam, R.M.. Towards zero industrial waste: Utilisation of brick dust waste in sustainable construction, Waste Management. Volume 31, Issue 8. 2011. pages 1867-1878.
[20] Kolias, S., Kasselouri-Rigopoulou, V., Karahalios, A. Stabilisation of clayey soils with high calcium fly ash and cement, Cement and Concrete Composites, Volume 27, Issue 2, February 2005, Pages 301-313.
[21] Kumar S. A perspective study on fly ash–lime–gypsum bricks and hollow blocks for low cost housing development, Construction and Building Materials, Volume 16, Issue 8, December 2002, Pages 519-525.

[22] Lin, K.L., Chiang, K.Y., Lin, C.Y. Hydration characteristics of waste sludge ash that is reused in eco-cement clinkers, Cement and Concrete Research, Volume 35, Issue 6, June 2005, Pages 1074-1081.

[23] Liu, Z., Chen, Q., Xie, X., Xue, G., Du, F., Ning, Q., Huang, L. Utilization of the sludge derived from dyestuff-making wastewater coagulation for unfired bricks, Construction and Building Materials, Volume 25, Issue 4, April 2011, Pages 1699-1706.

[24] Misra, A., Biswas, D., Upadhyaya, S., Physico-mechanical behavior of self-cementing class C fly ash-clay mixtures, Fuel, Volume 84, Issue 11, 2003 International Ash Utilization Symposium, August 2005, Pages 1410-1422.

[25] Mohamed, A.M.O. The role of clay minerals in marly soils on its stability, Engineering Geology, Volume 57, Issues 3-4, July 2000, Pages 193-203.

[26] McCarthy, M.J., Dhir, R.K. Towards maximising the use of fly ash as a binder, Fuel, Volume 78, Issue 2, January 1999, Pages 121-132.

[27] Nalbantoglu Z., Effectiveness of Class C fly ash as an expansive soil stabilizer, Construction and Building Materials, Volume 18, Issue 6, July 2004, Pages 377-381.

[28] Oti, J.E., Kinuthia, J.M., Bai, J. Engineering properties of unfired clay masonry bricks, Engineering Geology, Volume 107, Issues 3-4, 14 August 2009, Pages 130-139.

[29] Oti, J.E., Kinuthia, J.M., Bai, J. Design thermal values for unfired clay bricks, Materials & Design, Volume 31, Issue 1, January 2010, Pages 104-112.

[30] Ouhadi. V.R, Yong.R.N. Ettringite formation and behaviour in clay soils. Applied clay science. Volume 42, 2008, Pages 258-256.

[31] Ouhadi. V.R, Yong.R.N. The role of clay fractions of marly soils on their post stabilization failure. Engineering Geology. Volume 70, 2003. Pages 365-375.

[32] O'Rourke, B., McNally, C., Richardson, M. G. Development of calcium sulfate–ggbs–Portland cement binders, Construction and Building Materials, Volume 23, Issue 1, January 2009, Pages 340-346,

[33] Pacheco-Torgal, F., Castro-Gomes J., Jalali, S. Alkali-activated binders: A review: Part 1. Historical background, terminology, reaction mechanisms and hydration products, Construction and Building Materials, Volume 22, Issue 7, July 2008, Pages 1305-1314.

[34] Papayianni I., Stefanidou M. Strength–porosity relationships in lime–pozzolan mortars, Construction and Building Materials, Volume 20, Issue 9, November 2006, Pages 700-705.

[35] Poon, C-S., Chan, D., Effects of contaminants on the properties of concrete paving blocks prepared with recycled concrete aggregates, Construction and Building Materials, Volume 21, Issue 1, January 2007, Pages 164-175.

[36] Rajasekaran, G. Sulfate attack And ettringite formation in the lime and cement stabilized marine clays. Ocean Engineering Volume 32 (2005) Pages 1133-1159.

[37] Ravikumar, D., Peethamparan, S., Neithalath, N. Structure and strength of NaOH activated concretes containing fly ash or GGBFS as the sole binder, Cement and Concrete Composites, Volume 32, Issue 6, July 2010, Pages 399-410.

[38] Reddy, A.S., Pradhan, R.K., Chandra, S. Utilization of Basic Oxygen Furnace (BOF) slag in the production of a hydraulic cement binder. International journal of mineral processing. Volume 79 (2006) Pages 98-105.

[39] Rodríguez de Sensale, G. Effect of rice-husk ash on durability of cementitious materials, Cement and Concrete Composites, Volume 32, Issue 9, October 2010, Pages 718-725

[40] Samaras, P., Papadimitriou, C.A., Haritou, I, Zouboulis, A.I, Investigation of sewage sludge stabilization potential by the addition of fly ash and lime, Journal of Hazardous Materials, Volume 154, Issues 1-3, 15 June 2008, Pages 1052-1059.

[41] Seco, A., Ramírez, F.,Miqueleiz, L., García, B. Stabilization of expansive soils for use in construction, Applied Clay Science, Volume 51, Issue 3, February 2011a, Pages 348-352

[42] Seco, A., Ramírez, F., Miqueleiz, L., García, B., Prieto, E. The use of non-conventional additives in Marls stabilization, Applied Clay Science, Volume 51, Issue 4, March 2011b, Pages 419-423.

[43] Senol, A., Edil, T. B., Bin-Shafique, M., Acosta, H A,. Benson, C. H. Soft subgrades' stabilization by using various fly ashes, Resources, Conservation and Recycling, Volume 46, Issue 4, April 2006, Pages 365-376.

[44] Shen, W., Zhou, M., Zhao, Q. Study on lime–fly ash–phosphogypsum binder, Construction and Building Materials, Volume 21, Issue 7, July 2007, Pages 1480-1485.

[45] Singh, M., Garg, M. Cementitious binder from fly ash and other industrial wastes, Cement and Concrete Research, Volume 29, Issue 3, March 1999, Pages 309-314.

[46] Tay, J-H., Show, K-Y. Resource recovery of sludge as a building and construction material — A future trend in sludge management, Water Science and Technology, Volume 36, Issue 11, 1997, Pages 259-266,

[47] Taylor, M.R., Lydon, F.D., Barr, B.I.G. Mix proportions for high strength concrete, Construction and Building Materials, Volume 10, Issue 6, September 1996, Pages 445-450,

[48] Temimi, M., Rahal, M.A., Yahiaoui, M., Jauberthie, R. Recycling of fly ash in the consolidation of clay soils, Resources, Conservation and Recycling, Volume 24, Issue 1, October 1998, Pages 1-6.

[49] Thomas, M., Folliard, K., Drimalas, T., Ramlochan, T. Diagnosing delayed ettringite formation in concrete structures, Cement and Concrete Research, Volume 38, Issue 6, June 2008, Pages 841-847.

[50] U.S. Geological Survey. 2011. Mineral commodity summaries 2011: U.S. Geological Survey, 198 p.

[51] Wattanasiriwech, D., Saiton, A., Wattanasiriwech, S. Paving blocks from ceramic tile production waste, Journal of Cleaner Production, Volume 17, Issue 18, December 2009, Pages 1663-1668.

[52] Weng, C-H., Lin, D-F., Chiang, P-C. Utilization of sludge as brick materials, Advances in Environmental Research, Volume 7, Issue 3, May 2003, Pages 679-685.

[53] Wild, S., Kinuthia, J. M., Jones, G. I., Higgins, D. D. Effects of partial substitution of lime with ground granulated blast furnace slag (GGBS) on the strength properties of lime-stabilised sulphate-bearing clay soils, Engineering Geology, Volume 51, Issue 1, November 1998, Pages 37-53.

[54] Yarbasi, N., Kalkan, E., Akbulut, S., Modification of the geotechnical properties, as influenced by freeze-thaw, of granular soils with waste additives, Cold Regions Science and Technology, Volume 48, Issue 1, April 2007, Pages 44-54.

[55] Yong, R. N., Ouhadi, V.R., Experimental study on instability of bases on natural and lime/cement-stabilized clayey soils, Applied Clay Science, Volume 35, Issues 3-4, February 2007, Pages 238-249.

[56] Zerbino, R., Giaccio, G., Isaia, G.C. Concrete incorporating rice-husk ash without processing, Construction and Building Materials, Volume 25, Issue 1, January 2011, Pages 371-378.

# 8

# Mining or Our Heritage? Indigenous Local People's Views on Industrial Waste of Mines in Ghana

Samuel Awuah-Nyamekye[1,2] and Paul Sarfo-Mensah[3,4]
[1]*Department of Religion & Human Values, University of Cape Coast,*
[2]*Department of Theology & Religious Studies, University of Leeds,*
[3]*Bureau of Integrated Rural Development, Kwame Nkrumah University of Science & Technology (KNUST),*
[4]*University of Venda (UNIVEN),*
[1,3]*Ghana*
[2]*UK*
[4]*South Africa*

## 1. Introduction

The mining industry, particularly gold may be said to be as old as human civilisation. The history of the extraction of minerals in the various indigenous societies across the world will attest to this. In Ghana, the history of small scale gold mining dates as far back as the 4th century when indigenous craftsmen made use of gold in diverse ways (Hayford *et al.*, 2008). The search for gold took a higher dimension in 1471, when the Europeans arrived and the silent trade began. Commercial scale gold mining, however, is believed to have commenced in Ghana during the 19th century by the British (Tsikata, 1997; cited by Hayford *et al.*, 2008:1).

The industry in the country has expanded over the years, and mining currently includes the extraction of gold, bauxite, manganese and diamond in commercial quantities. Among these, gold mining is the most prominent and contributed to 93% of the exports made in 2004 (Minerals Commission, Ghana, 2004; cited by Hayford *et al.*, 2008:1). The country is one of the major gold producing countries; ranking 8$^{th}$ in the world and second in Africa. (Chamber of Mines of South Africa) The mining sector currently (2008) contributes about 7% of Ghana's total corporate tax earnings. It forms 41% of total export, 12% of Government revenue collected by the Internal Revenue Service and 5% of the total GDP (Hayford *et al.*, 2008).

The expansion of the industry in the country as elsewhere is characterised by improvement in the knowledge or technology in the extraction of the mines (Akabzaa and Darimini, 2001). This means that the volume of mining waste moves with its corresponding increase in the extraction of the minerals. However, the improvement in the technique in the extraction of the mineral has not corresponded with its waste management, rather very little investment has happened in that direction in Ghana due to lack of coordination among the relevant sector institutions (Akabzaa and Darimani, 2001:37). For example, several researches in the country as elsewhere have underscored that large and small scale mining processes impact negatively on the environment and the socio- economic status of the communities in which they operate (Hilson and Yakovleva, 2007; Hayford, *et al.*, 2008). And while emphasis is placed on the processing of the ore body and its waste product disposal method as major sources of environmental pollution, not much has been done to improve them (Hayford, *et al.*, 2008). Therefore, the typical waste product, tailings that consist of crushed ore and rock bodies after most of the needed metals have been removed, are retained in sedimentation ponds and piled up for future treatment, while the slime goes with the overflow into nearby streams. This practice normally leads to siltation of nearby streams, destroying aquatic fauna and flora (Hayford *et al.*, 2008:2). Other effluent that also ends up in nearby streams and rivers introduce dissolved toxic elements into them. The result is the increase in health problems, like severe intestinal upsets, keratosis and skin cancer, sleep disorders and salivation problems (Steinnes and Berg, 1998; Amankwah and Anim-Sackey, 2003).

The multinational mining corporations that operate in Ghana, as commonly found in other developing world tend to place much interest in their profit margins than the environmental hazards that their activities cause. For instance, in 2001, Goldfields Limited spilled cyanide into River Asuman, which serves as the drinking water for communities such as Abekoase, Samahu, Tebe, and Huniso. This also resulted in the death of hundreds of fish, crabs and birds (Anane, 2001). The statutory institutions including Environmental Protection Agency (EPA) that have the responsibility to oversee the operations of these companies are challenged in terms of human and material resources and are therefore unable to adequately deliver on their mandates (see for example, Akabzaa and Darimani 2001: 37). Thus, the management of industrial waste from the ever expanding mining sector in the country poses a real challenge to the country.

In this chapter we seek to examine industrial waste in Ghana with special reference to mining. Our focus is on local people's views on the impact of the industrial waste from mining on the local socio-economic environment. Evidence will be drawn from three regions in Ghana where mining related problems are more pronounced—Western, Ashanti and Brong Ahafo regions. For instance, greater percentage of Ghana's gold is mined in western and Ashanti Region. The chapter has been organised into five sections. The first section comprises the introduction. Section two presents a brief historical background of mining in Ghana and a profile of the study communities. The study methodology is presented in section three. Section four contains the results and discussion. In this section we present among others the conceptualization of industrial waste, types of mining waste and their impact as well as mitigation interventions. Finally, the conclusions and recommendations are presented in section five.

## 2. Brief historical background of the mining industry in Ghana

The mineral industry in Ghana as has been noted pre-dates the arrival of Europeans in the country in 1471(Akabzaa and Darimani, 2001). The industry was vibrant at the pre-independence period. For instance, Ghana was said to have accounted for 36 % of total world gold output (8,153,426 fine ounces) between 1493 and 1600 (Tsikata, 1997).

No wonder that Ghana was known at that time as 'Gold Coast' until her independence in 1957. It has been estimated that over 14.4 million ounces of gold were produced between

1471–1880 (Adadey, 1997). It has also been estimated that 2,488 metric tons (80 million ounces) of gold was produced in Ghana between 1493 and 1997 (Kesse, 1985; Ghana Chamber of Mines, 1998). It is on record that the first 'Gold Rush' was started by one Mr. Pierre Bonne, who was a French trader. He set up a company with several concessions in Ashanti region (Agra, 1997). However, mining as an organised industry, begun with gold production at the later part of the 19th century by the British and other foreign investors (Akabzaa and Darimani, 2001). Private Ghanaian gold miners were, however, banned from operating mines due to the promulgation of the Mercury Ordinance of 1932. This was the result of the Ghanaians preferring to work in their own mines rather than work for the Europeans (Akabzaa and Darimani, 2001).

The industry has now expanded tremendously. The country has 23 large-scale mining companies producing especially gold and other minerals such as diamonds, bauxite and manganese. Currently, Ghana has over 300 registered small scale mining groups and 90 mine support service companies. No wonder that Ghana is ranked Africa's second largest gold producer after South Africa and the 10th in the world (Hayford *et al.*, 2008).

## 3. Profiles of the study areas

Ghana comprises ten administrative regions which are all noted to be endowed with rich deposits of gold and other minerals. However, the Western, Ashanti and Brong Ahafo regions are rated among the most prominent in term of mining activities. Indeed, the Western and the Ashanti regions have the largest investments in mining; Brong Ahafo has gained prominence recently because of the activities of the Newmont Gold Ghana Limited (NGGL). This has also been attributed to the ubiquitous artisanal miners, known in Ghana as "Galamsey" many of whom have relocated from the Western and Ashanti regions to this emerging gold enclave, particularly within the Ahafo Kenyasi area where NGGL operates. In this section we give a profile of the study area; Western, Ashanti and Brong Ahafo regions. We will also provide an overview of the three main mining communities; Tarkwa, Obuasi and Ahafo Kenyasi, where we interacted with local people on industrial waste (see section on Methodology).

According to Ghana fact sheet (Modern Ghana Media, MGM, 2010), the Western region covers an area of 23,921 square kilometres which is about 10 per cent of Ghana's total land surface. It is located in the south-western part of Ghana, bordered by Ivory Coast on the west, Central Region on the east, Ashanti and Brong-Ahafo Regions on the north and on the south by 192 km of coastline of the Atlantic Ocean. The population of the region is 1,924,577, constituting about 10 per cent of the total population of the country. The region is endowed with considerable natural resources, which give it a significant economic importance within the context of national development. It is the largest producer of cocoa, rubber and coconut, and one of the major producers of oil palm. The rich tropical forest makes it one of the largest producers of raw and sawn timber as well as processed wood products. A wide variety of minerals, including gold, bauxite, iron, diamonds and manganese are either being exploited or are potentially exploitable. The region's total geological profile and mineral potential are yet to be fully determined. The Western Region is the largest producer of cocoa and timber, the second highest producer of gold, with the potential to become the highest producer of this commodity. There are five major gold mines, namely Teberebie and Iduaprem goldfields, both now owned by Ashanti goldfields, Prestea/Bogoso mines now owned by a South African company, Tarkwa goldfields, and Aboso goldfields located at Damang near Huni Valley. There are other proven but as yet unexploited ore deposits at Tarkwa, Aboso, Bondaye, and the forest reserve areas of Jomoro and Nzema East, Aowin-Suaman, Amenfi and Mpohor-Wassa-East districts.

Tarkwa, our study community in the Western region as indicated above, is a major gold mining centre not only in the region but Ghana as a whole. For example, Hayford *et al.* (2008) noted that although Anglo Gold Ashanti and Newmont, two of the major producers of gold in Ghana are in the Ashanti Region, the largest concentration of mining companies are, however, found in the Tarkwa-Wassaw District. Tarkwa our study community is the district capital. The 2010 Ghana Population Census estimated the community's population as 33, 466 (Ghana Statistical Service, 2010).

The community and its environs are characterised by an undulating terrain with a magnificent drainage system. It experiences the heaviest and most frequent rains in the country (Akbazaa and Darimani, 2001). The environmental challenges as result of industrial activities in the community and its environs are summarized by Akbazaa and Darimani (2001:30):

> *The heavy concentration of mining activities in the area has generated environmental and social issues in the area. The issues centre on resettlement and relocation, negotiation and compensation and environmental damage. The persistence of these socio-environmental problems accounts for the occasional and frequent resistance from the affected communities as well as clashes between them and the mining companies. The destruction of sources of livelihood and the spate of resistance and clashes have given rise to an environmentally conscious population from which local social movements are emerging.*

Obuasi, our second study community, is a major commercial and industrial centre in the Ashanti Region, the third largest of the 10 administrative regions in Ghana (MGM, 2010). The region occupies a total land surface of 24,389 square kilometres or 10.2 per cent of the total land area of Ghana. In terms of population, however, it is the most populated region with a population of 3,612,950 in 2000, accounting for 19.1 per cent of Ghana's total population (MGM, 2010). Kumasi the regional capital, is the second largest city in Ghana. Much of the region used to be forested but it has witnessed massive deforestation in the past few decades mainly as result of clearing of forest for agriculture which provides employment for more than half of its economically active population (MGM, 2010).

The Obuasi township is the administrative capital of the Obuasi Municipality Assembly (OMA). The community and its environs have been experiencing tremendous socioeconomic challenges, especially rapid population growth and environmental degradation due to several years of mining (Sarfo-Mensah *et al.*, 2010). The estimated population of the Municipality is 205,000 with an annual growth rate of 4%, making it one of the fastest growing districts in the country (Ghana Statistical Service, 2010). The inception of the Economic Recovery Programme (ERP) in 1984 and the subsequent expansion of mining activities, particularly production at AngloGold Ashanti (AGA), led to the establishment of several subsidiary companies, service and commercial activities which are either directly or indirectly related to mining. The resultant inflow of migrants into Obuasi township and its catchment areas in search of jobs has over the years had negative social, economic, cultural and environmental impacts such as illegal mining, high prevalence of prostitution and

HIV/AIDS, and high crime rates. Some attempts have been made by the OMA, NGOs, religious bodies, government agencies and AngloGold Ashanti to solve the unemployment problems in the municipality (Sarfo-Mensah *et al.*, 2010: 2). These attempts, however, have achieved little success. There has, therefore, been a steady increase in unemployment, particularly among the youth since 1984 which has been noted to mainly account for the high prevalence of illegal mining activities and exacerbation of the environmental conditions in the area (Sarfo-Mensah *et al.*, 2010:2).

The Brong Ahafo Region, the third study region covers an area of 39,557 square kilometres and shares boundaries with the Northern Region to the north, the Ashanti and Western Regions to the south, the Volta Region to the east, the Eastern Region to the southeast and La Cote d'Ivoire to the west. The region lies in the forest zone and is a major cocoa and timber producing area (MGM, 2010). The northern part of the region lies in the savannah zone and is a major grain and tuber producing region. The total population of the region is 1,815,408, representing 9.6 per cent of the country's population. The region is more populous than only four other regions though it is the second largest in terms of land area. The main occupation of the workforce of the region is in agriculture which employs about 68.6% of the economically active population (MGM, 2010).

The Newmont Gold Ghana Limited (NGGL) Ahafo gold property is based in the towns of Kenyasi and Ntotoroso in the Brong-Ahafo region. Production of gold in the two communities which started in 2006 was expected to be about 17,100 kilograms per year (kg/yr) (reported as 550,000 troy ounces) with a mine life estimated at more than 20 years (Bureau of Integrated Rural Development, BIRD, 2009a). The Kenyasi Township, our third study community has had its population increasing rapidly since inception of the operations of NGGL started in the communities in 2006. This has led to soaring prices of food and rent charges leading to high cost of living. The population increase has also brought about congestion in the use of some social facilities including water and sanitation (BIRD, 2009a). A massive inflow of illegal miners into the community and its environs is reportedly leading to the destruction of the farm lands and loss of employment of the youth whose livelihoods depend on farming. A major challenge in the community and its catchment area is environmental management due to mining and its related activities (BIRD, 2009a).

## 4. Methodology

The study methodology was participatory and mainly community-based. We scaled it up at the district and the regional levels to interact with local non-governmental organisations (NGOs), technocrats and government officials to triangulate views collected at the community level, and to collect secondary data. Thus, both primary and secondary data were collected for the study.

The primary data was mainly based on interaction with people in the following communities: Ahafo Kenyasi in the Brong-Ahafo region; Tarkwa in the Western region; and Obuasi in the Ashanti region (see Figure 1). These communities and their catchment areas are all prominent gold mining enclaves in Ghana. In the communities, participatory approaches were used to interact with the following groups: chiefs and opinion leaders; elderly males; elderly females; the youth (men and women groups); and children (between the ages of six and 18 years). Our groups were selected to ensure that gender views were

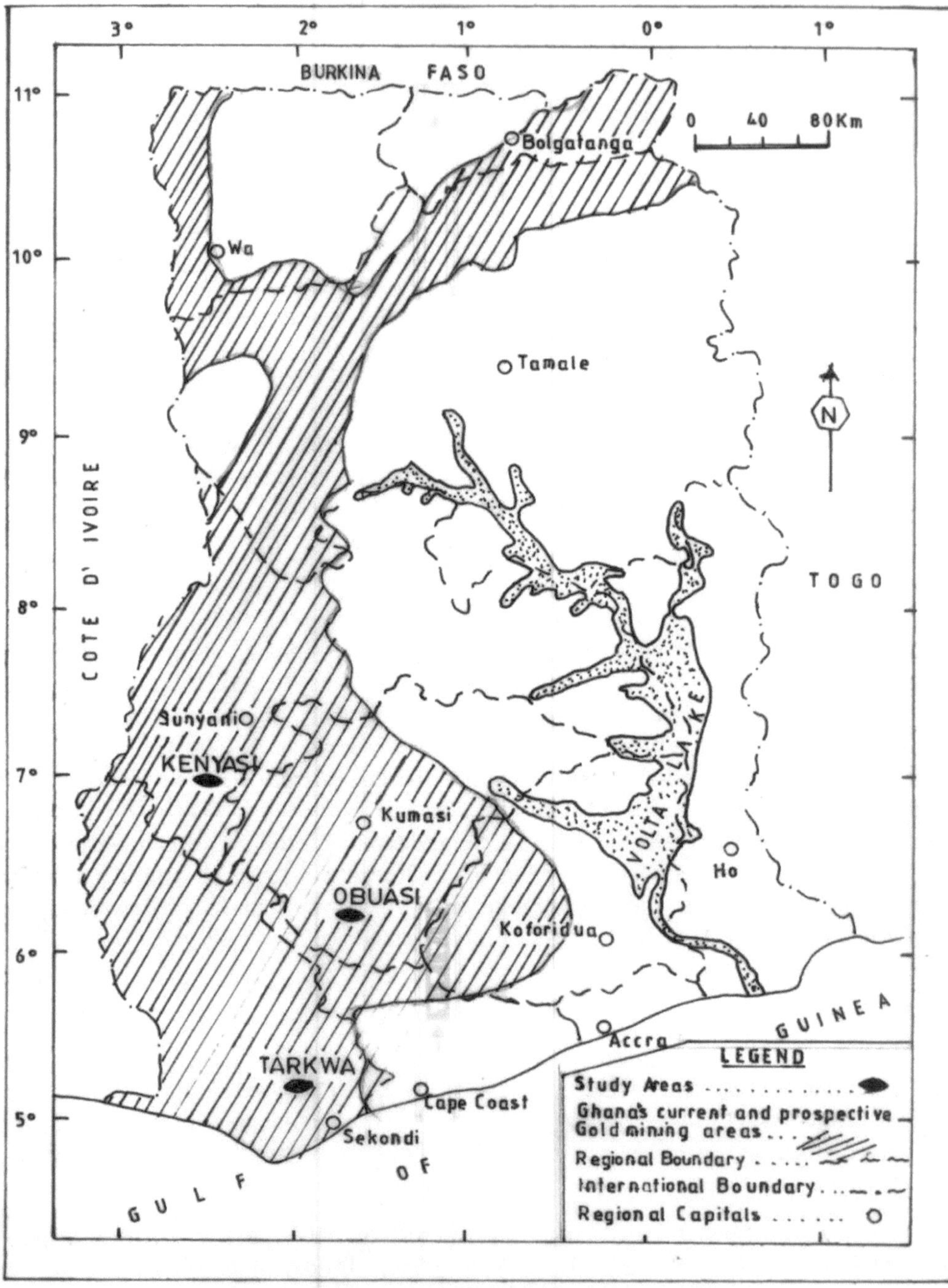


Source: the authors' construct 2011

Fig. 1. Map of Ghana showing Study Areas

captured. And, most importantly, children whose views are critical in society but often ignored in community engagement, was captured as well. And as indicated above, the participatory methodology was achieved by holding focus groups discussions with all our five carefully selected groups through the use of structured and semi-structured interviews. The focus of the discussions centred on the effects of mining especially mining waste on the lives of the people and the environment as a whole. Another key area of concentration was on the local people's environmental conservation strategies particularly on waste management. The possibility of integrating local knowledge with that of the modern means of waste management was also touched on.

The desk studies comprised mainly a review of relevant government documents and reports as well as reports produced by NGOs that operated in the study area. We also reviewed literature on industrial waste in the global, African and national context.

## 5. Industrial waste conceptualised

Industrial waste refers to waste resulting from industrial activity such as activities from factories, mills and mining and others. Some of these wastes may exist in liquid or solid form, some of which are hazardous -- those from automobile repair shops, petroleum and petrochemical industry, mining industry, etc. (see Books *et al.* 1976; Munson-McGee *et al.*, 1996; Chaudhary and Rachana, 2006; Moreno *et al.* 2009). That is, they are or pose potential threats to both human and the environmental health. Those harmful wastes are characterised by ignitability (waste oil), corrosivity (e.g. battery acid), reactivity, toxicity, infectious or pathogenic. There are, however, many industrial wastes that are neither dangerous nor harmful. For instance fibre resulting from agriculture and logging is neither hazardous nor toxic.

Since the 'Industrial Revolution'[1], industrial and mining operations have been accompanied by the problem of industrial waste, thus, making industrial waste a by-product of the Industrial Revolution. This waste is generated from the production process through the use and the disposal of manufactured products. Currently in Ghana, apart from the wastes generated from mining activities, the following are the main activities from which wastes are generated: "the textiles industries (Spinning, weaving, finishing, bleaching and dyeing, printing); food and beverages (fish processing, slaughter houses, breweries, soft drinks, fruit processing, oil processing, cocoa processing and flour mills); Petroleum and petrochemical industry; Wood processing industry (Sawmills, veneer processing, ply mills, furniture); Plastics and foam industry; paper, printing and publishing industry; Pharmaceutical industry; and the Paints and chemical industry" (see 18th Session of the United Nations Commission on Sustainable Development, National Report for Ghana Waste Management in Ghana). The basic concern of the Industrial Revolution has been the appropriate ways of managing industrial waste in order to avoid or mitigate their adverse environmental or human health impacts.

Managing industrial wastes has been a big source of worry to Ghana. The bane of the problem has been the lack of the requisite technical know-how, capital resources and the

[1]We refer to 'Industrial Revolution' as the major industrialisation that took place during the late 16th (1700s) and the early 17th centuries (1800s), which begun in the Great Britain was rapidly replicated throughout the world.

lack of political will by the various governments of Ghana (mostly paying lip-services to the issue since the policy, legal and institutional frameworks put in place are not strictly implemented). Although many of the industries have tried to do something by way of trying to recycle their waste products, their efforts have not been enough to contain the increasing volume of the current industrial waste level in the country. This chapter as has been pointed out examines the industrial waste in Ghana with special reference to mining waste.

## 6. Mining and the economy of Ghana

The contribution of the mineral industry to Ghana's economy is enormous. For instance, as back as 1991, the mining industry overtook cocoa as the largest foreign exchange earner for Ghana (Adadey, 1997). Minerals alone account for 40% of the Ghana's total gross foreign exchange earnings. Over 20,000 people are employed by the large scale mines whiles the small scale artisanal mining, especially gold and diamond, employ more than twice the number engaged by the large scale mining (Adadey, 1997). These figures appear to be conservative estimates. For example, according to Hilson (2001) an estimated 30,000 small scale miners work on registered plots while, 170,000+ are illegal *galamsey*. The contribution of small scale mining has been succinctly captured by Amankwah and Anim-Sackey (2003:133) as following:

> *The small-scale mining of these precious minerals has made significant socioeconomic impact on many individuals and communities since it provides both part- and fulltime jobs for the people and in some cases it is the only source of income available to the people. In the rural communities where mining takes place, the activity has reduced rural exodus, promoted local economic development and contributed towards poverty reduction. In addition, the mining operations are useful in basic skill development and contribute to the transformation of unskilled labour into semi-skilled and skilled workers. More importantly, due to the low barriers to entry in terms of capital needs and formal educational requirements, small-scale mining operations offer excellent opportunities for the evolution of indigenous entrepreneurs. In rural areas where other jobs are low paying or non-existent, small-scale mining appears as a valuable source of employment. The sector also provides raw materials for local industries.*

But as we discuss later, the small scale mining sector is a significant contributor to industrial waste in the mining industry in Ghana and generally regarded as the principal actor in land degradation due to their ubiquity and uncontrolled nature (Anane, 2001; Akabzaa and Darimani, 2001:26).

Gender related activities have also been highlighted in the mining sector, especially regarding females participation in the industry (Hilson 2001; Hinton et al., 2003). There are several activities, especially in artisanal mining, where women are said to be predominating, for example, in sieving, sorting, the transport of ore and water, in cooked food selling, petty trading and providing other essential support services (e.g. washing of the ore) to the male miners. According to Hilson (2001) quoted by Hinton et al., (2003:3-5) women in artisanal mining constitute the following: acting as licensed buyers (6%); concession holders (10%); and work group sponsors or participants (15-20%); women comprise approximately 15% of the legal small- scale metal mining labour force in Ghana; and their involvement in the illegal *galamsey* industry is up to 50%. And as noted by Hinton *et al.* (2003: 1) for many women, artisanal mining signifies an opportunity to relieve the strains of poverty.

In rural communities, where mining activities have seen an upsurge, the contribution of the industry to the local economies in terms of investment in people and infrastructure has been acknowledged by government and local people. These have ranged from housing and resettlement programmes, alternative livelihoods and direct training to employment opportunities for local people. Most mining companies are facilitating the management of comprehensive social responsibility programmes which fund these development projects. An example we found in our study area is the Newmont Ahafo Development Foundation of NGGL. As part of the company's social responsibility commitments, and through series of consultations and discussions with the relevant stakeholders including Traditional Authorities, District Assemblies, Regional Coordinating Council, Youth Groups, Women Groups, Farmers Groups and local NGOs, Newmont signed Social Responsibility Agreement with the communities around the Ahafo Mine in May 2008 (BIRD, 2009b). This agreement resulted in the setting up of Newmont Ahafo Development Foundation (NADeF) which is being supported by NGGL with $1 per oz of gold produced and 1% net profit. The fund is estimated to generate about $600,000 annually which will be used to support sustainable development projects and programmes in ten (10) communities namely; Kenyasi No.1, Kenyasi No.2, Ntotoroso, Gyedu, Wamahinso (all in Asutifi District), Yamfo, Susuanso, Terchire, Afrisipakrom and Adrobaa (all in Tano North District). The initial amount deposited by Newmont into the Fund is about $850,000 (BIRD, 2009b:36).

The Foundation is now a registered entity under the laws of Ghana with a Nine-Member Board of Trustees to manage the fund. The Foundation has set up a Secretariat to manage the day-to-day administration of the fund and a Tender Board to handle the tendering processes for projects under the Foundation. To ensure that the purpose for which the fund has been established is realized in a timely, cost effective, transparent and sustainable manner, the Agreement also requested the ten communities to establish Sustainable Development Committees in their respective communities to lead and manage the development processes (BIRD, 2009b: 36). Apart from the mainstream mining companies, it must also be noted that in rural communities, significant numbers of inhabitants are attracted to small-scale mining because the industry pays substantially higher wages than most other sectors of industry (Hilson, 2001: 8).

In major urban mining settlements such as Obuasi and Tarkwa, most social and economic activities revolve around the mines. In Obuasi, the entire local economy depends on AngloGold Ashanti. For instance, educational infrastructure, markets, roads and lorry stations have been built by the Municipality with funding support from the company (Sarfo-Mensah, *et al*, 2010). Business in the township is driven by incomes from miners. Indeed, responses from our respondents indicated that from children up to adults, everybody has something to do with the mining conglomerate. Although most of respondents were quick to add that because of the mines, cost of living is very high and several social vices are on the increase (see section 7 below). Socio-economic dependency on mining activities in communities where mining giants such as AngloGold Ashanti operate in Ghana is so pervasive that a slowdown in their businesses leads to virtual collapse of the local economies.

Mining has arguably made tremendous contributions to the socio-economic development of the country and its potential, as presented by the country's trade figures, is very enormous. In 2000 Ghana was ranked 8th in the world for gold production (Chamber of Mines of South Africa 2005). According to The Ghana Minerals Commission (2002), Ghana produced

2,241,125 ounces of gold and the revenue accruing from this was $646.00 million in 2002. Also, according to the Business Monitor (2011) gold production was up by 3.5% year on year (y-o-y), at 1.46mn oz by 2010 with gold export earnings standing at US$1.68bn. These figures indicate that Ghana derives substantial revenue from the mining sector, it is, however, estimated that only 30% of overall mining revenue remains in the country (Akabzaa 2009).

## 7. The negative impact of mining on Ghana

Mining as an industry has given society tremendous benefits. In terms of wealth creation, it is ranked amongst the highest income earners for many countries. However, its negative impact on society has been very significant. In this section, we discuss local people's views on the consequences of mining with emphasis on its industrial waste.

In our study communities, the negative impact of mining was defined severally perhaps due to the complexity of the problems that local people associated with mining. These were reflected in local terminologies such as *ekum asase* (land degradation), *esei kwae* (deforestation), *sika bonsam* (money that it generation drives people insane) and *ehyekro* (destroy community). Two major points are noteworthy from these definitions: first they were common in all the three major mining settlements and their catchment areas that we studied; second and most importantly they underlie three key areas that were recurrent during our interviews - environment, social and economic issues. Before we draw upon empirical data from our interviews to discuss these three domains and how they are driven by industrial waste from mining activities in the study area, we give a summary of a 56 year-old female respondent understanding of the negative impact of mining:

> *We have been made to believe since my infancy that citrus from our community is sweet and have even been given a label as "Obuasi ankaa wo" (to wit, the sweet as honey citrus of Obuasi). But little did we know that citrus and other agricultural products as well as our drinking water are deadly contaminated with poisonous substances that we cannot see with our naked eyes. Several diseases and sudden deaths which were attributed to witchcraft in the past, we now know have their bearings in the discharge of industrial waste from the mines in our community (Adwoa Agyeikumwaa, Obuasi, per comm.,2010)*

The above demonstrates the depth of local peoples' understanding of the threat industrial waste from mining activities poses to lives and livelihoods. However, we found from our study that most respondents considered the mining companies so powerful that they have resigned themselves to the situation (Babut, *et al.*, 2003; Hilson & Yakovleva, 2007; Bush, 2009). For example, we found across our study settlements when we asked our respondents of community members filing claims for compensation for deaths related to poisonous industrial waste from mining such as cyanide and mercury, most respondents said these were unheard of. But most of our respondents associated strange illnesses and some deaths in the communities with environmental pollution from mining activities, particularly the seepage of poisonous chemicals into water bodies (see for example Hayford et al., 2008:2). In the Tarkwa area and its environs for instance, many people had reportedly left villages where water quality has deteriorated to the extent that they no longer feel safe (Akbazaa & Darimani, 2001:46).

In addition to water pollution, negative environmental impact of industrial waste from mining was traced to loss of livelihoods. In our focus group discussions, many people, especially young men and women were critical of declining soil fertility in their communities which have negatively impacted on farming, the main occupation. They attributed this to massive land degradation by artisanal miners who they claimed used various chemicals which rendered the "soil dead". For instance, at Ahafo Kenyasi area, one of our respondents had this to say:

> *These people (artisanal miners) are everywhere and are "killing the land". Their activities are so destructive to the land that after they have left a piece of land it may take several generations before the land recovers. They do not care about our farms. In so far as they find gold on the land they will go for it. We are helpless because it appears some of our traditional leaders are in league with them. In addition, they are often armed and we cannot approach them, especially in those remote areas where we have our farms (Kwaku Agyapong, Ahafo Kenyasi, per comm.., 2010).*

Indeed agriculture, the main source of livelihood in mining communities, has been seriously affected due to land degradation (Akabzaa and Darimani, 2001; Hayford, *et al.*, 2008). For example, Akabzaa & Darimani (2001: 47) found that in most parts of Tarkwa, the environment is undergoing rapid degradation and its immense economic value is diminishing from year to year, due mainly to the heavy concentration of mining activities in the area. They also noted that agricultural lands are not only generally degraded, but the decrease in land for agricultural production has also led to a shortening of the fallow period from 10-15 years to 2-3 years. Thus in the Tarkwa area and our other two study communities, we found that land degradation has translated into many social and economic problems. For example, we found that many families were unable to adequately cater for their basic nutritional requirements either due to loss of their farm lands through government concessions to mining companies or indirectly through land degradation due to pollution from industrial waste of mining entities. This finding on industrial waste and land degradation has also been observed by Hilson (2002b: 54) with respect to small-scale mining as below:

> *....mercury, which is used in excessive quantities in mineral refining processes, has been released uncontrollably into natural ecosystems surrounding operations, accumulating to toxic levels in soils, water bodies and flora. Perhaps more important, however, is that small-scale gold mining, as a migratory activity, has caused a significant amount of land damage in Ghana.*

We found that in addition to land degradation which directly affects agriculture the main stay of the local economies in our study area, there has been recognisable impact in terms of income disparities which has been introduced into the communities by mining. For instance, many of our women respondents attributed the cost of living in their communities, which they claimed had increased more than fivefold, mainly to mining. They attributed this to money that artisanal miners have and are able to buy whatever at any price and also because of declining food production due to land degradation. High cost of living as a result of income disparities have been observed in several mining communities in Ghana including our study communities (see for example Akabzaa & Darimani, 2001:48). In the Ahafo Kenyasi area for example, we found from focus group discussions that many families have relocated outside the community. Some of the respondents said these people though benefitted from housing resettlement schemes have had to leave the communities because they no longer had adequate land to farm and the cost of living in the community had also become too high (Bureau of Integrated Rural Development, 2009a).

Results from our interviews in the communities also established that there was not much community-mining companies' engagement to address some of the issues raised above particularly regarding industrial waste management. Although we found that waste recycling technologies were being used by some mining companies to deal with the large volumes of industrial wastes they generate, there is still a long way to go in this direction. And with the ubiquitous artisanal miners we found no such inclination to deal with their waste.

Indeed as demonstrated above, the negative socio-economic consequences of mining waste in the study area and across Ghana can be tremendous. For example, literature exist on the metal pollution within the environment of mining communities in Ghana (Adimado and Amegbey 2003; Akabzaa et al. 2005; Carbo and Serfor - Armah 1997; Essumang et al 2007; Hilson 2002a; Manu et al 2004; Obiri 2007; Yidana et al 2008) but not much has been done on tackling the problem from the indigenous local people`s perspective, which is one of the key concerns of this chapter.

Literature on the mining industry in Ghana show that the corresponding damage to the environment and its effects on the people living in the mining communities far outweigh the benefits deriving from it . As at 1998, the government of Ghana had granted over 200 mining leases to companies. This means that about 30% of Ghana`s land surface is being mined, implying that the volume of mining related hazards particularly, wastes and its attendant environmental degradation is growing at an alarming rate. Interestingly, these mining activities are mostly found in the environmentally and socially vulnerable areas.

The Environmental Impact Assessment (EIA) is usually carried out as required by law in Ghana by the Environmental Protection Agency (EPA) before the mining companies are allowed to operate. Initially, these companies would present a very convincing arrangement to operate with the barest minimum negative impact on life and property. But as soon as they are granted permission to operate, they renege on their own control measures. This is exemplified by a number of cyanide spillages, damping mining waste into water bodies, destruction of farm lands and so on. For instance, a study has shown that Arsenic, Cadmium, and Mercury are found in food crops such as cocoyam (*Xanthosoma sagititolium*) and Watercocoyam (*Colocasia esculenta*) in Tarkwa, a prominent Mining Community in Ghana (Essumang *et al.*, 2007). There has been widespread tension between mining companies and the local people over the destruction of farmlands and the general disregard for the rights of the local people by the mining companies. More often than not, the military, police and other paramilitary groups are used by these mining companies to brutalise the protesters into submission. Deaths and serious injuries are usually reported by the local media on such inhuman behaviour of the mining companies.

There have been incessant complaints from the local people that these mining companies which are multinational in nature do not employ local people. For instance, our interaction with the people of the study areas revealed that the mining companies prefer labour from outside the mining areas. Meanwhile it is the local people who bear the brunt of their (mining companies) activities. For example, mining activities, particularly surface mining, have resulted in the alienation of large tracts of land from communities, depriving poor and marginalised communities of their land surface rights, and as a result depriving many communities of their sources of livelihood (Akabzaa, 2009). The appropriation of the land of local communities for mining has often engendered social upheavals and adversely impacted on the routine livelihood activities of these communities. Such social upheavals are commonplace in communities affected by mining projects in Ghana (Akabzaa, 2009).

Furthermore, we realised from our interactions with the local people that the mining companies are not paying adequate compensation to affected farmers based on the compensation principles in the Minerals and Mining Law (Act 703). The people from the study areas also emphasised that it is not only multinational companies whose actions are destroying the environment but also the activities of illegal miners whose method of mining has been tagged in Ghana as *galamsey*. This is a form of artisanal mining for gold and diamond. These illegal miners also carry out alluvial mining in some rivers. These miners troupe from all areas of Ghana into the mining areas to carry out their illegal activities. What is worrisome about this category of miners is that they lack the requisite technology and safety measures needed in mining. The open mining pits they leave behind have become death traps to people and livestock. They wash the ore directly in rivers and thus pollute these rivers with all sorts of chemicals particularly mercury. Because of the bad mining methods, mining related diseases such as tuberculosis, diarrhoea, buruli ulcer, pneumoconiosis, lung cancer and other occupational respiratory diseases are a common feature in the mining communities in Ghana. Generally, we found out that even though there exists the Mercury Law (PNDCL 217), Small-Scale Gold Mining Law (PNDCL 218), and Precious Minerals and Marketing Law (PNDCL 219) in 1989, the level of their enforcement leaves much to be desired.

## 8. Types of mining waste in Ghana and their impacts

Mining extraction in Ghana centre mainly on activities such as: Removal of overburdens and ore blasting; loading and hauling of ore to the processing plant; beneficiation (the processing of ore); disposal of waste generated from the ore processing; drainage of mine area and discharge of mine waters. Therefore, the mineral waste generated in Ghana result from the above activities. Specifically, the following are the main mining wastes in Ghana: waste rocks, open mine pits, disposal of mining waste usually these are damped into water bodies, atmospheric pollution (tailing from the mill and ore processing plants), chemical pollution of the land, and frequent cyanide spillage.

According to Hayford *et al.* (2008), the processing of the ore body and its waste product disposal method are the major source of environmental pollution. They indicate further that as typical of most mining operations in the country as elsewhere, the tailings that consist of crushed ore and rock bodies after most of the needed metals have been removed are often toxic and pose serious threats to human, animal and plant life (Hayford et al., 2008:2). Mercury has particularly been noted as a major pollutant, especially from the activities of small-scale miners (Amankwah & Sackey, 2002; Hayford *et al.*, 2008). Cyanide spillage has also become a major source of concern (Anane, 2001).

The sedimentation ponds containing trailing where future treatment is done has been noted to be major source of supply of the industrial waste from mining (Hayford et al., 2008:2). The sedimentation ponds contain sulphides (pyrites, arsenopyrites, etc.), which are oxidized in the tailings. The resultant acidic effluent water in the sedimentation ponds leach out elements such as mercury, arsenic, zinc, copper, vanadium, antimony and chromium (Hayford et al., 2008). The effluent subsequently ends up in nearby streams and rivers, introducing its dissolved toxic elements into them.

The environmental degradation from mining activities and its related industrial waste is evident in several forms. In Ghana those that are of great concern include atmospheric pollution, water pollution, land degradation and deforestation (Amankwah & Sackey, 2003:134). In our study communities as elsewhere in Ghana, the activities of small scale miners were noted by our respondents to be of major concern (see discussions on Negative Impact of Mining above). In Ghana, the environmental destruction caused by the unplanned and sometimes dangerous and irrational methods used by small-scale miners has been noted (Amankwah and Sackey, 2003).

Water pollution remains the major challenge in several mining communities countrywide. And as already noted in our study communities, this emanates from effluent from sedimentation ponds which subsequently ends up in nearby streams and rivers, introducing its dissolved toxic elements into them (Hayford, *et al*, 2008). This has resulted in increases in health problems, like severe intestinal upsets, keratosis and skin cancer (Steinnes & Berg, 1998; Akabzaa & Darimani, 2001; Hayford *et al.*, 2008). There have also been several instances of cyanide spillage in water bodies which has resulted in fatalities (Anane, 2001). Regarding atmospheric pollution, our respondents were quick to mention air pollution generated from dust from bulldozing and transport activities. Several of our respondents talked about coughs and chest problems which they attributed to dust particles.

Noise pollution from blast was also mentioned. In the latter, many of our respondents referred to cracks in the local buildings as partly to do with this operational activity of the mining companies. Literature points to the fact that mining communities in Ghana are victims of air, noise and water pollution as well as other forms of environmental degradation from mining companies (see for example, Akabzaa & Darimani, 2001:34). For example, Akabzaa and Darimani (2001) make reference to the release of airborne particulate matter into the environment, particularly minute dust particles of less than 10 microns which they emphasize as a serious health threat to the people of the Tarkwa, one of our study communities. All fine dust at a high level of exposure has the potential to cause respiratory diseases and disorders and can worsen the condition of people with asthma and arthritis. Dust arising from gold mining operations has a high silica content which has been responsible for silicosis and silico-tuberculosis in the area (Akabzaa and Darimani, 2001:55). The situation regarding industrial pollution and perhaps the apparent helpless situation of the mining communities emanates from what has been noted about mining projects that they generally have weak links with the rest of the host national economy, although they can have a decisive impact on the communities in which or near which the mines are located (Kwesi Anyemedu, 1992, cited by Akabzaa & Darimani 2001:35).

## 9. The role of indigenous people in mitigating the problem of mining waste in Ghana

Our previous studies (Sarfo-Mensah, 2002; Awuah-Nyamekye, 2009; Sarfo-Mensah and Oduro 2010) have shown that traditional Ghanaians` lifestyle is environmentally friendly. This is as the result of their worldview[2], which is hugely underpinned by their religion.

[2] By worldview of the local people, we mean what Mkhize (2004: 2-4) refer to as the set of basic assumptions of the people developed in order to explain reality and their place and purpose in this world. In other words, and to borrow Elkins' (1938:133, quoted in Rose, *et al*. 2003:59) words, worldview is " a view of nature and life, of the universe and man, which unites them with nature's activities and species' in bond 'of mutual life-giving."

Indigenous Ghanaians ascribe spiritual qualities to everything that constitutes nature. Thus, for them, forests, rivers and other natural endowments connote more than the eye can see. Behind these natural phenomena, is a socio-cultural dimension, particularly their spiritual connectedness to these resources that have as much value as their physical manifestations. Indeed, traditional Ghanaians` interpretation of nature (environment) is not different from culture (society) and they feel that these two mutually influence each other and the two underpin their heritage and very existence (Sarfo-Mensah and Oduro, 2010). This explains why indigenous Ghanaians will not pollute a river with dangerous chemicals. For instance, bathing or washing in a river is strictly prohibited. This is to prevent the chemical content of the soap from polluting the river and thereby endangering marine life (pers. comm. 3 July 2011). It is against this backdrop that African indigenous religion has been referred to as 'profoundly ecological' (Schoffeleers, 1978).

And within the mining sector, especially among artisanal miners, there is a lot of belief around precious minerals and local gods. This consequently impact on the conduct of their practices including how they treat rivers, sacred groves and other areas designated as the domain of local gods, such water heads, hills and mountainous areas which are potential areas for mining . For example, Addei and Amankwah (2011:249) relate that some major activities of artisanal and small scale gold miners in Ghana are to a large extent informed by superstitious beliefs and myths. They emphasize that each major activity such as prospecting/mining and processing of the precious metal are controlled by specific beliefs, and since mining activities take place around large rivers, sacred groves and forests, the belief that these bodies have inherent supernatural powers, demands that specific set of rules be observed (Addei and Amankwah, 2011:249). Indeed, the belief is not only limited to the artisanal mining but also relates to the formal big mining sector as Addei and Amankwah (2011:249) underscore below:

> *Oral traditions indicate that early miners in underground mines in the now AngloGold Ashanti Mine in Obuasi and the defunct Prestea Mine met small creatures purported to be dwarfs in underground tunnels and in some cases hens and chickens. These spirit beings saw the miners as intruding into their private spaces and miners had to pacify them in order to work safely.*

Another classic and interesting example found by Addei and Amankwah (2011) relates directly to beliefs concerning industrial waste. They claim that the first heap of waste built by the defunct Teberebie Goldfields could not be leached due to problems of lixiviate percolation. They contend that though the metallurgists thought that it had to do with the binder used, some opinion leaders in the community were of the view that it was because sacrifices had not yet been made to the gods of the land. To satisfy both schools of thought, a cow was sacrificed while the binder was changed to Portland cement. They further point out that on other mines, new earth moving equipment have been rendered irreparable after local people said unidentified white men were seen using them on a night shift (Addei and Amankwah, 2011: 249).

This implies that indigenous people`s beliefs and ecological knowledge, which is underpinned by their religion can be roped in when addressing environmental problems caused by mining activities, especially in the area of managing waste. But unfortunately, indigenous ecological knowledge and beliefs are often neglected by policy-decisions makers in Ghana.

It is regrettable, however, to note that traditional strategies and institutions to address environmental problems are under threat. A recent studies and elsewhere in Africa (Sarfo-Mensah and Oduro, 2010; Nwosu, 2010) in some Akan communities in the transitional agro-ecological zone of Ghana to investigate the spirituality of forests and conservation reveals that *tumi* (the traditional belief in super natural power suffused in nature by *Onyame,* the Supreme Creator Deity) and *suro* (the awesome reverence and fear) that were usually attached to nature are waning (see also Ntiamoa- Baidu 1995; Abayie Boateng 1998; Appiah-Opoku and Hyma 1999). This is partly attributed to the changes in the perceptions and attitudes of local people towards their worldview. For instance, our study revealed that certain forests which had been designated as sacred and thus their entry is limited to a few people have been cleared for mining.

Similarly, the 26th September edition of the national paper, *Daily Graphic* reports of the damping of waste from a wood treatment company in Takoradi into the *Butuah* lagoon which has resulted in the death of more than 40000 fishes in the lagoon. The report adds that those who ate the dead fishes suffered stomach running and dehydration and had to be sent to hospital (Daily Graphic 26 September 2011). This could not have happened when the day-to-day administration of the country was under the authority of the local chiefs who doubled as political and religious leaders.

Christianity is largely blamed to be responsible for this change in local perceptions and attitudes due to its sustained attack on African indigenous religion. Different studies on traditional people across Africa confirm this view (Parrinder 1961; Smith 1986:86; Nukunya 1986: 87; Juhe- Beaulaton 2008; Nwosu, 2010; Western Regional Directorate of CNC 2010; Teye 2010; Opoku-Ankomah *et al.* 2010). This seems to collaborates Lynn White`s 1967 view that Christianity is to blame for the world's current environmental crisis although Lynn's view has been challenged (Joranson and Butigan 1984; Harrison 1999; Johnson 2000).

But it should be explained that Christianity is being blamed not because it is seen as an anti environmental conservation per se, but the blame is in respect of its persistent attack on the indigenous religions which underpins the traditional conservation methods. Ghanaians and, for that matter, Africans are generally seen as 'incurably' religious (Parrinder 1974: 9). Consequently, their social and moral laws are believed to be divinely inspired and thus enforced through the agency of religion, so anything that affects their religion affects their laws. The only instance in Africa where a study's result seems to challenge the hypothesis that Christianity is to blame for loss of nature resources, particularly the forest, is that among the Muzarabani of Zimbabwe (Byers *et al.* 2001). The researchers, however, stated that about three-quarters of the people interviewed claimed to hold allegiance to both the local religion and Christianity. This indeed makes this challenge suspicious.

We also found among our respondents that most of them, as proven by other studies in Ghana (see for example, Awoonor, 1975; Adarkwa-Dadzie, 1998; Sarfo-Mensah, 2007; Opoku Ankomah, *et al.* 2010; Addei and Amankwah, 2011) were syncretic and that they were prone to be easily swayed by Christianity and Islam in their public conduct but were inherently superstitious and inclined to believe in traditional religion. But what we found worrisome, though, was the extent of deprivation in the study communities. It made many local people easy prey to economic motivations to become local collaborators for denigrating sacred areas and other natural resources including rivers and water bodies, actions which hitherto were considered taboo. Indeed, some local institutions were also found to be vulnerable to economic pressures. They sometimes either directly collaborated with mining concerns which were found to degrade the environment or turned blind eye because of inducements from miners, especially itinerant small-scale miners. For instance, Chieftaincy which is considered the bastion of culture and custodian of land and its related natural resources in the study area, was blamed by some respondents for the activities of recalcitrant artisanal miners.

## 10. The way forward

Based on the above findings, the following recommendations are made. There is the need for urgent amendment to the Mineral and Mining Laws in Ghana to be more binding on mining companies to comply with mining regulations. It is imperative that measures are put in place for a routine check on mining companies to make sure they comply with the environmental safety standards set especially in the area of illegal disposal of industrial wastes. This can effectively be done if there is a greater co-ordination in the implementation of waste management plans and programmes by the relevant agencies.

We also suggest that it is time for a comprehensive engagement of all stakeholders in the mining industry to evolve a strategy that will enhance mutual and harmonious co-existence between the formal mining sector (i.e. recognized big companies) and the small-scale sector which encourages the latter to adopt sound mining practices. This strategy should, among others, be geared towards achieving a regulatory tenure framework which will deal comprehensively with the incessant encroachment of land by artisanal miners. The overwhelming majority of small scale miners operate without security of tenure and any legal entitlement (Hilson and Potter, 2005).

There should be a constant sensitisation of the local people on their right to adequate compensation from mining companies and to keep 'eagle eye' on the mining companies particularly the way wastes from their operations are disposed off. This will enable them to report the bad activities of the mining companies to the appropriate quarters for disciplinary action.

It is also recommended that a framework is adopted which will enable or encourage more private companies to invest in waste management sector of the economy.

We further recommend that the Environmental Protection Agency (EPA) is adequately resourced to be able to do its work well and efficiently especially in its assessment and monitoring oversight role. This will also enable it to liaise well with the District Environmental Management Committees (DEMC). Additionally, EPA must put measures in place to enable it to constantly monitor and evaluate the activities of the mining companies especially with regard to their impact on the environment of the areas they operate. This can assist the EPA to revise its laws to meet the realities on the ground. Also in this direction, we suggest that EPA increases its activities to promote environmental improvement in small-scale gold mining operations which arguably have been noted to have caused disproportionately large share of environmental degradation (Hilson, 2002b).

It is furthermore recommended that policy-decisions makers set up special fund for the relevant agencies like the universities and other analogous institutions to conduct further research with the view to finding the most appropriate ways of dealing efficiently with the problem of industrial waste, particularly in the mining section in Ghana.

We also recommend that government begins a serious consideration for a comprehensive support package for the small-scale mining sector due to its huge employment creation capacity and its financial contribution to the economy. The support package, we suggest, should focus more on capacity development both in terms of environmental and ethical training as well as financial to enable them purchase relevant equipment that would make their operations environmentally sustainable.

## 11. Conclusion

The foregoing discussions have not only examined the concept of industrial wastes with particular reference to mining but have also focused on the complex long historical development background of the mining industry in Ghana. Related issues such as the effects of mining hazardous or toxic materials especially the wastes which pose real and potential danger to the lives of the people and the environment as a whole in Ghana, the local people`s environmental conservation strategies, as well as the role of indigenous people in mitigating the problem of mining waste in Ghana have been examined. The discussions have also pointed out the fact that over the years, the mining industry has contributed its quota to the development of the economy of Ghana. This notwithstanding, the poor manner in which the waste resulting from the industry especially the hazardous ones are managed is creating doubts in the minds of people especially environmentalists and eco-friendly minded-people about the actual benefits of the industry to local people and the country as a whole.

It is imperative to stress that the study also is not by the picture thus far painted, suggesting the banning of mining in Ghana, but rather, advocating the application of sustainable mining techniques to reduce risks posed by mining wastes to human life, the environment and also enable Ghanaians to enjoy the socio-economic benefits of the industry and at the same time conserving the natural resources -- our heritage. To be able to do this effectively, there is the urgent need to re-examine the mining industry by taking the above recommendations into consideration.

## 12. References

Abayie Boaten, A. 1998. Traditional conservation practices: Ghana's example. *Institute of African Studies Research Review,* Vol. 14, No 1, pp. 42-51.

Adadey, E. (1997).The role of the mining industry in the economy of Ghana: The mining industry and the Environment, Proceedings of a National Symposium, UST/IDRC Environmental Research Group: Kumasi, Ghana, April, 14 &15.

Adarkwa-Dadzie, A. 1998. 'The Contribution of Ghanaian Traditional Beliefs to Biodiversity Conservation', in Amlalo D.S., L. D. Atsiatorme and C. Fiati (eds.), Proceedings Of The Third UNESCO MAB Regional Seminar On Biodiversity Conservation And Sustainable Development In Anglophone Africa (BRAAF), Cape Coast, 9-12th March 1997. Accra: Environmental Protection Agency (EPA).

Addei, C. and Amankwah, R. K. (2011). Myths and Superstition in the Small Scale Gold Mining Industry of Ghana *Research Journal of Environmental and Earth Sciences, Vol.* 3 No. 3 pp. 249-253.

Adimado A., A.; Amegbey N., A. (2003). Incidents of cyanide spillage in Ghana, Mineral Processing and Extractive Metallurgy. (Trans. IMMC) 112, 2.

Agra, V. (1997). Current environmental practice in the mining industry: The mining industry and the Environment, Proceedings of a National Symposium, UST/IDRC Environmental Research Group: Kumasi, Ghana, April, 14 &15.

Akabzaa, T. M., Banoeng - Yakubo, B. K. And Seyire, J. S. (2005): Impact of Mining Activities on Water in the Vicinity of the Obuasi Mine.

Akabzaa, T. M. and Darimani, A. (2001). Impact of Mining Sector Investment In Ghana: A Study Of The Tarkwa Mining Region. A Draft Report Prepared For SAPRI.

Amankwah, R.K. and Anim-Sackey, C. (2003). Strategies for sustainable development of the small-scale gold and diamond mining industry of Ghana. *Resources Policy,* 29 (2003): 131–138.

Amevor, S. (2006) Ghana: Mining is killing Agricultural sector says Wacam (*Public Agenda,* 6 June, 2008).

Anane, M. (2001). Another Cyanide Spillage in Ghana. MAC: Mines and Communities, In: FIAN International. Accessed on 28 September, 2011. Available from: <http://www.minesandcommunities.org/article.php?a=1077>.

Appiah-Opoku, S. and B. Hyma. 1999. Indigenous institutions and resource management in Ghana'. *Indigenous Knowledge and Development Monitor,* vol. 7 no. 3 pp. 15-17.

Awoonor, Kofi. (1975). The breast of the Earth: A survey of the history, culture, and literature of Africa south of the Sahara. Anchor Press: Garden City, N.Y., 387 p.

Awuah-Nyamekye, S. (2009a). Teaching sustainable development from the perspective of indigenous spiritualities of Ghana, In: *religion and sustainable development opportunities and challenges for higher education,* Cathrien de Pater & Irene Dankelman (eds.), pp. 22- 39, Lit Verlag Fresnostr, ISSBN 978-3-643-90017-3, Berlin.

Awuah-Nyamekye, S. (2009b). Salvaging nature: the Akan religio-cultural perspective, *World Views: Global Religions, Culture, and Ecology,* Vol. 13, No. 3. pp. 251- 282, ISSN 1363-5247.

BIRD. (2009) a. Newmont Ghana Gold Limited Ahafo Mine, External Stakeholder Perception Study, Final Report prepared by Bureau of Integrated Rural Development (BIRD), Kwame Nkrumah University of Science and Technology, Kumasi, Ghana.

BIRD, (2009) b. Newmont Ahafo Development Foundation Mine, Training of Sustainable Development Committees from ten communities in the Newmont Ahafo Mine Area, Final Report prepared by Bureau of Integrated Rural Development (BIRD), Kwame Nkrumah University of Science and Technology, Kumasi, Ghana.

Babut, M. R. et al. (2003). Improving the environmental management of small-scale gold mining in Ghana: a case study of Dumasi, *Journal of Cleaner Production,* 11 (2003), pp. 215–221.

Bush, R. (2009).Soon there will be no-one left to take the corpses to the morgue': Accumulation and abjection in Ghana's mining communities, *Resources Policy,* Vol. 34, no. 1-2, March-June 2009, pp. 57-63.

Brooks, R.R, et al. (1976). Mercury and other heavy metals in trout of central North Island, New Zealand. New Zealand Journal of Marine and Freshwater Research, 10, 233-244.

Byers, B. A. *et al.* (2001). Linking the conservation of culture and nature: A case study of sacred forests in Zimbabwe' in *Human Ecology,* Vol. 29, No. 2, pp. 187-218.

Carboo, D., Serfor-Armah, Y. (1997). Arsenic in stream and sediments in Obuasi area. Proceeding of the symposium on the mining industry and the environment KNUST/IDRC 1997, 114 - 119.

Coakley, G. J. (2003). The Mineral Industry of Ghana. United Sates Geological Survey Minerals Yearbook—2003. Washington DC. Accessed 27 September, 2011 Available from: < http://minerals.usgs.gov/minerals/pubs/country/2003/ghmyb03.pdf>.

Chaudhary R., Rachana M., (2006). Factors affecting hazardous waste solidification/stabilization: A Review. In: Journal of Hazardous Materials B137 pp. 267–276.

Essumang, D., K. Et al. (2007). Arsenic, Cadmium, and Mercury in Cocoyam (Xanthosoma sagititolium) and Watercocoyam (Colocasia esculenta) in Tarkwa, a Mining Community. Bulletin of Environmental Contamination and Toxicology 79:377-379.

Ghana Statistical Services. (2010). Ghana Population and Housing Census. Government of Ghana, Accra.

Harrison, Peter.1999. Subduing the earth: Genesis1, early modern science, and the exploitation of nature, *in The Journal of Religion, Vol. 79, no.1, pp* 86-- .

Hayford E. K. *et al* . (2008). Impact of Gold Mining on Soil and some Staple Foods Collected from Selected Mining Communities in and around Tarkwa-Prestea Area, *West African Journal of Applied Ecology,* (14): 1-12.

Hilson, G. (2001). A Contextual Review of the Ghanaian Small-scale Mining Industry. Report commissioned by the (MMSD) project of IIED.

---------- (2002a). An overview of land use conflicts in mining communities. Land Use Policy 19(1): 65-73.

----------(2002b). Promoting sustainable development in Ghanaian small-scale gold mining operations. *The Environmentalist, 22, 55-57.*

Hilson, G. and Potter, C. (2005). Structural Adjustment and Subsistence Industry: Artisanal Gold Mining in Ghana Development and Change 36(1): 103–131 (2005).

Hinton J. J. , Veiga, M. M. and Beinhoff, C. (2003). Women and Artisanal Mining: Gender Roles and the Road Ahead, In ed. Hilson, G. The Socio-Economic Impacts of Artisanal and Small-Scale Mining in Developing Countries. A. A. Balkema, Swets Publishers, Netherlands, Chapter 11, pp. 52

Jehu-Beaulaton, D. (2008). Sacred Forests and the Global Challenge of Biodiversity Conservation: The Case of Benin and Togo, In: *Journal for the study of Religion, Nature and Culture, Vol. 2 no.3, pp.* 351-372, ISSN 1363-7320.

Johnson, Philip N. and Butigan, Ken (eds.). 1984. *Cry of the environment: Rebuilding the Christian creation tradition.* Santa Fe, NM: Bear and Company.

Johnson, William T. (2000). The Bible on Environment Conservation: A 21st Century Perspective, In: *Quodlibet Journal,* Vol. 2 No. 4. Accessed 29 December 2010. Available from: <http:www.quodlibet.net/articles/Johnson-environment.shtml>.

Kesse, G. O. (1985). The Mineral and Rock Resources of Ghana. Journal of African Earth Sciences (7): 601-610.

Leonard, J. and Robinson, G., Eds. (2009). *Managing Hazardous Materials.* Institute of Hazardous Materials Management.

Manu, A. et al. (2004). Application of Remote Sensing and GIS Technologies to Assess the Impact of Surface Mining at Tarkwa, Ghana. Geoscience and Remote Sensing Symposium IGARSS '04 Proceedings. IEEE International 1:572- 574.

MGM ( 2010). Regional profiles of Ghana. Modern Ghana Media. Accessed 28 September, 2011. Available from:
<http://www.modernghana.com/GhanaHome/regions/ashanti.asp?menu_id=6&men u_id2=14&sub_menu_id=131&gender=>.

Minerals Commission (2004). Ghana Minerals Commission, 2004. Annual Report for 2003 (Self Published).

Mkhize, N. (2004). Psychology: An African perspective. In ed. Ratele, K., N. Duncan, D. Hook, N. Mkhize, P. Kiguwa and A. Collins. *Self, Community and Psychology.* Lansdwone: UCT Press, Republic of South Africa., Chapter 4.

Moreno, F.N, et al. 2009. Analysis of mercury-rich plants and mine tailing using the hydride-generation AAS method.

Munson-McGee S.H., Parsa, J., Steiner, R., 1996. Stabilization/solidification of hazardous wastes using fly ash. In: Journal of Environmental Engineering, 122 (10), pp. 935-940.

Ntiamoa-Baidu, Yaa. (1995). Indigenous versus introduced biodiversity conservation strategies; The case of protected areas in Ghana. Biodiversity support program (Issue in African Biodiversity No. 1) Washington, DC.

Nwosu, P.U. 2010. 'The Role of Okonko Society in Preserving Igbo Environment' in *Journal of Human Ecology,* Vol. 31, no. 1, (2010), pp. 59-64.

Obiri, S. (2007). Determination of heavy metals in boreholes in Dumasi in the Wassa West District of Western Region of the Republic of Ghana. Environmental monitoring and assessment 130: 455-463.

Opoku-Ankomah, Y, *et al.* 2010. 'Research Findings in Water bodies: The human and Cultural factors and their impact on the Environment', in Report of the National Conference on Culture and Water organised by the National Commission on Culture at Akosombo, 24- 26th February 2010.

Orloff, Kenneth, and Henry Falk. 2003. An international perspective on hazardous waste practices. International Journal of Hygiene and Environmental Health 206 (4-5):291-302.

Owusu-Koranteng, H. 2008. Report WACAM-DKA collaboration: Preliminary review of the Alternative Livelihood Project (ALP) of Goldfields Ghana Limited.

Parrinder, E. Geoffrey. 1961. *West African Religions: A Study of the Beliefs and Practices of Akan, Ewe, Yoruba, Ibo and Kindred peoples.* London: The Epworth Press.

Parrinder, Geoffrey. 1(974). *African Traditional Religion 3rd ed.* London: Sheldon Press.

Regional Directorate of Centre for National Culture, Western Region. 2010. Culture as a vehicle for Wealth creation and Socio-economic Development', in Report of the National Conference on Culture and Water organised by the National Commission on Culture at Akosombo, 24- 26th February 2010.

Rose, D., James, D and Watson, C. (2003). Indigenous kinship with the Natural World in New South Wale. New South Wales National Park and Wildlife Services.

Sarfo-Mensah, P and Oduro, W. 2007. Traditional Natural Resources Management Practices and Biodiversity Conservation in Ghana: A Review of Local Concepts and Issues on Change and Sustainability. NOTA DI LAVORO 90.2007 Available on line at:

http://www.feem.it/Feem/Pub/Publications/WPapers/default.htm. Last accessed 20 Aug, 2011.

Sarfo-Mensah, P. et al. (2010)a Traditional Representations of the natural environment and biodiversity conservation: Sacred groves in Ghana.. In: Fondazione Eni Enrico Mattei Working Papers, Accessed 20 November 2010, Available from: < http://www.bepress.com/feem/papers398>.

Sarfo-Mensah, P. et al. (2010)b. Youth Unemployment Challenges in Mining Areas of Ghana. *NOTA DI LAVORO, 122.2009, GLOBAL CHALLENGES Series* Fondazione Eni Enrico Mattei, web site: www.feem.it, e- mail: working.papers@feem.it.

Schoffeleers, J. M. (ed.). 1978. Guardians of the Land: Essays on Central African Territorial Cults. Harare, Zimbabwe: Mambo Press.

Steinnes E. and Berg T. (1998). Significance of Long-Range Atmospheric Transport for Heavy Metal and Radionuclide Contamination of Terrestrial Environment. Am. Mineral. 80: 429–443.

Teye, Joseph K. 2010. 'Network Management of Water Resources in Ghana : A Framework for Integrating Traditional and Scientific Methods', in Report of the National Conference on Culture and Water organised by the National Commission on Culture at Akosombo, 24- 26th February 2010.

The Ghana Chamber of Mines (2005). "The impact of mining on local economy". Annual report of the Chamber mines. Accra, Ghana.

The UNC. (2010). 18th Session of the United Nations Commission on Sustainable Development, National Report for Ghana Waste Management in Ghana).

Tsikata F. S. (1997). The vicissitude of mineral policy in Ghana. *Resource. Pol.* Vol. 23, no.1/2, pp.9- 14.

White, Lynn. 1967. The historical roots of our ecological crisis. Science 155 (3767): 1205-1207.

Yidana, S.M. et al (2007). A multivariate statistical analysis of surface water chemistry data in The Ankobra Basin, Ghana. Journal of Environmental Management 86: 80-87.

# 9

# Use of Phosphate Waste as a Building Material

Mouhamadou Bassir Diop
*Université Cheikh Anta Diop de Dakar, Sénégal*

## 1. Introduction

A phosphate mining waste, called Feral is produced during the processing of a phosphate rich alumina (25wt% $P_2O_5$, 27wt% $Al_2O_3$, 9.53 wt% $Fe_2O_3$) which makes up the Lam-Lam deposit in western Senegal (Fig. 1). Lam-Lam is one of the few aluminium phosphate deposits in the world. These unweathered Al-phosphates are mined by the Société Sénégalaise des Phosphates de Taiba (SSPT) near the village of Lam Lam, northwest of Thies. These phosphates occur as 7 m thick layers under a thick iron crust. Proven reserves of this deposit are 4 million tones of marketable product with an average grade of 33% $P_2O_5$. Only 1.5 million tones of these reserves have an overburden of less than 24 m (McClellan and Notholt 1986).

During processing, 30% of the phosphates are disposed of as waste (0/5mm in size). The waste is called Feral due to its high iron and aluminum oxide content: 10 and 27 wt %; respectively. Lam lam deposit is only a few kilometres flying distance from the seashore.

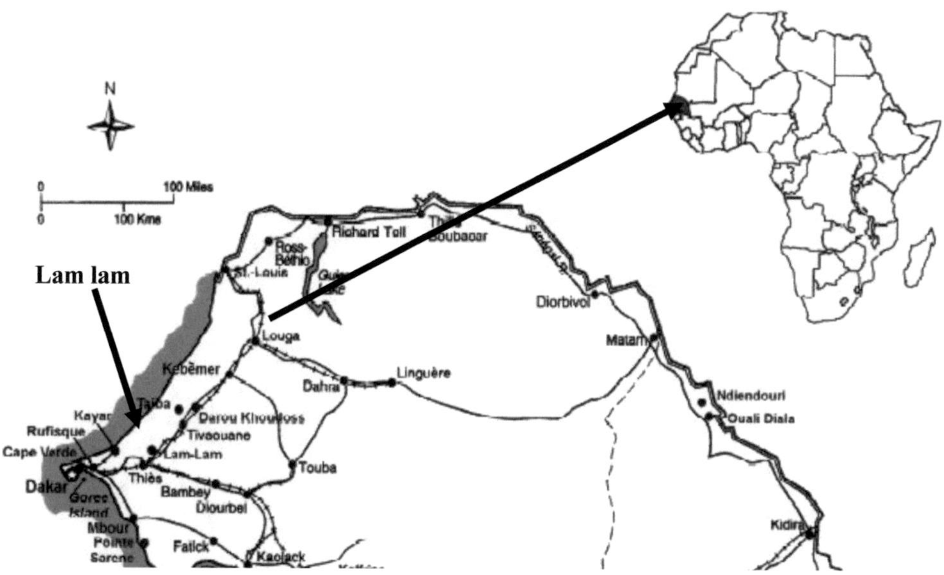


Fig. 1. Map of the African continent showing the location of Senegal and the Lam-Lam phosphate deposit.

It is anticipated that approximately 15 million tonnes of Feral will be produced in the coming years. This is a significant amount of waste material. Unfortunately Senegal, a developing country in West Africa, does not have the money or technology to recycle these wastes. Thus they languish in lakes and ponds where they have the potential to impact the environment in a negative fashion. In this note, we explore the possibility of using it as a building material and in road construction.

Table 1 gives the chemical composition of Feral in weight percent. The major oxides are: $Al_2O_3$, $P_2O_5$, $SiO_2$ and $Fe_2O_3$. The position of Feral in the double triangle diagram shows that it's possible by adding caustic solution to form geopolymer and $AlPO_4$ zeolite (Dyer, 1999).

| $SiO_2$ | CaO | $Al_2O_3$ | $Fe_2O_3$ | MgO | $K_2O$ | $Na_2O$ | $TiO_2$ | MnO | $P_2O_5$ | BaO | SrO | LI |
|---|---|---|---|---|---|---|---|---|---|---|---|---|
| 12.5 | 6.76 | 27.0 | 9.53 | 0.08 | 0.13 | 1.27 | 2.01 | 0.03 | 24.9 | 0.09 | 0.42 | 15.28 |

Table 1. Chemical composition of Feral (wt %)

Figure 2 shows the positions of the Feral deposit both on the $SiO_2$, $Na_2O$ and $Al_2O_3$ triangle and the $P_2O_5$, CaO and $Al_2O_3$ triangle.

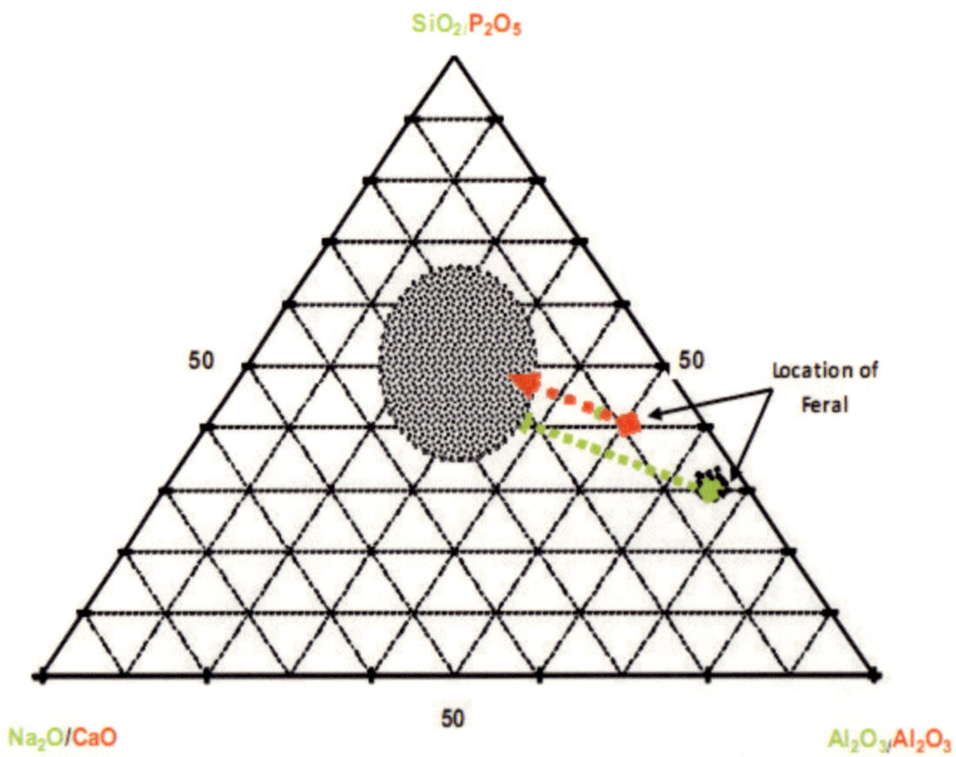


Fig. 2. Positions of the Lam-Lam phosphate deposit on the $SiO_2$, $Na_2O$ and $Al_2O_3$ and $P_2O_5$, CaO and $Al_2O_3$ triangles and the location of the zeolite domain

## 2. Available raw materials

Varieties of clay, highly weathered rock (laterite, tuff) and mining waste (aluminum phosphate, calcium phosphate, phosphogypsum) are materials available in Senegal for making bricks.

The substratum of the Senegalese territory is made up of two major geological domains: the the shallow-dipping Upper Cretaceous to Quaternary sediments in most of the central and western parts of Senegal, which occupies more than 75% of the territory, and the Precambrian basement, and in the east by the Palaeoproterozoic volcano-sedimentary sequences of the Kedougou- Kenieba inlier.

The Precambrian basement formations are constituted at the west by the Mauritanides range bordering the eastern part of the Sedimentary Basin and in the east by the Palaeoproterozoic volcano-sedimentary sequences of the Kedougou- Kenieba inlier. The Kedougou-Kenieba inlier is limited to the west by the Mauritanides chain and on all other sides by the Upper Proterozoic and Cambrian sediments of the Basin of Taoudenni. The Kedougou-Kenieba inlier is interpreted as an accretion of north-easterly trending Birimian age volcanic terrains. Geological studies suggest that mineralisation in the prospective Sabodala volcano sedimentary belt and the Senegal-Malian shear zone is associated with an altered and sulphidised gabbro, which has intruded along the main structure, and a typical shear zone, hosted, where a structure has developed at the contact between a package of volcaniclastics and sediments. A lapilli tuff acts as a prominent marker horizon in the hanging wall of mineralisation.

The inlier is divided into three main stratigraphic units from west to east: the Mako Supergroup, the Diale Supergroup and the Daléma Supergroup.

- The Mako supergroup include basalt flows; often carbonate alterations and minor volcaniclastic intercalations, magnesium basalt or komatiites, ultramafic sub-volcanic intrusions (pyroxenites) and numerous massive biotite and amphibole granitoids.
- The Diale Supergroup, located between the Mako Supergroup and the western edge of the Saraya granite consist of shale, greywacke, quartzite and volcanodétritic rocks.
- The Dalema Supergroup, is composed of volcano-sedimentary schist and grauwacke rocks.

In addition, there are large marble and other ornamental rocks deposits, but also non metallic indices and deposits of barytes, kaolin, asbestos etc.

The Senegal Basin occupies the central part of the Northwest African Coastal Basin, which extends from the Reguibat ridge at the north end of the Guinean fault.

Most of the outcrops of the basin are composed of recent sandy covers. The Secondary formations include Palaeocene zoogenic limestone exploited at Bandia and Pout by cement plants and aggregates producers. They include also Maestrichian sands, clays and sandstones.

Tertiary formations hold into the Eocene compartment, significant resources of phosphates, limestone, attapulgite, clay and ceramics, solid fuels, etc. A major part of the basin is covered with superficial Quaternary formations, which in the middle and recent parts are characterised by fixed red sand dunes, semi-fixed or alive yellow and white dunes. These dunes, often exploited as building materials around urban centres, constitute also important reservoirs of heavy minerals.

## 3. Procedure

### 3.1 Identification of Feral

This particular Feral has been characterized using a variety of techniques including chemical and physical analyses. X-ray diffraction was used for mineralogical analyses.

The California bearing ratio (CBR) is a penetration test for evaluation of the mechanical strength of road sub grades and base courses.

The test is performed by measuring the pressure required to penetrate a soil sample with a plunger of standard area. The measured pressure is then divided by the pressure required to achieve an equal penetration on a standard crushed rock material. The CBR test is described in ASTM Standards D1883-05 (for laboratory-prepared samples) and D4429 (for soils in place in field), and AASHTO T193. The CBR test is fully described in BS 1377: Soils for civil engineering purposes: Part 4, Compaction related tests.

The CBR rating was developed for measuring the load-bearing capacity of soils used for building roads. The CBR can also be used for measuring the load-bearing capacity of unimproved airstrips or for soils under paved airstrips. The harder the surface, the higher the CBR rating. A CBR of 3 equates to tilled farmland, a CBR of 4.75 equates to turf or moist clay, while moist sand may have a CBR of 10. High quality crushed rock has a CBR over 80. The standard material for this test is crushed California limestone which has a value of 100.

### 3.2 Treatment of Feral by cement Portland

Some materials used in road geotechnics have to be treated by binders in order to have their geotechnical characteristics conform to standards.

The aims of these treatments are to improve their geotechnical properties:

- aptitude to compactness;
- decrease of water sensibility;
- increase of strength and load bearing capacity.

The cement used here is a Portland cement fabricated by SOCOCIM-Industries, the first cement plant in Senegal. The composition of the cement used in this study and fabricated by SOCOCIM-Industries is summarized on table 2. The cement have around 1wt% of $P_2O_5$.

| $SiO_2$ | $Al_2O_3$ | $Fe_2O_3$ | CaO | MgO | $P_2O_5$ | $K_2O$ | $Na_2O$ | $SO_3$ | Free CaO |
|---|---|---|---|---|---|---|---|---|---|
| 21.8 | 5.10 | 3.40 | 66.18 | 0.68 | 0.82 | 0.33 | 0.05 | 0.61 | 2.04 |

Table 2. Composition cement fabricated by SOCOCIM-Industries (wt%)

For aggregate used in Senegal for road construction, the CBR must be greater than 80 % for substructure and 30% for sub base (Table 3).

| **Lateritic aggregate** | **substructure** | **sub-base.** |
|---|---|---|
| CBR after 4days immersion (%) | ≥80 | ≥30 |

Table 3. Criteria of use of lateritic aggregate in Senegal

Table 4 represents the criteria of use of a soil-cement in Senegal based mainly on the CBR value and the compressive strength.

| **Soil-cement** | **substructure** | **sub-base** |
|---|---|---|
| CBR after 3days in air and 4days immersion (%) | 6-80 | ≥ 160 |
| CBR after 7days in air (%) | 80-120 | - |
| Compressive strength after 3days in air and 4days immersion (bars) | 25-5 | ≥5 |
| Compressive strength after 7days in (bars) | 5-10 | 18-30 |

Table 4. Criteria of use of a soil cement in Senegal

Cement content used for the treatment of soil are low for cost reasons, they vary generally between 0 and 3% as represented in Table 5.

| Improvement of Feral waste | | | | | |
|---|---|---|---|---|---|
| % of cement | 1 | 1.5 | 2 | 2.5 | 3 |

Table 5. Cement content (wt %) used to treat Feral

### 3.3 Treatment of Feral by alkali activation

#### 3.3.1 Mixes

Geopolymers are chains or networks of aluminosilicate mineral molecules linked with co-valent bonds. Different geopolymers can be distinguished following their molecular units (Davidovits, 1989; Cioffi et al., 2003):

(-Si-O-Al-O-) polysialate with $SiO_2/Al_2O_3$ ratio equal to 2,
(-Si-O-Al-O-Si-O-) polysialatesiloxo with $SiO_2/Al_2O_3 = 4$,
(-Si-O-Al-O-Si-O-Si-O-) polysialatedisoloxo with $SiO_2/Al_2O_3 = 4$.

The aluminosilicate kaolinite reacts with NaOH at 100–150°C and polycondenses into hydrosodalite-based geopolymer ($SiO_2/Al_2O_3=2$). A polysialatesiloxo ($SiO_2/Al_2O_3 = 4$) is obtained from metakaolin and NaOH. These inorganic polymers have a chemical composition somewhat similar to zeolite A ($Na_{12}$ $[Al_{12}Si_{12}O_{48}]$ $27H_2O$) (Breek, 1974; Dyer, 1988) but exist as amorphous solids, rather than having a crystalline microstructure.

Figure 2 shows the location of Feral in a $SiO_2$ - $Na_2O$ - $Al_2O_3$ compositional diagram. The $SiO_2/Al_2O_3$ is equal to 4.3. It suggests that geopolymer can be obtained from this Feral by adding NaOH. Different processes can be proposed. In the first one, Feral is mixed with sodium hydroxide and allowed to react in large heated boiling-water vessels. The water-rich slurries are stirred until geopolymer formation and the completion of the reaction. In the second one, monolithic samples can also be produced, but with a lower water content. In this last case, stoichiometric amounts of 4 to 12M NaOH are mixed with the dry ingredients to form a thick putty-like paste. This paste is then molded and cured at elevated temperatures. Monoliths made in this way are generally very strong and highly insoluble. (Berg et al., 1965; Palomo et al., 1999b; Bao et al., 2004; Grutzeck et al., 2004; Bao et al., 2005, Bao and Grutzeck, 2006). It is this last process which used in this study.

The work reported here is an outgrowth of our previous work (Diop, 2005, 2007, 2008). In this study we point out that this process using the Feral is possible.

### 3.3.2 Samples

Feral sample were mixed with different alkali concentrations (4, 8 and 12 molar NaOH) to form thick pastes (Table 6). The $D^{TM}$ sodium silicate used in the formulation 11 is composed of silicic acid, sodium salt; sodium silicate which constitute 44.1wt% and water. In the $D^{TM}$, the ratio $SiO_2/Na_2O$ in weight percent is equal to 2.

| Samples | Feral (wt %) | NaOH (wt %) | Cure Temp (°C) | Cure time (days) |
|---|---|---|---|---|
| 1 | | | | 0.25 |
| 2 | 80 | 4Molar (20) | 120 | 0.5 |
| 3 | | | | 1 |
| 4 | | | | 0.25 |
| 5 | 80 | 8Molar (20) | 120 | 0.5 |
| 6 | | | | 1 |
| 7 | | | | 0.25 |
| 8 | 80 | 12Molar (20) | 120 | 0.5 |
| 9 | | | | 1 |
| 10 | 80 | 15Molar (20) | 120 | 0.5 |
| 11 | 80 | Mix Molar (20)* | 120 | 0.5 |

*(100g 8Molar NaOH + 250g " $D^{TM}$ " sodium silicate)

Table 6. Formulations studied for bricks manufacturing

The entire eleven samples tested are composed of 80%wt Feral and 20% of solution. After vigorous hand-mixing, the treated Feral was statically compacted in a 2.5cm diameter cylinder. The compaction was carried out by a hand-operated hydraulic press. Pressure was applied until water began to be squeezed out of the sample. Pressures were typically in the 10 MPa range. The cylinders were trimmed to 5.0 cm in length and then allowed to sit overnight at room temperature before being cured at 120 C. This so called "soaking" is typically used to allow time for dissolution and for geopolymer precursors to form (Breek, 1974; Dyer, 1988). The 40°C samples were cured in a "walk in" chamber that was maintained at 60% relative humidity. The 120°C samples were cured in sealed Parr type vessels fitted with Teflon liners at 120 °C for varying periods of time.

Concentrations used for road construction (table 7) are lower.

| Quantity of NaOH (g) | Water quantity (litre) | Concentration (g/l) | Molarity |
|---|---|---|---|
| 100 | 5 | 20 | 0,5M |
| 200 | 5 | 40 | 1M |
| 400 | 5 | 80 | 2M |
| 800 | 5 | 160 | 4M |

Table 7. Composition of solutions used in alkali activation of Feral for road construction

### 3.4 Mechanical tests

The Californian Bearing Ratio (CBR) test done on Feral gives a value of 13%. This value is obtained after 3days conservation in air and 4days immersion in water of the compacted sample. Or the minimum requirements CBR for a material to be use in road construction is 80% for substructure and 30% for sub-base.

Then, we explore a treatment of Feral with cement. For soil treated with cement, the minimum requirements for CBR are 160% for substructure and 6-80% for sub-base.

#### 3.4.1 Compressive strength

After different periods of curing, the mechanical behavior of the cylinders was tested. The compressive strength values were measured after 6, 12 and 24 h for the samples cured at 120°C.

### 3.5 Durability tests

In order to test durability, pieces of three samples cured at 120°C for 12 h (#s 2, 9, 16) were ground to sizes less than 150 μm and more than 75 μm and then dried at 105 °C. One gram of each of the powdered specimens was placed in 10 mL deionized water and held at 90°C for 1 and 7 days in a sealed Teflon container. This test is a modified product consistency test (PCT) designed to test glass leaching (ASTM C1285, 2008). These samples were chosen because it was assumed that reaction has reached the greatest degree at 120°C and leaching of these samples would better reflect what would happen to all bricks that had been cured for a longer time once it was used to build a house (some years).

## 4. Results

### 4.1 Physical and mechanical properties of Feral

The physical and mechanical properties of Feral are summarized on table 8 (Sy, 2000). Feral has an important fine content (24 % < 80μm). 13% of these fines are of clay size (≤ 2 μm). The CBR of Feral (13%) is low. According to table 4, the Feral must be treated for its use in the sub-base of road construction.

| Grain size | |
|---|---|
| % > 2mm | 10 |
| % < 80μm | 24 (26 after CBR) |
| % < 2μm | 13 |
| $C_u$ | 600 |
| $C_c$ | 32 |
| **Sand equivalent (SE)** | |
| piston | 34 |
| observed | 30 |
| **Atterberg limit** | |
| $W_l$ (%) | 32.6 |
| $W_p$ (%) | 29.3 |
| $I_p$ (%) | 3.3 (4.2 after CBR) |
| Activity (A) | 0.25 |
| $W_{OPM}$ (%) | 20.4 |
| $\gamma_{dmax}$ (kN/m$^3$) | 18.3 |
| CBR at 95% OPM | 13 |
| Swelling (%) | 0 |
| $\gamma_s$ (kN/m$^3$) | 27.6 |
| $\gamma_{app}$ (kN/m$^3$) | 13 |

Table 8. Physical and mechanical properties of Feral.

### 4.2 Feral treated for road geotechnics

#### 4.2.1 CBR of Feral treated with Portland cement

By adding cement in Feral (table 9), its bearing capacity is gradually improved enabling its use in sub-base.

| | | | | | |
|---|---|---|---|---|---|
| CBR' (%) | 94 | - | 110 | 164 | 257 |
| CBR (%) | 19 | 39 | 60 | 70 | 87 |
| % of cement | 1 | 1.5 | 2 | 2.5 | 3 |

CBR after 3days in air and 4days immersion in water - CBR' after 7days in air

Table 9. Evolution of Californian Bearing Ration with cement addition

We notice for CBR and CBR' an increase of bearing capacity with cement (Figure 3). But it shows also that the Feral is sensitive to water. This result can be explained:

- the material has been compacted at the modified optimum proctor and 4days of immersion will cause an excess of water,
- the infiltration of water after conservation at air (cause shrinkage) may provocate the weakness of the material due to cracks.

2 and 3% cement content give CBR conform to CEBTP standards for use of Feral in road substructure.

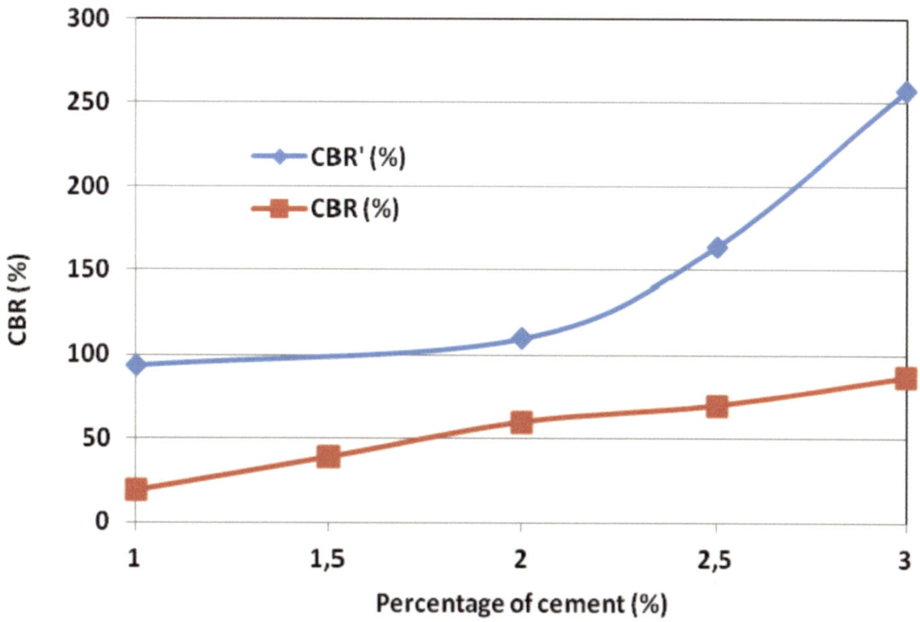


Fig. 3. Evolution of CBR with cement content

Figure 4 represent the evolution of strength with age for different cement content varying between 1 to 3%. No matter what the cement content, strength decrease with age.

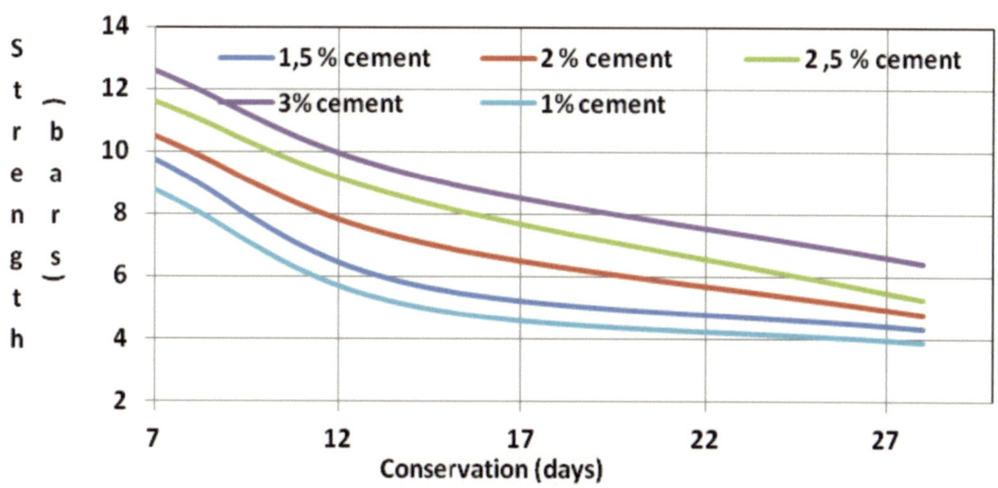


Fig. 4. Evolution of strength with age for different cement content

Figure 5 represents the evolution of CBR/CBR′ with different cement content. The CBR/CBR′ is an indication of the sensitivity of a material to water.

The values of cement between 2 to 3% give a soil-cement in conformity with CEBTP standard for a use in sub-base. However, compressive strength measurement is necessary to fully appreciate the aptitude of the material for road construction.

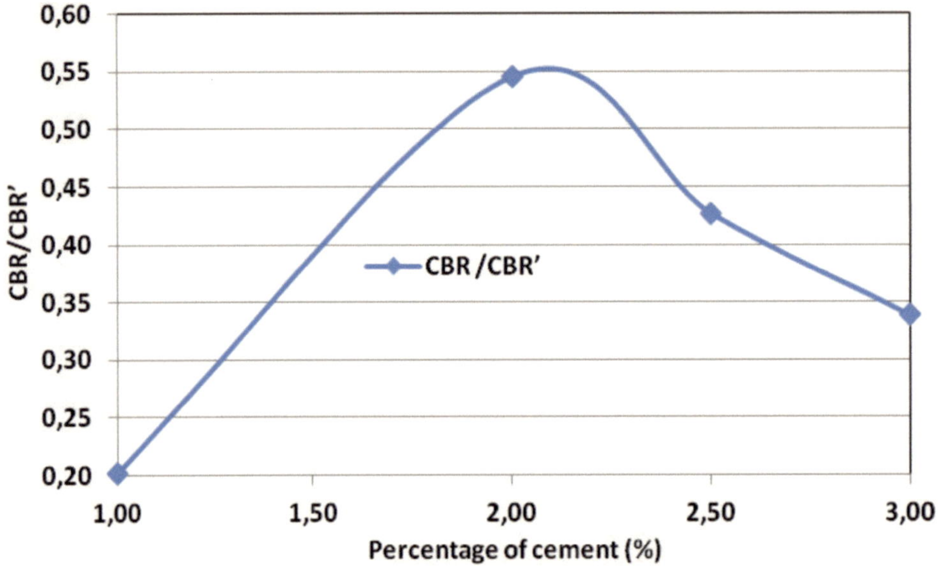


Fig. 5. Evolution of CBR/CBR' with different cement content

| Improvement of Feral waste | | | | | |
|---|---|---|---|---|---|
| Compressive strength after 7days $R_{c7}$ (bars) | 8.76 | 9.74 | 10.51 | 11.59 | 12.58 |
| Compressive strength after 14days $R_{c14}$ (bars) | 5.06 | 5.74 | 7.15 | 8.45 | 9.25 |
| Compressive strength after 28days $R_{c28}$ (bars) | 3.89 | 4.32 | 4.75 | 5.24 | 6.41 |
| $R_{c7}/ R_{c28}$ | 0.44 | 0.44 | 0.45 | 0.45 | 0.51 |
| Percentage of cement (wt) | 1 | 1.50 | 2.00 | 2.50 | 3.00 |

Table 12. Variation of Compressive strength functions of the age and cement content

Figure 6 represents the variation of compressive strength as a function of the age and cement content.

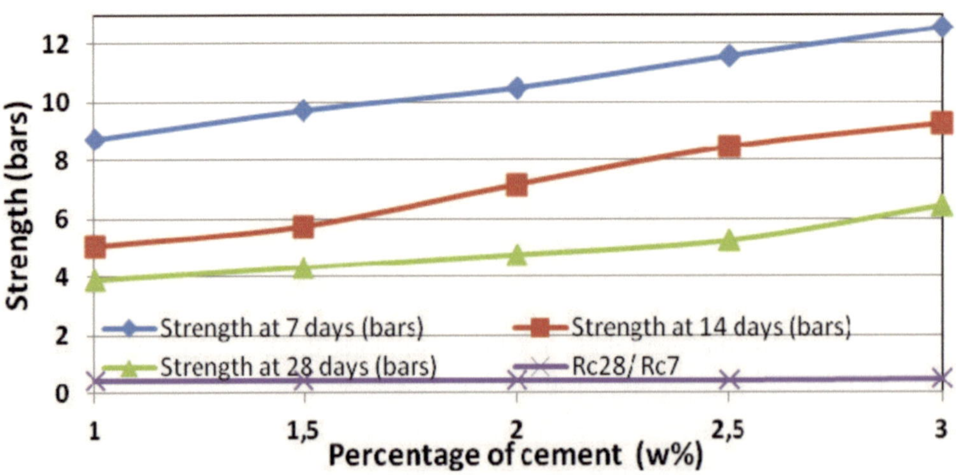


Fig. 6. Variation of Compressive strength function of the age and cement content

We notice for all cement content a decrease of compressive strength with age. The compressive strength after 28days is at least 45% less than the compressive strength after 7days. Hence, an alkali activation of Feral with concentrated caustic solution (NaOH and Na-Si) is tested.

### 4.2.2 CBR of alkali activated Feral

It was found that caustic addition caused zeolite minerals with aluminium-phosphorus-oxygen frameworks ($AlPO_{4s}$) to form (Dyer, 1999). In the related aluminophosphates ($AlPO_4$), each negatively charged $AlO_4$ tetrahedron is balanced by a positively charged $PO_4$ tetrahedron, and non framework cations are not needed (Sherman, 1999). Still other variants include the silicoaluminophosphate (SAPO) structures in which Si substitutes some P in the $AlPO_4$ framework; each added Si needs a non framework cation to balance the charge on the framework. The results of alkali activation of Feral with concentrated caustic solution will be exposed in the paper.

| Molarity | CBR'(%) | CBR(%) | Compressive strength after 3days in air and 4days immersion (bars) | Compressive strength after 7days in (bars) |
|---|---|---|---|---|
| 0,5M | 54,0 | 73 | 28 | 59 |
| 1M | 115 | 55 | 31 | 65 |
| 2M | 98 | 11 | 35 | 73 |

*CBR after 3days in air and 4days immersion in water - CBR' after 7days in air

Table 13. CBR* tests on alkali activated Feral

Either activated with 0.5M or 1M NaOH Feral can be used in road substructure. The same conclusion can be done for Feral Activated with 2M NaOH. Feral activated with these concentrations can't be used in road sub-base (CBR ≥ 160).

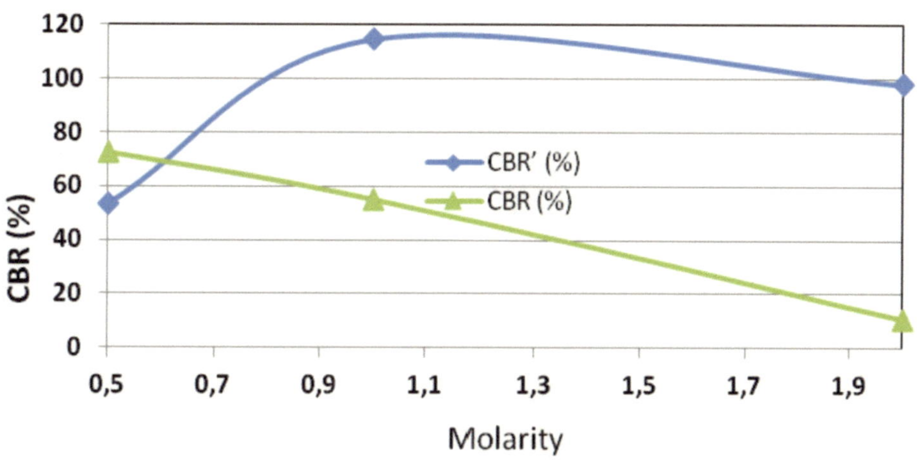


CBR* after 3days in air and 4days immersion in water - CBR' after 7days in air

Fig. 7. CBR* tests on alkali activated Feral

### 4.3 Feral treated for the manufacture of bricks

#### 4.3.1 Mechanical tests

Typical stress strain curves for Feral samples made with 4, 8 and 12 M NaOH solutions and cured at 6h, 12 h and 24h at 120 °C are given in figure 8. The deformation curve of the cylinders shows that the rupture is progressive. Breaking of the 2.5 by 5.0 mm cylinders during compression was good; all breaks exhibited a typical double pyramidal shape.

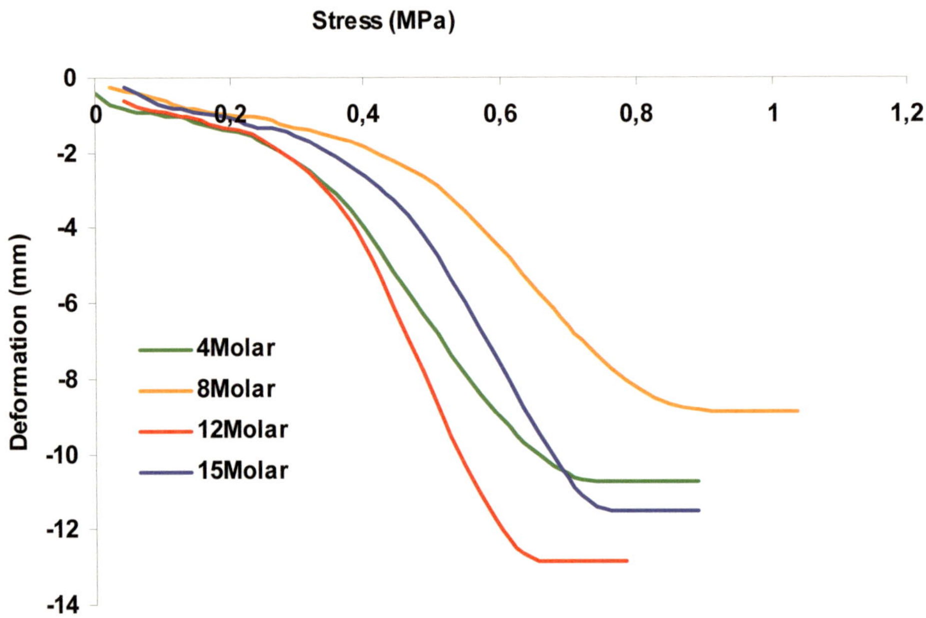


Fig. 8. Stress strain curves for samples made with 4, 8 and 12 M NaOH solutions as a thick paste and then cured at 120 °C.

A summary of the strength data is given in figure 8.

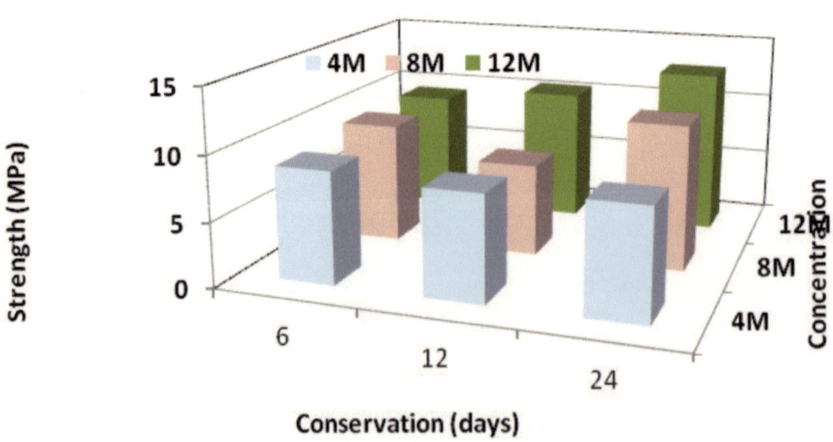


Fig. 9. Summary of compressive strengths of samples cured at 120 °C.

### 4.3.2 Physical and chemical analysis

Figure 10 represents a Micrograph microstructure of Feral treated with 12 M NaOH solution and cured at 120°C for 12 hours. The micro-structure is more columned-like texture. Based upon X-ray data in figure 11, it is proposed that these crystals are millisite.

Figure 11 represents the X-ray diffraction patterns for the 120 °C samples cured for 12 h made with different molar NaOH solution. The tabular crystals are the most common and based upon X-ray diffraction data presented are probably millisite. There is not very much difference between the patterns. There is a general broadening of peaks with increasing concentrations. Millisite ((Na, K) $CaAl_6(PO_4)4(OH)_9$ $3H_2O$) a zeolite phosphate is the most representative of forming minerals; then Huangite (Ca Al $6(SO)_4(OH)_{12}$. Quartz ($SiO_2$) and Dickite ($Al_2Si_2O_5$ $(OH)_4$) are new silicate minerals. It is proposed that other zeolites are also present because of the reduction in the amount of starting materials, but that they are probably sub microscopic and thus not able to diffract X-rays in a coherent fashion.

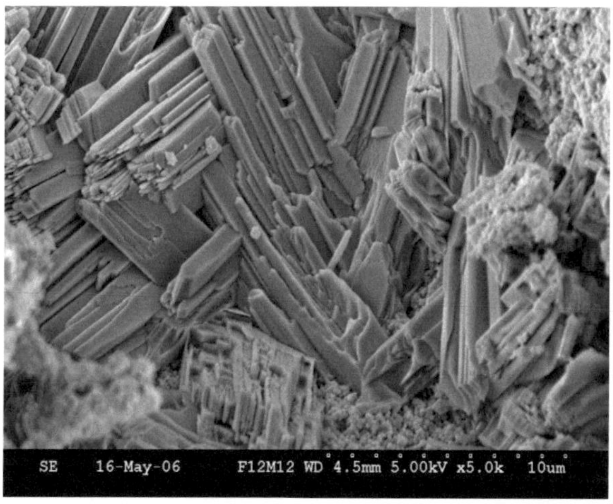


Fig. 10. View represents the microstructure of the Feral treated with 12 M NaOH solutions and cured at 12 h.

Samples fabricated from Feral and NaOH solutions attain compressive strengths that range from 6 to more than 16 MPa (Fig. 9). These strengths are in the range of similar samples made with metakaolin. Strength development is dependent on temperature and length of curing, alkali concentration, fineness, and composition of the raw materials. The best mechanical performance of the samples depends of the concentration of NaOH solution used to make the sample. For 4 molar concentrations the greatest compressive strength is obtained after 12hours curing. For 8 and 12 molar concentration the greatest compressive strength are obtained after one day curing. The strength given by the Feral treated with 15Molar concentration and cured for 12 hours (8.35 MPa) is less than for the 12Molars concentration and cured for the same duration (12.84 MPa). By using a mixture between "D" sodium silicate and 8Molar NaOH solution (100g 8Molar NaOH + 250g "D" sodium silicate), the compressive strength is neatly increased (16 MPa). This may be due to the fact that with "D" sodium silicate, we have both the formation of zeolite silicates and zeolite phosphates. The deformation curve of the cylinders shows that the rupture is progressive. Breaking of the 2.5 by 5.0 mm cylinders during compression was good; all breaks exhibited a typical double pyramidal shape.

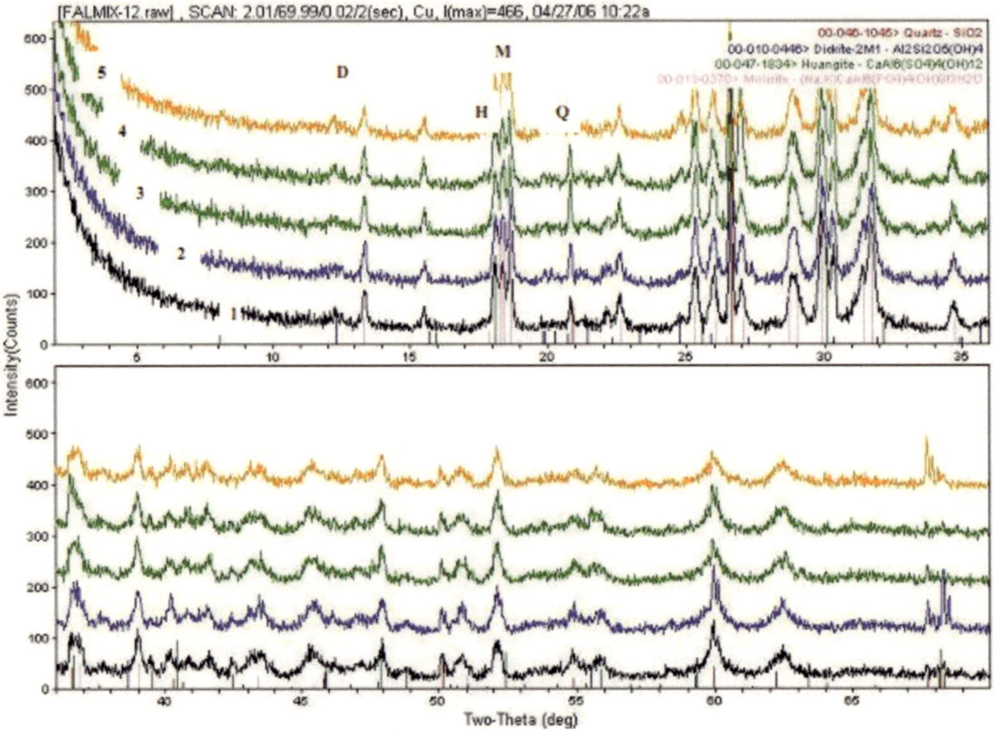


Fig. 11. Comparison of X ray diffractometer powder of Feral (curve n°1) and samples made with 4 (curve n°2), 8 (curve n°3), 12 (curve n°4) and 15Molar (curve n°5) NaOH concentration reacting with Feral after 12hours conservation. D = dickite, H = huangite, M = millisite, Q = quartz

Solubility tests were performed with Senegalese Feral reacting with caustic ASTM C1285 - 02(2008). Samples were ground to size less than 250 µm and 1 gram placed in 10 mL deionized water and held at 90 °C for 7 days. After 24 h, the leaching tests of the 120 °C samples showed very low conductivities no matter what concentration of alkali used to make the brick. The values of conductivities increased with time (Table 14), the one day conductivities for all three samples are lower than they are at 7 days. There is a kinetic process (possibly diffusion controlled) that limits the buildup of Na in solution. There is little Al or Si present in the solution, because these species are essentially insoluble; it is the sodium which accounts for the conductivity. For example a standard solution of NaOH in water with a conductivity of 1 mS/ cm contains 200 ppm NaOH (Bao and al 2005).

The 4 M and 8 M samples have the lowest overall conductivities, whereas the 12 M samples are almost twice as high. This suggests that the amount of Na in the 4 and 8 M brick reacts nearly completely, whereas the 12 M sample may contain excess NaOH or soluble sodium silicate which washes out giving it a higher conductivity and pH. The moderate pH confirms the formation of silicate minerals of some type. These values are in line with those one obtains when natural zeolites are put in water. Sodium does not leach from these minerals to any great extent. Nevertheless, all conductivity values are reasonably low, proving that reactions are occurring during curing and that zeolite-like mineral(s) are probably forming (Bao and al 2006). Solubility is low. Durability should be better.

| Concentration (molar) | Measurement at 1 day | | Measurement at 7 days | |
|---|---|---|---|---|
| | Conductivity (mS/cm) | pH | Conductivity (mS/cm) | pH |
| 4 | 0.7 | 10 | 1.00 | 10 |
| 8 | 0.6 | 10 | 0.90 | 10 |
| 12 | 1.5 | 11 | 2.10 | 11 |

Table 14. Results of leaching test of bricks cured at 120 °C

## 5. Discussion

The treatment of Feral with cement using different percentage has permitted to appreciate the mechanical behavior of this material and its sensitivity to water.

The treatment of Feral with cement improves its bearing capacity but for the long term (7days, 14 days and 28days), strength decrease no matter what the cement content. This phenomenon may be caused by the high content in $P_2O_5$ (0.82) of the Portland cement used. The Feral itself coming from phosphate mineral treatment contain around 25wt% of $P_2O_5$. According to several studies, among them, those of Lafarge laboratories (Cochet, 1995) stating that $P_2O_5$ is a strong retarding agent for the setting and hardening of mortar. Percentage higher than 0.5 will causes a decrease of initial resistance and an increase of the setting time.

| | Resistance of mortar (French standards) | | |
|---|---|---|---|
| $P_2O_5$ (%) | 7days | 14days | 28days |
| 0.05 - 0.55 | * | * | c |
| 0.05 - 1.10 | ** | * | ◘ |
| 0.05 - 1.55 | *** | * | ◘ |

* = decrease; ** = high decrease; *** = very high decrease; c = constant; ◘ = increase; ; ◘ = high increase

Table 15. Impact of $P_2O_5$ on the characteristics of cements

So other means of treatment should be explored.

Alkali activation of Feral present the main advantage in connection with its composition: $SiO_2$ (13wt %), $Al_2O_3$ (27wt %), $P_2O_5$ (25wt %), CaO (7wt %) and $Na_2O$ (1.3wt %). The attack of Feral with caustic solution (sodium hydroxide and/or sodium silicate) will enable the formation of geopolymers and zeolite minerals with aluminium-phosphorus-oxygen frameworks ($AlPO_{4s}$). Forming mineral may include other variants like silicoaluminophosphate (SAPO).

The treatment of Feral with alkali makes it possible to use them in road substructure. However tests must be performed to determine the optimum concentration.

For bricks, Feral is treated with varying alkali concentration: between 4M, 8M and 12M.

For samples cured at 120 °C, regardless of the concentration of NaOH used, 12 h of curing gives the best compressive strength. This can be explained by the fact that with temperature, the reaction between the alkali solution and the Feral has taken place. After 12 h of curing, the hydrated phase that forms are $AlPO_4$ type zeolitic, because the bulk composition of the starting material falls within a compositional range typical for zeolites. Although the zeolites minerals that form are not as "stable" as kaolinite, they are able to bond to each other and form a solid that is much more resistant to softening and deformation during annual wet/dry cycles. All zeolites are metastable to some degree (Sherman, 1999; Diop and Grutzeck, 2008). The ones that form first, in the presence of abundant water, will become less hydrated and undergo phase transitions as a result of diagenesis. If placed in an aggressive environment such as one with an acidic pH, they will also dissolve. In neutral and alkaline environments however they are very insoluble. Zeolites will change with geological time if buried, but if exposed to heat and humidity at the surface, this change is nearly imperceptible. The type of zeolite that forms during the manufacture of the aforementioned brick is somewhat temperature dependant. What will happen on occasion is that an initially formed zeolite will change into another one, one that is more stable. This is a problem associated with nucleation and growth from supersaturated solutions and is similar to what happens during diagenesis. Early formed zeolites are often the least stable converting to a more stable form after a few hours or days of curing. This makes it mandatory for the person making brick using this method (especially at 120 °C) to test strength versus time to see if a disruptive phase change occurs that could reduce performance by introducing shrinkage/expansion cracks. If these occur, samples should not be cured longer than necessary to achieve initial maximum strength.

For all the samples tested, those mixed with 12 molar NaOH had the highest strengths. This is what one might expect, because as one increases the concentration of the alkali solution more Feral will dissolve in the solution and more sodium aluminophosphate precursors will form that have the ideal ratio to form $AlPO_4$, i.e. Na:Al:Si P = 1:1:1). However, increasing the concentration of NaOH indefinitely (e.g. 15 M) will not cause a continuous increase in the strength; rather it will cause more sodium rich phases to form that may not be as insoluble as zeolites. This will increase solubility and possibly have a negative effect on durability.

The strengths and leachabilities of the samples are similar to those for alkali activated metakaolinite samples which usually run about 3 MPa, have a pH of about 10 and a conductivity of 2-3 mS/cm (Breck, 1974) Bao et al, 2004; Bao and Gritzeck. The ability to make a brick with these characteristics is rather exciting because of its implications and potential impact on the nature of sun dried brick making in developing countries.

## 6. Conclusions

Zeolites are very insoluble. They are currently forming at the bottom of the World's oceans. However, the Na ions in a zeolite are mobile, which accounts for a zeolite's ability to exchange cations with other substances in solution. In this case the measurement of conductivity used here is actually measuring two things: the degree of reaction that the sample has undergone prior to being tested, and the mobility of the Na ion in the zeolitic-matrix as it exchanges with protons in the water (Bao et al, 2005). If the conductivity of the solution is low this suggests that the NaOH that was used to make the sample has reacted with the Feral and has been "tied up" in a tectosilicate matrix. Conductivity reflects the degree of fixation (effectiveness of the recipe to do what it is meant to do) and the magnitude of the cation exchange of $Na^+$ for $H_3O^+$ that takes place. It is safe to say that if no NaOH had reacted it would dissolve in the leaching solution and conductivity would be in the 20-30 mS/cm range. Because we are in the 1-3 mS/cm range, this suggests that zeolites are forming and they are very much part of the structure even though their presence may not be evident in SEM or X-ray diffraction scans. Based on the low conductivity numbers of 120 °C samples, it is predicted that durability of the long term room temperature cured alkali activated Feral brick should be better than that of conventional sun dried clay brick.

The development of a quick temperature process to create durable bricks can be accomplished according to the needs of the community. The process does not generate chemical pollutants like fired clay bricks. It can use by-product materials like industrial waste enabling brick makers to solve environmental problems. A wooden mold and a mix of granular material and seawater (if commercially produced silicates are not available) are enough to fabricate good quality block with this technique. Locally available materials can be tested with different amounts of NaOH as mixing solution and cured as a function of temperature to determine the optimum concentrations of NaOH to use. At this time, we recommend 8 M NaOH because it seems to be a compromise between strength and cost. If a stronger brick is needed, one can use 12 M NaOH because strength was significantly higher in this case. This technology appears to be a solution for African developing countries like Senegal, but for the US too, if clean technology is desired. It seems possible that villagers could start their own businesses providing income and at the same time upgrading the brick used to build houses on a village to village basis. It is also proposed that a simple manual press might be used to make full sized bricks.

## 7. References


ASTM C1285 - 02(2008), Standard Test Methods for Determining Chemical Durability of Nuclear, Hazardous, and Mixed Waste Glasses and Multiphase Glass Ceramics: the Product Consistency Test (PCT), American Society for the Testing of Materials ASTM International, 100 Barr Harbor Drive, PO Box C700, West Conshohocken, PA, 19428-2959, USA.

Bao Y, Grutzeck MW. Solidification of sodium bearing waste using hydroceramic and portland cement binders. Ceram Trans 2005; 168:243–52. Environmental Issues and Waste Management Technologies in the Ceramic and Nuclear Industries X.

Bao Y, Grutzeck MW. General recipe and properties of a four inch hydroceramic waste form. In:Ceramic Transactions, vol. 176 (Environmental Issues and Waste Management Technologies in the Ceramic and Nuclear Industries XI), Am Ceram Soc, Westerville, OH; 2006. p. 63–74.

Bao Y, Grutzeck MW, Jantzen CM. Preparation and properties of hydroceramic waste forms made with simulated Hanford low-activity waste. J Amer Ceram Soc 2005;88(12):3287–302.

Bao Y, Kwan S, Siemer DD, Grutzeck MW. Binders for radioactive waste forms made from pretreated calcined sodium bearing waste (SBW). J Mater Sci 2004;39(2):481–8.

Bao Y, Kwan S, Siemer DD, Grutzeck MW. Binders for radioactive waste forms made from pretreated calcined sodium bearing waste (SBW). J Mater Sci 2004;39(2):481–8.

Boa Y, Grutzeck MW. Solidification of sodium bearing waste using hydroceramic and portland cement binders. Ceram Trans 2005; 168:243–52. Environmental Issues and Waste Management Technologies in the Ceramic and Nuclear Industries X.

Bao Y, Grutzeck MW. General recipe and properties of a four inch hydroceramic waste form. In:Ceramic Transactions, vol. 176 (Environmental Issues and Waste Management Technologies in the Ceramic and Nuclear Industries XI), Am Ceram Soc, Westerville, OH; 2006. p. 63–74.

Bao Y, Grutzeck MW, Jantzen CM. Preparation and properties of hydroceramic waste forms made with simulated Hanford low-activity waste. J Amer Ceram Soc 2005;88(12):3287–302.

Breck D. W. Zeolite molecular sieves. New York: John Wiley & Sons; 1974.

Cochet G., 1995, rapport inédit, Lafarge CTI - France.

Davidovits, J., Geopolymers and Geopolymeric Materi- als, J. Therm. Anal., 1989, vol. 35, pp. 429–441. 15. Rowles, M. and Connor, B.O.

Diop M. B. and Grutzeck M. W. "Sodium silicate activated clay brick" Bull Eng Geol Environ, DOI 10.1007/s10064-008-0160-3

Diop M. B. and Grutzeck Michael W. "Low temperature process to create brick ", Construction and Building Materials 22 (2008) 1114–1121

Dyer A. An introduction to zeolite molecular sieves. New York: John Wiley & Sons; 1988.

Grutzeck MW, Siemer DD. Zeolites synthetized from Class F fly ash and sodium aluminate slurry. J Amer Ceram Soc 1997; 80(9):2449–53.

McClellan GH and AJG Notholt 1986. Phosphate deposits of sub-Saharan Africa. In: Mokwunye AU and PLG Vlek (eds.) Management of nitrogen and phosphorus fertilizers in sub-Saharan Africa. Martinus Nijhoff, Dordrecht, Netherlands:173-224.

Palomo A, Grutzeck MW, Blanco MT. Alkali-activated fly ashes A cement for the future. Cem Concrete Res 1999; 29(8):1323–9.

Palomo A, Blanco MT, Granizo ML, Puertas F, Vasquez T, Grutzeck MW. Chemical stability of cementious materials based on metakaolin. Cem Concr Res 1999;29(7):997–1004.

Sherman J. D. "Synthetic zeolites and other microporous oxide molecular sieves" Proceeding of the National Academy of Sciences of the USA; PNAS Marsh 30 1999, vol 96 7 3471 -3478

Siemer DD, Grutzeck MW, Scheetz BE. Comparison of materials for making hydroceramic waste forms. Ceram Trans 2000; 107:161–7. Environmental Issues and Waste Management Technologies in the Ceramic and Nuclear Industries V.

Sy Papa Amadou. Caracterisation et performances d'un matériau de type nouveau en corps de chaussée: les résidus des phosphates de Lam lam (produits secodaires de la chaine de traitement de la SSPT). Mémoire de fin d'études d'ingnéieur de l'Institut des Sciences de la Terre (IST), Faculté des Sciences et Tchniques, Université Cheikh Anta Diop de Dakar, Sénégal.

10

# Utilization of Coal Combustion By-Products and Green Materials for Production of Hydraulic Cement

James Hicks
*CeraTech, Inc.,*
*USA*

## 1. Introduction

### 1.1 Condition of highways

The condition of highways is reaching conditions such that immense expenditures are required just to remediate existing conditions. (Floris & Hicks, 2009)

According to the 2005 report card by the American Society of Civil Engineers (ASCE), the state of America's infrastructure has reached alarmingly unacceptable levels which threaten the current lifestyle and standard of living. The average grade for America's infrastructure was D (Poor), requiring an estimated US$1.6 trillion just to return the infrastructure systems back to a serviceable condition.

Another fact is that US$9.4 billion a year for 20 years is required to eliminate the deficiencies in the nation's 600,000 bridges, suggests serious systemic problems exist in the construction industry. Society is slowly coming to realize that initial predictions of cement durability may have been excessive and that concrete buildings or structures that last for only 50–70 years may not be cost effective. (Phair, 2006)

Volumes of materials required for the repair of deficiencies are further exacerbated by any new construction required. Much of this will require use of portland cement or other hydraulic cements.

### 1.2 Greenhouse gasses emitted

A large producer of $CO_2$ emissions is portland cement kilns. For instance, the use of fly ash (a by-product of coal burning in power generation and most common Coal Combustion Products (CCP) in the cement-making process) could reduce substantial amounts of $CO_2$ emitted by a cement kiln. Worldwide, the production of portland cement alone accounts for 6 to 8 percent of all human generated $CO_2$ greenhouse gases (Huntzinger, Deborah N. and Eatmon, Thomas D., 2009). Portland cement production is not only a source of combustion-related CO2 emissions, but it is also one of the largest sources of industrial process-related emissions in the United States. Combustion related emissions from the U.S. [portland] cement industry were estimated at approximately 36 Tg of $CO_2$ accounting for approximately 3.7 percent of combustion-related emissions in the U.S. industrial sector in 2001 (USGS, 2002).

Extensive research is underway to find more economically feasible alternatives for carbon dioxide capture and storage (CCS). However, until financially and environmentally sustainable alternatives are in use, the byproducts of pulverized coal-based power generation (conventional) will be an issue for decades to come. A successful $CO_2$ mitigation process presently lies in private-public strategy that combines existing power plants (revamped to capture, geologically store and/or enhance oil recovery), and new ones using more advanced coal power generation technologies like the Integrated Coal Gasification Combined Cycle (IGCC). These approaches alongside proactive regulation could build a relatively sustainable alternative for the future (Floris, Vinio, 2009).

In 2009, a survey of US coal-fired power plants showed production of 122.2 million metric tonnes (t), (134.7 million short tons (st)) of CCP's (US Energy Administration, 2010). Of this amount, only 22.4 million t (24.7 million st) were used as fly ash in concrete or as a concrete product. The survey reported that more than 71.7 million t (79 million st) of fly ash is still being disposed of in US landfills annually. (ACAA, 2009) Clearly, the use of otherwise waste materials for beneficial use can reduce the need for more landfills and the amount of CO2 produced.

Coal has been the primary fuel used worldwide in recent times (Figure 1). This has been forecast to continue well into the future.

Not all current power plants will be refurbished as mentioned; yet CCP's from plants that do not convert to CCS technologies could still be put to good use through innovative ways in order to assist in the decrease of greenhouse gases.

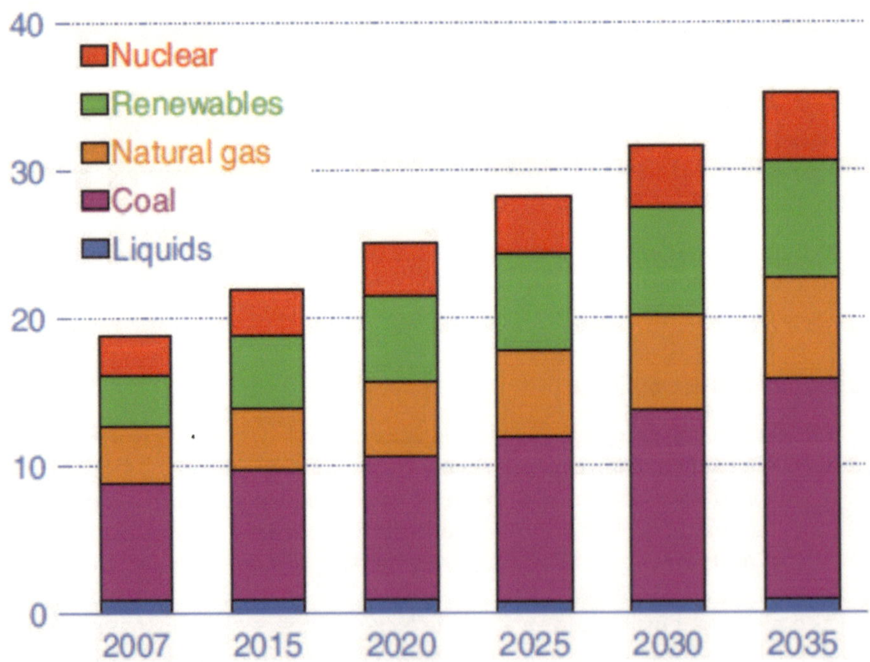


Fig. 1. World net Electricity generation by fuel, 2007-2035 (trillion kilowatt hours) (U. S Energy Administration, 2010)

For instance, fly ash (a by-product that largely ends up in landfills) can substitute for portland cement and improve structural changes to the end product (concrete and others). The use of one unit of fly ash reduces approximately one unit of $CO_2$ emitted by a cement kiln. Fly ash and other CCP's in the cement-making process could also avoid the use of the high energy requirement and significantly reduce the volume of useable material taken to waste management sites (Floris and Hicks, 2009). The amount of fly ash beneficially used in the United States is illustrated in Figure 2.

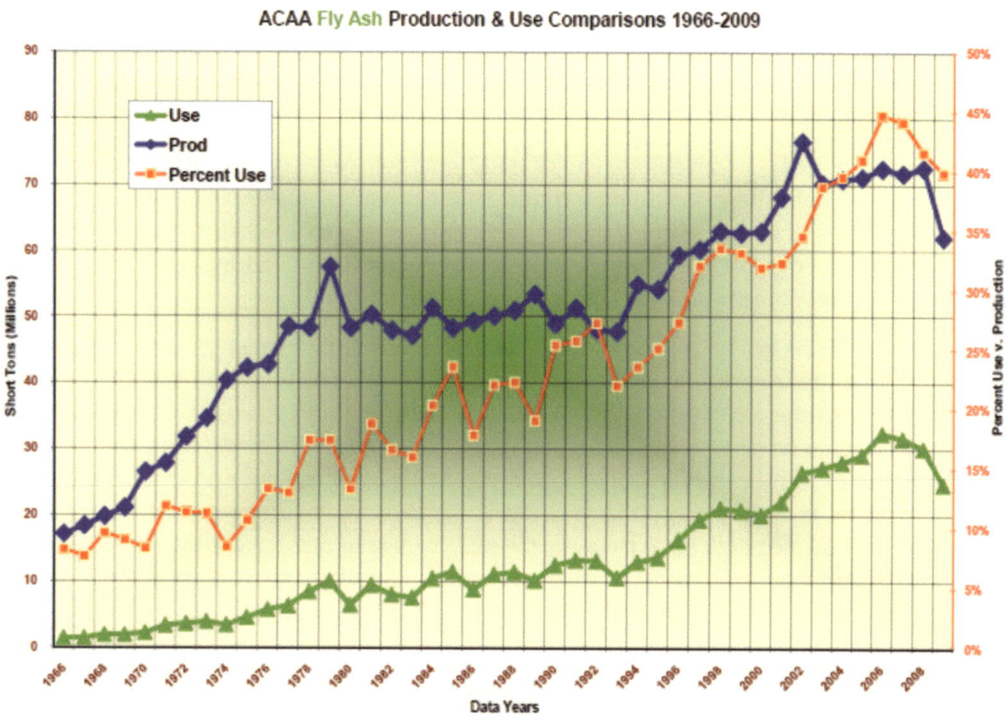


Fig. 2. CCP Beneficial Use versus production (American Coal Ash Association, 2010)

## 2. Waste materials converted into hydraulic cements

This section provides a brief overview of concretes in which the binder phase is made of materials other than portland cement. The non-portland cement binders discussed here are alkali activated slag and fly ash cements, calcium aluminate cements, and calcium sulfo-aluminate cements. A brief introduction of the recycled glass-based cements and concretes is also provided. It should be mentioned that most of these binder systems have been used for specific applications across the world for several decades. However, recent concerns about the environmental impact of portland cement production have created a renewed interest in these less energy- and $CO_2$-intensive binder systems. With energy, resource and infrastructure demands growing in the foreseeable future, a sustainable concrete that is more durable and costs less to maintain, is becoming increasingly desirable (Touzo & Espinosa, 2010).

### 2.1 Alkali activated fly ash concretes

Alkali activated fly ash is a potential substitute for portland cement. Fly ash is mainly composed of glassy alumina ($Al_2O_3$) and silicate ($SiO_2$) phases which can be activated (i.e., dissolved) by a concentrated aqueous alkali hydroxide (e.g., NaOH, KOH), alkali silicate (e.g., $Na_2SiO_3$), or a combination of these solutions. As the concentration of alumina and silicate species approach saturation in the pore fluid, an amorphous to semi-crystalline inorganic polymer is formed which is called by a common name geopolymer (Davidovits, 1984). This inorganic geopolymer is mainly composed of an alkali (e.g., Na)-alumino-silicate-hydrate gel (in short, NASH) which creates the binding phase; in contrast with calcium-silicate-hydrate (CSH) gel which is the main constituent and binding phase of portland cement concretes (Fernández-Jiménez & Palomo, 2009). In addition to fly ash (both class F and C), geopolymers can be produced using metakaolin.

Alkali activated products can compete with traditional concrete in terms of performance in a wide range of applications including variations in temperatures in use. Depending on the curing conditions and the concentration of activator, alkali activated fly ash can exhibit a wide variety of properties including high compressive strength, low creep and drying shrinkage, acid resistance and fire resistance (Su, M., et. al, 1997).

The spherical shape of fly ash particles is beneficial for reducing water demand and for maximizing particle packing to reduce porosity. On the other hand, alkali activated Class-F fly ash either requires a highly concentrated alkali solution (highly caustic) for activation or high temperature curing to gain sufficient strength. Also, variably in the composition of fly ash can result in significant variability in the performance of concrete produced unless the ash composition is continuously monitored and changes in mixture proportions are made accordingly. These cements have been engineered for use in fast track concrete repairs and construction, conventional paving, walls and concrete block masonry, new construction and repair projects.

Developments in fly ash based cements offers the user a unique set of mechanical and dimensional properties. In addition, they are now competitive in cost to current cementitious product offerings.

These newly developed activated fly ash based products leave virtually no carbon footprint. Updated cementitious binder technology eliminates approximately 0.9 t (1 st) of $CO_2$ emitted into the atmosphere per ton of portland cement produced.

Concrete is the most widely used man-made material in the world. In 2008 nearly 2.6 billion t (3 billion st) of portland and hydraulic cement was produced worldwide (PCA, 2009). Cement production generates carbon-dioxide emissions because it requires fossil fuels to heat the powdered mixture of limestone, clay, ferrous and siliceous materials to temperatures of 1,500°C (2,700°F). Limestone (Calcium Carbonate - $CaCO_3$) is the principle ingredient of cement. During the portland cement clinker calcining process, $CaCO_3$ is changed to CaO. This conversion releases one mole of $CO_2$ (carbon dioxide) for every mole of $CaCO_3$ consumed in the production process. Approximately one ton of $CO_2$ is released in the production of one ton of portland cement. In the United States, portland cement production alone constitutes about 2-3 percent of $CO_2$ gasses generated annually. Given the impact portland cement production has on the environment, it is incumbent on concrete manufacturers to actively pursue immediate programs and/or practices that reduce the generation of $CO_2$ emissions. The concrete industry shouldn't consider this obligation a negative, however, because this responsibility also brings the opportunity to develop innovative technological advances in both material and a production processes. The cements thusly produced develop Calcium Silicate Hydrates (CSH) as shown in Figure 3.

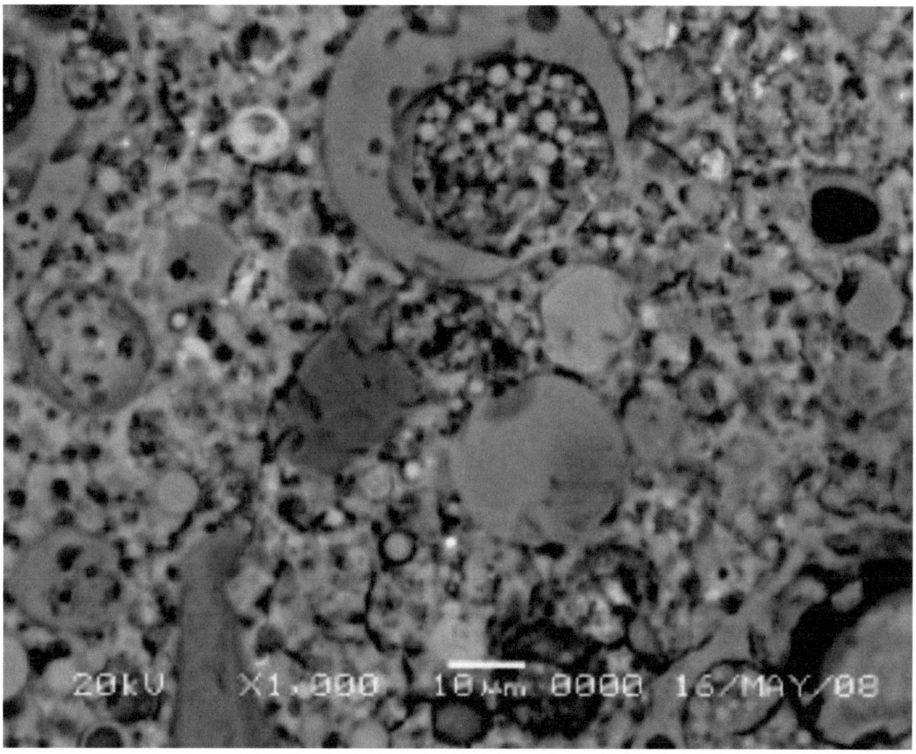


Fig. 3. Micrograph of CSH Formation in Activated Fly Ash Cement, sample age is 14 months from addition of water. (James K. Hicks, et. al., 2009).

Portland cement has long been used in standard building materials. Over the years, various modifiers have been developed for cement formulations to provide particular properties or advantages, such as more rapid curing, compatibility with and resistance to certain materials, and varying strengths, etc. In the past, at times the modified formulations have worked at cross purposes, so that a cement formulation that initially cures more rapidly results in a final product with a lower ultimate strength, while the higher late strength portland cement formulations frequently cannot be demolded for substantial periods of time because there is not sufficient early strength.

Over the past thirty years, scientists have pursued various methods to produce a class of fly ash based cement known as geo-polymers. These early precursors to present products were found - even though mineral in composition - to provide many of the properties of molding resins, such as epoxies and polyurethanes.

Some geopolymeric cementitious products are in used still today in various parts of the world. Such geopolymers are described and claimed, for example, US Patents. (Davidovits,

1982). These geo-polymers are primarily composed of silicas and aluminas, mixed and reacted in particular ways to provide the desired structure. While, in general, these geopolymers are perfectly adequate for the purposes intended, as such, they do not always provide the types of strengths sought in a concrete composition. Furthermore, geopolymers typically require post reaction thermal processing for up to 24 hours in order to achieve desirable strengths.

Below is a recent historical summary of earlier versions of pozzolan based cements:

- Alkali activation of solid, non-portland cement precursors (usually high-calcium slags) was first demonstrated in reasonably modern times by Purdon in 1940, and was developed on a larger scale primarily in Eastern Europe in the succeeding decades, (vanDeventer, Jannie S.J. et. al, 2010)
- 1970's: Geo-polymers from fly ash, cements high in Al-Si. J. Davidovits makes references to their use in historical construction techniques.
- 1980's: Activated fly ashes blended with cement, e.g. mostly two step mixes unconditionally require addition of the activator at the jobsite.
- 1990's through mid-decade beginning in 2000: development of one step mixes, activator in product package or cement. The cementitious compositions typically consisted of harsh acids and bases such as citric acids (pH~2.2) and alkali metal activators including alkali hydroxides (pH~12-14) and metal carbonates (pH~11.6). These included patents by Gravitt, Kirkpatrick, Styron, Hicks and others. There were some drawbacks to these materials. The prior art required acid -base reactions. These reactions sometimes were non-uniform and difficult to control.

The art has needed and continued to seek a hydraulic cement composition, which provides for utilization in standard situations, while providing both a high early strength and an ultimate, very high strength. In particular, compositions having a minimum strength of 28 MPa (4,000 psi) at 4 hours, the release strength necessary for prestress work, have been sought.

### 2.1.1 The new generation of cement technology

This new generation of fly ash based cements offers the user a unique set of mechanical and dimensional properties competitive in cost to current cementitious product offerings, providing the user with a value added alternative solution for today's most challenging construction cementitious repair, product and paving applications. The technology is built around a highly flexible chemistry that allows for the inclusion of a wide array of waste materials as part of its binder matrix, establishing it as a truly green sustainable construction material with unique performance and application advantages.

This new green cement technology is based upon an all fly ash cement design that requires no portland cement in its matrix. Through a detailed study of various types of chemistry and reactive fly ash-based cement pastes, key aspects of the mineralogy have been identified for determining the usefulness of various fly ash sources as high performance cements, including non acid-alkali activated cements.

Key to green cement development was creating a material matrix that had a very dense crystal structure eliminating the movement of water and other chemicals through the material matrix; water being the catalyst for many of the reactions that occur in the concrete matrix.

This is accomplished through the simultaneous dissolution and retardation of the Calcium Oxide phase to solubilize both the silicate and aluminate amorphous phases. The minerals recombine to the desired structure providing desired mechanical and dimensional properties. Thusly, pozzolanic materials are modified with chemicals to produce the desired structure. They are characterized by a very dense crystal structure exhibiting the optimum chemical ratio of calcium to silicates to aluminates. The micro pore structure is very small, greatly limiting the movement of liquids within the material matrix.

The crystal structure of portland cement is dominated by Tricalcium Silicate ($C_3S$) and Dicalcium Silicate ($C_2S$) components producing a crystal structure that is not as dense leading to relatively a large voids structure within the material matrix. The chemical and mineralogical improvements, coupled with the much higher fineness of pozzolan based cements ground leads to much lower porosity in the concrete. The lower porosity provides for very low water to cementitious ratios and improved durability factors.

Having developed a technique to "fingerprint" raw materials as well as a "road map" of good fly ash sources, the new approach is able to maintain quality assurance on product lines using a broad array of fly ash sources, and blends of sources.

The improved activated hydraulic cement technology is the principal backbone chemistry for a range of product offerings from small area repair packaged goods to new construction concretes. Products from the non acid-alkali activated cements were developed specifically to satisfy user or application performance requirements. Each product is water mixed, single component activated, turnkey concrete, mortar or grout with flexible working times from 15 minutes to three hours. The products were engineered to allow for mixing, hauling, placing and finishing using standard industry equipment and practices. The products were designed for applications where speed, strength and durability were desirable performance characteristics. Compressive strengths of more than 17 MPa (2,500 psi) in as little as 60 minutes supported by bond strengths of over 21 MPa (3,000 psi) and flexural strengths over 10 MPa (1,500 psi) in 7 days frame the technology's mechanical properties. Dimensional stability is highlighted by shrinkage of less than 0.04% length change in 28 days.

Principle benefits of this new class of products include:

- Non-shrink.
- Exceptional sustained bond strengths (slant shear and direct tension).
- Low coefficient of thermal expansion.
- Modulus of elasticity consistent with Portland cement concrete.
- Low permeability.
- High resistance to freezing and thawing.
- High resistance to scaling. High resistance to sulfate and chemical attack.
- Exceptional durability.
- Placement temperature tolerant.
- No epoxy resins are contained.

Specific areas of products developed meeting objective criteria fall into several areas:

- Rapid Repair
- Ready Mix including paving
- Volumetric mixer concrete and mortar
- Concrete block/grout/mortar
- Precast
- High Temperature Resistant Materials such as that being placed in Figure 4.
- Chemical Resistant Materials

Fig. 4. Construction of base for a heat treating facility in Houston, Texas. Cycling of temperatures caused severe breakdown of conventional concrete, (James K. Hicks, et. al., 2009).

Some of the more specific examples descriptions are:

- Rapid repair products all have cementitious components greater than 90 percent coal ash, and contain no portland cement. Based upon the size of the repair, products range in working time from 15 to 45 minutes, offering return to service ranging from 1 to 4 hours (See Table 1). All products can be mixed with conventional mixing equipment and placed like portland cement products, however without the requirement of bond coats.
- Ready-mix truck delivery. For large placements such as roadway slabs, ash-based pozzolanic cements have been adapted to ready-mix batch plant/transit truck mixing and placement. These products are able to be site activated (up to 4 hours transit time),

and adjusted to placement times from 1 to 3 hours. Return to service can be achieved in as little as 6 to 12 hours (See Table 2). Slump control can be adjusted to range from roller-compacted concrete (RCC) to a self-consolidating concrete (SCC).

| Property | As Packaged 4 in. x 8 in. cylinders | Test Method |
|---|---|---|
| **Compressive** Strengths, **psi** (MPa) | | |
| **2 hours** | **2820** (19.4) | **ASTM C 39** |
| **1 day - 24 hours** | **6115** (42.2) | **ASTM C 39** |
| **7 days** | **9345** (64.4) | **ASTM C 39** |
| **28 days** | **10,510** (72.5) | **ASTM C 39** |
| **Flexural** Strength, **psi** (MPa) | | |
| **1 day - 24 hours** | **690** (4.8) | **ASTM C 78** |
| **7 days** | **945** (6.5) | **ASTM C 78** |
| **28 days** | **1405** (9.7) | **ASTM C 78** |
| **Splitting Tensile** Strength, **psi** (MPa) | | |
| **28 days** | **590** (4.0) | **ASTM C 496** |
| **Bond** Strength, **psi** (MPa) | | |
| **1 day - 24 hours** | **1960** (13.6) | **ASTM C 882** |
| **7 days** | **2745** (18.9) | **ASTM C 882** |
| **Rapid Freeze Thaw** Resistance (Durability Factor - Retained percentage of Dynamic Modulus) | | |
| **300 cycles** | **100%** | **ASTM C 666A** |
| **Scaling** Resistance, **lbs/ft²** (kg/m²) | | |
| **50 cycles** | **0** | **ASTM C 672** |
| **Modulus** of Elasticity, **msi** (GPa) | | |
| **28 days** | **5.2** (35.1) | **ASTM C 469** |
| **Coefficient** of Thermal Expansion, **in/in/°F** | | |
| **28 days** | **1.32** | **AASHTO TP 60** |
| **Length** Change, **% of total length** | | |
| **28 days soak / 28 days dry** | **-0.052 /-0.041** | **ASTM C 157** |

Table 1. Characteristics of an activated pozzolan cement fast return to service ready mixed concrete. Source, CeraTech, Inc.

| Property | [1]Rapid Set | [1]Standard Set | Test Method |
|---|---|---|---|
| **Compressive** Strengths, **psi** (MPa) 4 in. x 8 in. cylinders | | | |
| **6 hours** | **3500** (24.1) | **NA** | **ASTM C 39** |
| **24 hours** | **3604** (24.9) | **2497** (17.2) | **ASTM C 39** |
| **3 day - 72 hour** | **4502** (31.0) | **4193** (29.0) | **ASTM C 39** |
| **7 days** | **6487** (44.7) | **5998** (41.3) | **ASTM C 39** |
| **28 days** | **8511** (58.7) | **8502** (58.6) | **ASTM C 39** |
| **Flexural** Strength, **psi** (MPa) | | | |
| **7 days** | **510** (3.5) | **485** (3.3) | **ASTM C 78** |
| **28 days** | **650** (4.5) | **630** (4.3) | **ASTM C 78** |
| **Splitting Tensile** Strength, **psi** (MPa) | | | |
| **28 days** | **720** (5.0) | | **ASTM C 496** |
| **Rapid Freeze Thaw** Resistance (Durability Factor - Retained percentage of Dynamic Modulus) | | | |
| **300 cycles** | **100%** | | **ASTM C 666A** |
| **Scaling** Resistance, **lbs/ft²** (kg/m²) | | | |
| **50 cycles** | **0** | | **ASTM C 672** |
| **Abrasion** Resistance, Depth of wear, millimeters @ 28 day | **0.14** | | **ASTM C 944 (2005)** |
| **Modulus** of Elasticity, **msi** (GPa) | | | |
| **28 days** | **5.00** (34.0 ) | | **ASTM C 469** |
| **Coefficient** of Thermal Expansion, **in/in/°F** | | | |
| **28 days** | **4.6** | | **AASHTO TP 60** |
| **Length** Change, **% of total length** | | | |
| **14 days** | **0.04** | | **ASTM C 157** |
| **Creep (365 days)** (μ Strain / psi) **Creep** Coefficient | **1.91** | | **ASTM C 512** |

Table 2. Characteristics of an activated pozzolan cement fast return to service ready mixed concrete. Source, CeraTech, Inc.

- Volumetric mobile mixer use. The volumetric pozzolanic product utilizes the same backbone chemistry as the rapid repair products. For larger placements that also require fast return to service, the pozzolans have been adapted to work in a volumetric mixer, allowing from 20 to 50 minutes of placement time, with return to service in as little as 1 hour depending upon the user requirements. With DOT and DOD applications, the principal benefit of volumetric placement is the ability to place larger volumes while still taking advantage of the quick return to service. One version of this product can be used as a flowable grout capable of providing up to three hours of working time, yet providing up to 35 MPa (5000 psi) in compressive strength in 24 hours.

Fig. 5. Marine Corps Engineers training for fast track construction and repair prior to deployment. Marine Corps Base Concrete Installation, (James K. Hicks, et. al., 2009).

Among the general construction and precast benefits are:

- For vertical construction markets, including columns, flooring, and tilt-up construction, ash-based pozzolanic cements have been adapted to perform as self-consolidating concrete (SCC). These products permit easy pumping and long working times, yet can suspend aggregate, provide sufficient placement time, and offer early return to service. These are placed with a conventional batching and mixer system.
- Precast. Additional benefits of non acid-alkali activated ash-based pozzolanic cements also extend to precast concrete applications. Higher strength precast components can be developed, offering the ability to strip molds much earlier than with cement based concrete. This ability permits faster turn-around and throughput to the manufacturer.
- High Temperature Resistant Materials. A unique benefit of ash-based pozzolanic cements is their high temperature resistance capabilities. Ash-based cements are naturally refractory given their amorphous glass chemistry. Coupled with other high-temperature admixtures, these products are the only materials that have passed Mach 1 shock testing at 1700°F (927 °C) for 300 cycles. This result has qualified the material for use as a run-up and takeoff pad for current emerging vertical takeoff aircraft (VTOL) including the AV-8, V-22 Osprey, and the Joint Strike Fighter.

- Armor and Protective Materials. Non acid-alkali activated fly ash-based pozzolanic cements are not only able to achieve high-early strength, but very high strengths overall. In one development area, a class of cements has been developed capable of achieving over 69 MPa (10,000 psi) in 24 hours, and up to 152 MPa (22,000 psi) within 28 days. These products are in development with the US Army Corps of Engineers as a field emplaced armor material capable of withstanding both blast and fragment penetration.
- Concrete block/grout/mortar. The non acid-alkali activated fly-ash based pozzolanic cements have also been optimized to product both normal strength and high strength concrete masonry units (CMUs). Products have been able to achieve strengths ranging from over 14 to 69 MPa (2,000 to 10,000 psi) using conventional concrete block manufacturing facilities, techniques, and cement percentages equal to those used by conventional cement. (Hicks, et. al., 2009)

Notes:

1. Strength development and working times can be adjusted by varying the cement ratio and by use of various proprietary activator admixtures.

- Rapid Set Test results based on 846 lbs. of cement per cubic yard mix design and Fast Set Activator.
- Standard Set Test results based on 564 lbs. of cement per cubic yard mix design and Fast Set Activator.

### 2.2 Alkali Activated Slag cements

An alkali activated slag (AAS) concrete is one in which the binder phase is made of ground granulated blast furnace slag (GGBFS), water, and an alkali activator which triggers the chemical reactions involving dissolution of slag and polymerization of calcium-silicate and aluminum-silicate phases which serve as the binder. These systems contain no portland cement. Although slag is cementitious and can self-activate, the reaction is typically slow and requires external activators to enhance reaction rates and form stronger products. Sufficient alkali content is also necessary for the development of significant strength. Since slag is deficient in alkalis, these have to be supplied externally. These alkali-activated slag systems also appear in the literature with a variety of names such as alkali-activated cements (Palomo & López dela Fuente, 2003), and alkali-slag cements (Roy, D.M. 1999).

The most commonly used alkaline activating agents are sodium hydroxide (NaOH) and sodium silicate ($Na_2SiO_3$), also known as water glass, or combinations of these (Shi & Day, 1996), (Krizan & Zivanovic, 2002). A variety of industrial by-products containing alkalis and sulfates such as cement kiln dusts are also equally effective in activating slag (Chaunsali & Peethamparan, 2010) . Among the activators used, water glass is widely reported to give rise to rapid hardening and high compressive strengths (Oh et. al. 2010). The main hydration product in AAS systems is calcium silicate hydrate (C-S-H) with a low Ca/Si ratio regardless of the type of activator used (Song, et. al., 2000). The morphology of the hydration products changes with the activator and other hydration conditions. Hydrotalcite (magnesium-calcium aluminate hydrate: $(M,C)_4AH_{13}$) and minor amounts of strätlingite ($C_2ASH_8$) are also found in AAS.

Alkali-activated slag displays very good strength (Atis, et. al, 2000), durability (Fernández-Jiménez & Puertas, 2002), and a variety of other potentially valuable characteristics such as fire and waste water resistance. However, shorter setting time, increased shrinkage, and difficulties with safe handling of the liquid water glass outside the laboratory have limited its use (Shi, et. al., 2006). Sodium silicate with higher modulus (i.e., silica-to-alkali ratio) can often alleviate some of these problems.

Supersulfated slag cement is an example of a cement that has used ettringite in combination with CSH as the main binding phase to a relative degree of success. Supersulfated slag cements consist mostly of granulated blast-furnace slag, gypsum or anhydrite with a small amount of lime or Portland cement to catalyse the chemical reaction. Although, they exhibit slow strength development and may undergo deleterious carbonation reactions, they still have significant application potential.

More research is needed to modify the early age properties of the alkali activated slag concrete. The low $CO_2$ emission rate and the low embodied energy of AAS, as slag being a commercial by-product, compared to traditional portland cement are primarily attributed to the absence of high temperature clinkering processes involving a large amount of fossil fuel-derived energy. While the use of alkaline hydroxide or silicate activating solutions rather than water does reintroduce some greenhouse costs, the overall beneficial effects due to widespread use of alkali activated slag is expected to be highly significant.

### 2.3 Calcium Aluminate Cements

Calcium aluminate cements (CAC's) are a special class of cement containing primarily aluminates and calcium. Small amounts of ferrite and silica are also typically present. CAC's were invented in the early 1900's to resist sulfate attack. CAC's are inherently rapid hardening and can be rapid setting, adjustable with appropriate chemical admixtures. These cements are often used in refractory applications, building chemistry and rapid repair, rehabilitation and construction of concrete flatwork (e.g. sidewalks, overlays and full-depth pavement construction). The rapid hardening properties, resistance to sulfate attack and alkali-aggregate reaction and abrasion resistance make these cements desirable in a wide-range of special applications (Scrivener & Capmas, 1998, Juenger, et.al. 2010). The manufacturing process of CAC's generates significantly less $CO_2$ than the ordinary portland cement (OPC); roughly on the order of 50% less (Juenger, et.al. 2010), Gartner, E. (2004). However there is slightly more grinding energy required (than OPC) due to the increased strength of the clinker less (Juenger, et.al. 2010, Scrivener & Capmas, 1998).

The most widely discussed and controversial aspect of CAC's is a process referred to as conversion. Conversion is a process where metastable hydrates ($CAH_{10}$ and $C_2AH_8$) formed at low and moderate temperatures (T=5 to ~70°C) convert to stable hydrates ($C_3AH_6$) formed at high temperatures (T>70°C). This process leads to an increase in porosity and subsequent decrease in strength. Conversion is an inevitable process and must be accounted for when designing the concrete mixture. The hydration process of CAC's is thus highly temperature dependant and the time spent during hydration within specific temperature ranges will dictate the type and amount of metastable and/or stable hydrates formed (Scrivener & Capmas, 1998).

Several building collapses in the 1970s were attributed to CAC conversion, and many structural codes subsequently banned the use of this material. Since this time, intensive research has provided a greater understanding of CAC chemistry and behavior. Furthermore, a report by the Concrete Society (Palomo & López dela Fuente, 2003) revisited these landmark collapses and revealed that improper structural detailing, a lack of understanding about CAC properties and not following manufacturers' guidance had lead to the majority of the structural failures. Improved guidance for predicting long-term properties and a better scientific understanding of the material has seen resurgence in interest and use of this alternative cementitious binder (Juenger, et.al. 2010).

There are several recommendations that have been employed to keep the strength lost during conversion to a minimum. This includes keeping the w/cm below 0.40 and including a minimum cement content of 400 $kg/m^3$ ( Concrete Society Technical, 1997, Fentiman, C. et.al. , 2008).

### 2.4 Calcium sulfoaluminate (CSA) cements

Calcium Sulfo-Aluminate Cements or ettringite cements are a type of high alumina cement that first came to prominence in the 1970s. To form calcium sulfoaluminate clinker, limestone, bauxite, and gypsum are mixed and heat fired in rotary calcining kiln. Commercial sulfoaluminate clinkers developed by the Chinese predominantly consist of $C_4A_3\bar{S}$ (55–75%) (also known as Klein's compound) and a-$C_2S$ (15–25%). The remaining phases present are $C1_2A_7$, $C_4AF$ and CaO, but $C_2A\bar{S}$ and $C\bar{S}$ are considered deleterious and, therefore, undesirable. Belite ($C_2S$)-rich sulfoaluminate cements are preferred to alite ($C_3S$)-rich, since belite-based cements can be formed at around 1200 °C, as opposed to 1400 °C for the alite cements. This equates to an energy savings of 20% during manufacture( Popescu, et.al. ) and results in less $CO_2$ being generated from the reaction of formation of $C_2S$ compared to $C_3S$. Belite-based sulfoaluminate cements are also preferred for certain performance reasons, since cements containing larger quantities of $C_2S$ than $C_3S$ are less permeable as well as being more resistant to chemical attacks and smaller drying shrinkage. On the other hand, large proportions of $C_2S$ do reduce the rate of strength evolution and setting point but the presence of calcium sulfate and calcium sulfoaluminate more than compensate for this deficiency.

Sulfoaluminate cements may be defined by the compositional system of CaO–$Al_2O_3$–$SiO_2$–Fe2$O_3$–$SO_3$. Strength and material performance rely heavily on the specific composition and phases present. The bonding phases within sulfoaluminate cements consist predominantly of ettringite, monocalcium sulfo-aluminate hydrate, ferrite and alumina gel. Calcium aluminates such as $C_3AH_6$ and $C_4AH_{13}$ have been observed in some instances.

To date, studies on the durability of these cements have been promising. Typically, sulfoaluminate cements are used where rapid setting, early strength or shrinkage compensation is required. They also have the advantage that their long-term strength and durability can exceed that of Portland cement. These cements have seen widespread and high volume use as bridge decks, airport runways, patching roadways etc. where rapid setting is required.

Approximately 81,000 t of CSA Cement was produced in 1996 in the US alone, but it has been used even more in Japan and China to construct bridges and buildings. Production now exceeds 106 tonnes per year. However, rapid setting and expansive cements are inappropriate for certain building engineering applications. Since the hydration of belite-sulfoaluminate cements at ambient temperatures leads to the formation of ettringite as the major phase, the cements may be susceptible to degradative problems associated with carbonation, delayed ettringite formation as well as thaumasite attack. So if the setting time is to be properly regulated, special activators must be added to reduce the setting rate. Further research is ongoing in this area to widen the potential use of sulfoaluminate cements.

Significant environmental advantages exist in using sulfoaluminate cements. The calcium sulfoaluminate clinker ($C_4A_3S$) generates the least amount of $CO_2$ per g of raw material as by-product of its reaction of formation. This makes it extremely attractive. Furthermore, the calcining of the raw materials for clinker formation occurs at temperatures (1160–1200°C) much lower than those used for firing Portland cement clinker (1450°C). Another environmental advantage of the manufacturing process of belite-rich sulfoaluminate clinkers is that it can utilize industrial by-products with high sulfate content. For example, fluidized bed combustor (FBC) fly ash, blast furnace slag, low-calcium fly ash or flue gas desulfurization (FGD) sludge can be utilized to manufacture belite-rich sulfoaluminate clinkers, whereas they could not be used in the manufacture of Portland cement clinker, which limits the $SO_3$ content to 3.5% dry weight. Energy savings also occur in the grinding of the clinker compared to Portland cement, since the low firing temperatures result in a clinker, which is generally softer.

Another advantage of calcium sulfoaluminate cements is that it can use gypsum to form hydration products which have not undergone any heat treatment in a kiln. This results in a considerable energy saving.

Utilization of sulfoaluminate cements in both China and Japan is now becoming so widespread that they may almost be referred to as a commodity material, particularly in China. In Europe and North America, its utilization has largely been restricted to high early strength, self-stressing or shrinkage applications. Sulfoaluminate cements could potentially become a major large-scale alternative to Portland cement, however their classification has largely been restricted to certain geographic regions with different climatic and economic conditions. As for alkali-activated cements, a significant disadvantage of sulfoaluminate cements has been the confusing nomenclature and lack of international consensus as to standards and names.

Calcium sulfoaluminate (CSA) cements are receiving increasing examination from the cement industry and researchers as a lower-energy, lower-$CO_2$ alternative to portland cement. Such cements are not new; they contain as a primary phase Ye'elimite ($Ca_4Al_6SO_{16}$; or $C_4A_3$ S) in cement chemistry notation), which was used by Alexander Klein in the 1960s as an expansive additive to portland cement and is sometimes called Klein's compound (Klein, A. 1966) . The Chinese have been industrially producing these cements since the 1970s and their annual production may exceed 1 million tons (Glasser & Zhang 2001). CSA cements are used for some structural applications, but they are especially well-suited for precast and cold-weather applications which take advantage of the rapid strength gain of these materials (Quillin, 2001). Their mechanical properties and durability are reported to rival portland cement (Mehta, 1980), but much work is needed to fully vet the long-term material performance.

CSA cements lack standardized composition or performance criteria. They are distinguished by the presence of $C_4A_3S$ as the primary cementing phase, belite ($C_2S$) as a secondary phase, and the absence of alite ($C_3S$). Other phases that are commonly present include $C_4AF$, $C_2AS$, CA, and CS, and gypsum is added for hydration control as in portland cement. Because ettringite is the primary reaction product of $C_4A_3S$ with gypsum and water, these cements can exhibit rapid setting, rapid strength gains, and expansion. However, these properties can be controlled through manipulating the composition of the cement and through the addition of chemical retarders. Mechanical properties and dimensional stability similar to portland cement can be achieved (Juenger & Chen, 2011, Sharp, 1999).

The reputation of CSA cements as environmentally-friendly comes from reductions in energy use and $CO_2$ emissions during manufacturing. CSA clinkers can be made at lower kiln temperatures than portland cement clinkers (1250°C instead of 1350-1450°C), and are more porous and friable, thereby requiring less energy for grinding (Shi, et. al, 2005, Rajabipour, et. al, 2010). The lime (CaO) content of $C_4A_3S$ is low (36.7 wt.%) compared to that of $C_3S$ (73.7 wt.%); therefore much less limestone is calcined in the production of CSA cements, leading to lower $CO_2$ emissions. CSA cements may be able to reduce limestone use by 40% and energy by 25% compared to OPC (Sharp, 1999). One of the primary challenges facing the widespread adoption of CSA cements, aside from further knowledge of its material properties, is the cost of raw materials. The high alumina content demands the use of bauxite as a raw material, which is not widely available and is expensive. Alternatively, the high sulfur and iron (if $C_4AF$ is included) contents allow for the use of many types of industrial by products as raw materials, reducing the cost of the clinker, both in terms of the economy and the environment (Palou & Majling, 1995, Phair 2006).

### 2.5 Magnesia based cements

Magnesia cements are a range of cements based on magnesium oxide (MgO) as the key reactive ingredient. The first type of magnesia cement was developed by Frenchman Stanislas Sorel in 1867 and is now referred to as Sorel, magnesite or magnesium oxychloride cement (Phair, J.W. 2006). Sorel cement was produced initially by the combination of magnesium oxide with concentrated aqueous magnesium chloride. This results in a hardened cement paste consisting of four main bonding phases: $2Mg(OH)_2{\bullet}MgCl_2{\bullet}4H_2O$, $3Mg(OH)_2{\bullet}MgCl_2{\bullet}8H2O$, $5Mg(OH)_2{\bullet}MgCl_2{\bullet}5H2O$ and $9Mg(OH)_2{\bullet}MgCl_2{\bullet}H2O$. It was soon discovered, however, that the magnesium oxychloride phases are not stable after prolonged exposure to water, and will leach out in the form of magnesium chloride or magnesium hydroxide. As a consequence, Sorel cement has found limited application in the building and construction industry despite demonstrating other properties that are far superior to those of Portland cement. For instance, it has high fire resistance, low thermal conductivity, high abrasion resistance, high transverse and crushing strengths and does not require wet curing. Various additives are being investigated to improve its water resistance although no spectacular discovery in this area has been made yet.

With time, a variety of other magnesia cements have been developed based on permutations of magnesium oxide as the binding phase with varying levels of success. Magnesium oxysulfate cements, formed by the reaction of a magnesium sulfate solution with magnesium oxide, have similar properties to magnesium oxychloride cements. Again, they have good binding properties and combine well with a variety of inorganic and organic aggregates such as sand, marble flour, gravel, saw dust and wood flour to produce a cement with high early strength. Poor weathering resistance, however, is again the main drawback for this type of cement.

Another analogue is magnesia phosphate cements, which can be synthesized by reacting magnesium oxide with a soluble phosphate (e.g. ammonium phosphate). In essence, this is an acid–base reaction between the phosphate acid and the magnesium oxide to form an initial gel that crystallizes into an insoluble phosphate, mostly in the form of magnesium ammonium phosphate hexahydrate ($NH_4MgPO_4 \bullet 6H_2O$).

Magnesia phosphate cements are characterized by very high early strength and rapid setting, which makes them useful as a rapid patching mortar. It can also bind well to a wide variety of aggregates and substrates. Unlike magnesium oxychloride and oxysulfate cements, this cement has good water and freeze-thaw resistance and is, therefore, amenable to a wide variety of applications. A major drawback, however, is the expensiveness of phosphate, which confines its application to niche areas.

More promising magnesia cements that have recently attracted considerable interest from industry have been magnesium carbonate or magnesite cements (Harrison, 2003). The bonding phase for these cements includes magnesium hydroxide and magnesium carbonate. In the presence of accelerating additives (such as $KNO_3$, $Fe_2SO_4 \bullet xH_2O$, $CaCl_2$, etc.) magnesium oxide will undergo hydration to form magnesium hydroxide, or brucite, which is considerably less soluble than portlandite ($Ca(OH)_2$), the calcium analogue. Brucite can then react in the presence of sufficient carbon dioxide and a calcium source to form magnesite, hydromagnesite and calcium magnesium carbonate or dolomite ($CaMg(CO_3)_2$). These carbonates add strength to the binding phase, and in being relatively insoluble and fibrous in structure, have the ability to form a better network nanostructure than observed for calcium carbonates.

The presence of the accelerator also serves to increase the setting time making the long-term strength gains comparable to that of Portland cement.(Harrison, 2003).Typically, magnesia carbonate cements can be formed when magnesia is in the presence of considerable quantities of waste materials or traditional pozzolans, provided sufficient brucite, carbon dioxide and accelerators are present.131 Such waste materials include fly ash, slag, silica fume, silica flour, mine tailings, sewerage ash etc.

The possibility also exists for adding magnesium oxide to ordinary portland cement, yielding more durable composite cement. This may be achieved by promoting the reaction of portlandite with pozzolans, thereby allowing brucite to replace it and sequester excess alkali. This is an advantageous scenario since brucite is some 100–1000 times less soluble than portlandite. ($Ca(OH)_2$ $K_{SP}$ = 5.5 6 x $10^{-6}$, $Mg(OH)_2$ $K_{SP}$ = 1.8 x $10^{-11}$ at 25°C). The main environmental advantage associated with magnesia carbonate cements is that the starting material, magnesia, is readily obtained from the calcining of magnesite. The process of magnesite decomposition occurs at temperatures around 400°C less than that of limestone so it requires considerably less energy to manufacture. However, the reaction for formation of magnesia results in the emission of considerable amounts of CO2, which slightly offsets the gains achieved in the lower energy requirements for processing magnesia. Nevertheless, the carbonation reaction resulting in the formation of the main bonding phase from magnesia can consume atmospheric CO2. This suggests that such cements could in fact act as carbon sinks, acting to favorably reduce greenhouse gas emissions. However, this reaction is slow since it is diffusion controlled, so the useful lifetime of the cement would probably be exceeded before complete carbonation could occur. Theoretically, since magnesium carbonate is the main binding phase, these cements are also readily recyclable.

As opposed to portland cements, which cannot be easily reheated to regenerate the original starting materials (calcium carbonate and clinker), the formation of magnesia from the magnesium carbonates should be much easier. The utilization of secondary materials such as fly ash and slag is routine in the cement industry but it is the chemistry of the bonding phases that is more important. Significantly, magnesium hydroxide is less alkaline than calcium hydroxide and, therefore, has a considerable reduction in potential problems later on. It has been reported that the addition of the accelerator allows the cement to overcome many of the problems associated with magnesia oxychloride cements. As a result, magnesium carbonate cements exhibit greater resistance to acids and are a less corrosive substrate that is stable in moist environments.

While magnesia carbonate cements are still in the developmental stages, there are a few downsides. For instance, despite the fact that filler materials such as industrial by-products are reasonably cheap and inexpensive to obtain, acquiring magnesium carbonate is generally more expensive than calcium carbonate and harder to find. With increasing demand, however, it is possible that the prices could dramatically drop. The technology, to some civil engineers, may also be regarded as unproven. A range of concerns still surround the material such as the load-bearing performance, long-term dimensional stability, freeze-thaw resistance, creep under load as well as a host of basic characteristics such as heat output during hydration, porosity/permeability relationships and fire resistance. However, magnesia carbonates and OPC/ magnesia carbonate composite cements are the most likely large-scale magnesia cement alternatives to portland cement.

### 2.6 Recycled glass-based cements and concretes

The main incentive behind using post-consumer recycled glass (e.g., bottles, windows) in concrete is the prohibitive costs of shipping recycled glass from collection points to remelting facilities that manufacture new glass products. As of 2003 in the U.S., about 600,000 tons/year of "recycled" glass is landfilled or stockpiled in hopes that future technologies would allow a profitable use of this material (Reindl, 2003). Pulverized glass can be used in concrete as fine aggregates or as cement replacement. Aggregate applications are more developed and have been commercialized in the U.S., U.K., and Australia. Desirable strength and workability of concretes containing glass sand have been achieved by proper mixture proportioning (Polley, et.al, 1998). The main challenge on the use of these materials is the alkali-silica reaction (ASR) of glass particles leading to cracking and deterioration of concrete. However, ASR has been successfully controlled by (a) use of fly ash, (b) use of glass finer than #50 sieve (0.3 mm) (Jin, et.al., 2000), or (c) annealing crushed glass before use in concrete (Rajabipour, , et. al., 2010).

In comparison, the use of glass powder as a cementitious material is in its infancy. Due to a high concentration of amorphous silica (~70% wt.) soda-lime glass can react pozzolanically with portlandite in a glass-portland cement system and produce low Ca/Si C-S-H. At moderate (up to 30%wt.) replacement levels of OPC, glass powder has been found to improve compressive strength beyond 28 days; however, early strengths can be reduced when using the same w/cm

Shao, et. al., 2000, Shi, et. al, 2005). Fineness of glass powder has a significant impact on its reactivity; glass finer than 38μm satisfies the strength activity index of ASTM C 618 (SAI>75% at 7days) and can be classified as a Type N pozzolan (Byars, et. al., 2004, Shao, et. al., 2000). By further increasing glass fineness and/or heat curing, concretes with 3-day strengths surpassing that of OPC concrete can be prepared (Shi, et. al, 2005). In addition, fine glass powder (<10μm) can mitigate the alkali-silica reaction generated by glass aggregates or other natural reactive aggregates ( Shayan & Xu, 2006).

## 3. Conclusions

Despite all global warming concerns and being in the midst of a financial crisis, an approximate growth of 30-50 percent of coal power generation is expected between Years 2007 to 2025. The installed capacity would jump to approximately 2.1 million MW. Initial estimates were even higher but the US and Europe are scaling back due to strong environmental pressures. China alone would add approximately 350,000 MW during this period while India would follow with more than 100,000 MW.

It is essential to emphasize that the use of CCP could make key reductions in $CO_2$ emissions by using byproducts to make cement, substituting for portland cement in concrete, and reducing energy given the energy-efficient nature of concrete structures. It is important to point out that the CCP option is only available for pulverized coal plants. IGCC units follow a different technique and do not produce any cementitious materials as by-products.

These cutting-edge, next-generation "green" non acid-alkali activated fly-ash based pozzolanic cements provide the construction a value added alternative to traditional cement product offers. The extent of engineering that has been done with the product offers widest range of end-use applications from any pozzolan, removing it from its previous limited use as a short-life rapid repair product only. Moreover, the amount of research that has been conducted on understanding fly-ash chemistry and mineralogy has extended the ability to use a much wider range of high calcium coal ash while maintaining predictable product performance. These truly green building materials are comprised largely of renewable, recyclable or reusable resources. They are the only cements in the world whose chemical matrix is comprised of more than 95% waste materials. See Figure 5.

This new generation of all ash-based pozzolanic cements also furthers the ability to utilize green building technology for the widest range of end-use markets, including most DOT, DOD, and building construction market applications while meeting International Building Code and ASTM Standards.

It is important to note that although the environmental and even economic benefits from using CCP's are apparent, they are still under-utilized. The American Coal Ash Association reported that less than 40 percent of CCP are used. The Association only reports affiliated utilities. The authors estimate that less than those amounts are currently used and end up in landfills, creating a burden to the environment and the economy of different enterprises.

We all need to understand that we must adopt sustainable energy policies to avoid endangering energy security and control carbon emissions. Without any intervention, $CO_2$ could increase 42.4 gigatons in 2035 from 29.7 in 2007 (EIA, 2010). This increase is a real and immense challenge that has to be managed promptly.

## 4. References

"Calcium aluminate cements in construction: a re-assessment", *Concrete Society Technical Report TR 46, 1997*

A. J. W. Harrison, Reactive magnesium oxide cements, US Pat. 2003/0041785A1, 2003.

American Coal Ash Association (ACAA), *2009 Fly Ash Survey,* (2010).

Atis, C.D., Bilim, C., Celik, O., Karahan, O. (2000) "Influence of activator on the strength and drying shrinkage of alkali-activated slag mortar", *Construction and Building Materials*, 23, 548-555.

British Petroleum, *BP Statistical Review of World Energy,* (2009)

Byars, E.A., Morales, B., Zhu, H.Y. (2004) "Conglasscrete II", Report published by University of Sheffield.

C. D. Popescu, M. Muntean and J. H. Sharp, *Cem. Concrete Compos.*, 2003, 25, 689.

*Calcium Aluminate Cements: Proceedings of the Centenary Conference*, Fentiman, C. et al. Eds, IHS BRE Press, Avignon, France, 2008.

California Environmental Protection Agency, Air US Coal Supply and Demand, 2007 Review, www.eia.doe.gov.coal, 2007.

Carr, M. and Morales, A. 2009. "World CO2-Emissions Growth Keeps Focus on Coal", *Bloomberg Business Week*, June 10, 2009 http://www.bloomberg.com/apps/news

Chaunsali, P., Peethamparan, S. (2010) "Evolution of strength, microstructure and mineralogical composition of a CKD-GGBFS binder", *Cement and Concrete Research,* oi:10.1016/j.cemconres.2010.11.010.

Clodic L., Patterson J., Holland T., 2010. "Next generation paving materials using mineralized $CO_2$ captured from flue gas," CD-Proceedings, FHWA International Conference on Sustainable Concrete Pavements., Sacramento, California, September.

Davidovits, J. (1991) "Geopolymers: Inorganic polymeric new materials", *Journal of Thermal Analysis*, 37(8), 1633.

Davidovits, Joseph, 1984. United States Patent Office (USPTO), U.S. Pat. Nos. 4,349,386 (1982) and 4,472,199 www.USPTO.gov.

Davidovits, Joseph, United States Patent Office (USPTO), U.S. Pat. Nos. 4,349,386(1982) and 4,472,199 (1984), www.USPTO.gov.

EIA International Energy Outlook, EIA predicts surge in world coal consumption, www.im-mining.com/ 2009/06/07/eia-predicts-surge-in-world-coal-consumption/2009.

EIA International Energy Outlook, U. S Energy Administration 2011 CO2 Forecast Highlights. Report Number: DOE/EIA-0484(2010), http://205.254.135.24/forecasts/ieo/.

Fernández-Jiménez, A., Palomo, A. (2009), "Nanostructure/microstructure of fly ash geopolymers", in: *Geopolymers: Structure, Processing, Properties, and Industrial Applications*, J.L. Provis, J.S.J. Deventer Eds., CRC Press, Boca Raton, USA, pp. 89-117.

Fernández-Jiménez, A., Puertas, F. (2002) "The alkali–silica reaction in alkali-activated granulated slag mortars with reactive aggregate", *Cement and Concrete Research*, 32, 1019–1024.

Floris, Vinio, Challenges and Achievements of Coal-Based Power Generation: Moving Towards a Holistic and Sustainable Approach. *Proceedings of the International Association of Energy Economics*, Santiago, Chile (2009).

Floris, Vinio, Hicks James K., "Environmental Benefits of Coal Combustion Products", *Pittsburgh Coal Conference*, University of Pittsburg, PA (2009).

Glasser, F.P., Zhang, L. (2001) "High-performance cement matrices based on calcium sulfoaluminate-belite compositions", *Cement and Concrete Research*, 31, 1881-1886.

Guetzow, G. and Rapoport, J., 2010."New Architectural Masonry Units Lower $CO_2$, Energy Consumption" *Masonry Edge/The Storypole,* Vol. 5, No.1, pp.1-3.

Hicks, J. (2010) "Durable "Green" Concrete from Activated Pozzolan Cement", *Proceedings of ASCE Green Streets and Highways Conference*, Denver, CO.

Hicks, James K. and Scott, Ryan M., 2007. United States Patent Office (USPTO), U. S Patent 7,288,148 B2, Rapid Hardening Hydraulic Cement from Subbituminous Fly Ash and Products thereof, October 23, 2007.

Hicks, James K., *CCP Beneficial Use versus production,* American Coal Ash Association, 2009, www.acaa-usa.org/.

Huntzinger, Deborah N and Eatmon, Thomas D., *Journal of Cleaner Production,* Elsevier, (2009) 668–675.

James K. Hicks, P. E., Mike Riley, Glenn Schumacher, Raj Patel and Paul Sampson, 2009, Utilization of Recycled Materials for High Quality Cements and Products, World of Coal Ash Conference, Lexington, KY, USA.

Jin W., Meyer C., Baxter S. (2000) "Glascrete - concrete with glass aggregate", *ACI Materials Journal*, 97(2) 208-213

Juenger, M., Chen, I. (2011) "Composition-property relationships in calcium sulfoaluminate cements", *Proceedings of the 13th International Congress on the Chemistry of Cement,* Madrid, Spain.

Juenger, M.C.G., Winnefeld, F., Provis, J.L., Ideker, J.H. (2010) "Advances in alternative cementitious binders", *Cement and Concrete Research,* doi:10.1016/j.cemconres.2010.11.012.

Klein, A. (1966) "Expansive and shrinkage-compensated cements", US Patent 3251701.

Krizan, D., Zivanovic, B. (2002) "Effect of dosage and modulus of water glass on early hydration of alkali-slag cements", *Cement and Concrete Research,* 32, 1181–1188.

Lea, F. M., *The Chemistry of Cement and Concrete,* Arnold, 1970.

Mehta, P.K. (1980) "Investigations on energy-saving cements", *World Cement Technology,* 11 (4) 166-177

*Minerals Yearbook, Vol. 1. Metals and Minerals, 2002. U.S. Geological Survey. U.S. Department of the Interior.* Phair, J.W. (2006) E. Gartner, Cem. Concr. Res., 2004, 34, 1489.

Oh J.E., Monteiro, P.J.M., Jun S.S., Choi, S., Clark, S.M., (2010) "The evolution of strength and crystalline phases for alkali activated ground blast furnace slag and fly ash-based geopolymers", *Cement and Concrete Research,* 40, 189-196.

Palomo, A., López dela Fuente, J.I. (2003) "Alkali-activated cementitious materials: alternative matrices for the immobilization of hazardous wastes. Part I. Stabilization of boron", *Cement and Concrete Research,* 33, 285–295.

Palou, M., Majling, J. (1995) "Preparation of the high iron sulfoaluminate belite cements from raw mixtures incorporating industrial wastes", *Ceramics – Silikáty,* 39 (2) 41-80

Phair, J.W. (2006) "Green chemistry for sustainable cement production and use", Green Chemistry, 8, 763-780

Polley, C., Cramer, S.M., De La Cruz, R.V. (1998) "Potential for using waste glass in portland cement concrete", *ASCE Journal of Materials in Civil Engineering,* 10, 210-219.

Portland Cement Association (PCA), *Economic Research, Market Data Reports,* www.cement.org (2009).

Quillin, K. (2001) "Performance of belite-sulfoaluminate cements", *Cement and Concrete Research,* 31, 1341-1349

Rajabipour, F., Maraghechi, H., Fischer, G. (2010) "Investigating the alkali silica reaction of recycled glass aggregates in concrete materials", *ASCE Journal of Materials in Civil Engineering,* 22(12) 1201-1208.

Reindl, J. (2003) "Reuse/recycling of glass cullet for non-container uses", http://www.glassonline.com/infoserv/Glass_recycle_reuse/Glass_reuse_recycling_doc1.pdf, Accessed on Nov. 30, 2010.

Roy, D.M. (1999) "Alkali-activated cements: Opportunities and challenges", *Cement and Concrete Research*, 29, 249–254.

Scrivener, K.L., Capmas, A. (1998) "Calcium aluminate cements", in: P.C. Hewlett (Ed.), *Lea's Chemistry of Cement and Concrete*, Elsevier, Ltd., Oxford, UK, pp. 713–782.

Shao, Y.X., Leforta, T., Morasa, S., Rodriguez D. (2000) "Studies on concrete containing ground waste glass", *Cement and Concrete Research*, 30, 91-100

Sharp, J.H., Lawrence, C.D., Yang, R. (1999) "Calcium sulfoaluminate cements - low-energy cements, special cements, or what?", *Advances in Cement Research*, 11 (1) 3-13.

Shayan, A., Xu, A.M. (2006) "Performance of glass powder as a pozzolanic material in concrete: A field trial on concrete slabs", *Cement and Concrete Research*, 36, 457-468

Shi, C., Day, R.L. (1996) "Some factors affecting early hydration of alkali-slag cements", *Cement and Concrete Research*, 26, 439–447.

Shi, C., Krivenko, P.V., Roy, D. (2006), *Alkali-Activated Cements and Concretes*, Taylor and Francis, New York.

Shi, C.J., Wua, Y., Rieflerb, C., Wang, H. (2005) "Characteristics and pozzolanic reactivity of glass powders", *Cement and Concrete Research*, 35, 987-993.

Song, S., Sohn, D., Jennings, H.M., Mason T.O. (2000) "Hydration of alkali-activated ground granulated blast furnace slag", *Journal of Materials Science*, 35, 249-257.

Su, M., Wang, Y., Zhang, L., Li, D. (1997) "Preliminary study on the durability of sulfo/ferroaluminate cements", p. 4iv029 in: *Proceedings of the 10th International Congress on the Chemistry of Cement*, Gothenburg, Sweden, H. Justnes (Ed.).

Touzo B and Espinosa B., Glasser Symposium, http://www.abdn.ac.uk/chemistry/cement-symposium/abstracts, 2009 (accessed 01/02/2010).

U. S. Energy Information Administration, *2010 CO2 Forecast Highlights*, www.eia.doe.gov/international (2010)

United States Environmental Protection Agency (USEPA), Recent Trends in U.S. Greenhouse Gas, Climate Change, Emissions, 2010 Inventory of Greenhouse Gas Emissions and Sinks, www.epa.gov/climatechange/emissions.

US Environmental Protection Agency. 2008. "Inventory of U.S. Greenhouse Gas Emissions and Sinks: 1990-2006", Report No. 430-R-0805.

USGS. 2003. "Green chemistry for sustainable cement production and use", *Green Chemistry*, 8, 763-780.

van Deventer, Jannie S.J., et. al, "The role of research in the commercial development of geopolymer concrete", *International Cement Microscopy Association, March 2010 Conference*, www.cemmicro.org.

Van Ost, Hendrik G, *United States Geological Survey, Cement*, 2002.

World CO2-Emissions Growth Keeps Focus on Coal, *Bloomberg Business Week*, June 10, 2009

www.epa.gov/climatechange/emissions/usgginventory.htmlUS Energy Information Administration. www.eia.doe.gov 2007Schlesinger, Richard. *Energybiz*, Coal Ash Piles Up (April/March 2009).

# 11

# Polyoptimal Multiperiodic Control of Complex Systems with Inventory Couplings Via the Ideal Point Evolutionary Algorithm

Marek Skowron and Krystyn Styczeń
*Wrocław University of Technology*
*Poland*

## 1. Introduction

The paper is devoted to the polyoptimal control of a complex system with inventory couplings, which transfer the by-products of some subsystems to other subsystems as their input components or energy carriers. The cooperation of the subsystems on the recycling may enhance desired ecological features of complex production processes reducing the waste stream endangering the natural environment. The consideration of such systems is connected with the tendency of the rearrangement of complex industrial production systems from an open loop form with many waste products to a closed loop form guaranteeing their beneficial utilization (Ignatenko et al., 2007; Salmiaton & Garforth, 2007; Tan et al., 2008; Tatara et al., 2007; Yi & Luyben, 1996). The networks of interconnected chemical or biochemical reactors can be mentioned as the examples of systems discussed (Diaconescu et al., 2002; Russo et al., 2006; Smith & Waltman, 1995). The recycling problem is analyzed for various operation modes of the networks. Because of the flexible couplings the subsystems have high autonomy degree and can be operated in their own mode. In particular the following three nested operation kinds of the flexibly coupled network can be distinguished (Skowron & Styczeń, 2009): the steady state process (the low intensity production process), the periodic process (the increased intensity production process with the same operation period for all subsystems), and the multiperiodic process (the high intensity production process with different operation periods for the subsystems adjusted to their particular dynamic properties).

Since each of the subsystems has its own objective function composed of the product value, the recycled loop cost, and the waste neutralization cost, the polyoptimal (multiobjective) formulation of the control problem for the complex dynamic network comes to mind. The importance of the search of a compromise solution for a set of conflicting objectives has been widely emphasized in the literature (Huang & Yang, 2001; Sanchis et al., 2008; Sawaragi et al., 1985; Zitzler & Thiele, 1999). The extension of the admissible control processes may yield an essential improvement of the optimal objective function. To this end the periodic dynamic processes of the subsystems (the cycles) are represented by the finite-dimensional vectors encompassing their periods, their initial states, their local controls, and their inventory interactions. The three nested control problems are considered, namely the polyoptimal steady-state control problem, the polyoptimal periodic control problem, and the polyoptimal multiperiodic control problem. The evolutionary optimization algorithm is proposed, which finds the approximations of the polyoptimal ideal point for steady-state, periodic, and multiperiodic processes. The proper dominance for such three polyoptimal points in the objective function space is analyzed. The algorithm is generalized to the case of the improper dominance of the approximated nested ideal points. It uses increased aspiration levels of the objective functions for suitably chosen subsystems.

The application of the evolutionary algorithm to the nested polyoptimization has the advantage of the searching for a globally optimal solutions on each nested stage of the system operation. The easiness of the incorporation of various side constraints including the stability conditions for the optimized control process should also be emphasized (Skowron & Styczeń, 2006). On the other hand the algorithm proposed is time consuming. It deals with dynamic interconnected processes, it evaluates the objective function of the complex recycled systems implementing the globalized Gauss-Newton method for the finding of periodic control processes of the subsystems, and it reconstructs both the averaged control constraints as well as the interaction constraints. This complicates its application for advanced polyoptimal approaches requiring a broad scanning of the Pareto set, the niching technique or the nondominated sorting technique (Audet et al., 2008; Sarkar & Modak, 2005; Tarafder et al., 2005; 2007; Zhang & Li, 2007). In this context the ideal point method shows the advantage of the moderate extent of the computations necessary for the determination of a nested polyoptimal solution.

The theoretical and algorithmic developments are illustrated by the illustrative example of the nested polyoptimization of production processes performed in systems of cross-recycled chemical reactors.

## 2. Polyoptimal multiperiodic control problem for recycled systems

Consider the following polyoptimal multiperiodic control (POMC) problem for systems composed of $N$ subsystems with the inventory couplings (IC): minimize the vector objective function

$$G(z) \doteq (G_1(z), G_2(z), ..., G_N(z)) \tag{1}$$

composed of the $\tau_i$-averaged objective functions of the particular subsystems

$$G_i(z_i) = \frac{1}{\tau_i}\int_0^{\tau_i} g_i(x_i(t), u_i(t), v_i(t))dt \quad (i = 1, 2, ..., N), \tag{2}$$

and subject for $i = 1, 2, ..., N$ to the $\tau_i$-periodic state equations of the subsystems

$$\dot{x}_i(t) = f_i(x_i(t), u_i(t), v_i(t)), \;\; t \in [0, \tau_i], \;\; x_i(\tau_i) = x_i(0), \tag{3}$$

to the resource constraints

$$\frac{1}{\tau_i}\int_0^{\tau_i} u_i(t)dt = b_i, \tag{4}$$

to the stability constraints

$$|s_i(\Phi_i(z_i))|_\infty \leq \alpha_i, \tag{5}$$

to the box constraints

$$\tau_i \in \mathcal{T}_i, \;\; x_i(t) \in X_i, \;\; u_i(t) \in U_i, \;\; v_i(t) \in V_i, \;\; t \in [0, \tau_i], \tag{6}$$

and to the averaged inventory interaction constraints

$$\frac{1}{\tau_i}\int_0^{\tau_i} v_i(t)dt \leq \sum_{j=1}^{N} \frac{1}{\tau_j}\int_0^{\tau_j} h_{ij}(x_j(t), u_j(t), v_j(t))dt, \tag{7}$$

where $\tau_i \in R_+$ is the operation period of the $i$-th subsystem, $x_i \in W_\infty^{1,n_i}(0,\tau_i)$ is its state trajectory, $u_i(t) \in L_\infty^{m_i}(0,\tau_i)$ is its control, $v_i(t) \in L_\infty^{r_i}(0,\tau_i)$ is its inventory interaction, $z_i \doteq (\tau_i, x_i, u_i, v_i)$ is its control process, $b_i \in R^{m_i}$ is its averaged level of the resource availability, and $\mathcal{T}_i \doteq [\tau_i^-, \tau_i^+]$, $X_i \doteq [x_i^-, x_i^+]$, $U_i \doteq [u_i^-, u_i^+]$ and $V_i \doteq [v_i^-, v_i^+]$ are the box sets with the bounds $\tau_i^\pm \in R_+$, $x_i^\pm \in R^{n_i}$, $u_i^\pm \in R^{m_i}$ and $v_i^\pm \in R^{r_i}$, and $s_i \doteq (s_{ij})_{j=1}^{n_i}$ is the vector of the Floquet's multipliers of the state equation (3) i.e. the vector of the eigenvalues of its monodromy matrix $\Phi_i(z_i)$ endowed with the norm $|s_i|_\infty \doteq \max_{j=1,2,...,n_i} |s_{ij}|$, and $\alpha_i \in R_+$ is the local stability F-level of the $i$-th subsystem, and

$$g_i : R^{n_i} \times R^{m_i} \times R^{r_i} \to R, \;\; f_i : R^{n_i} \times R^{m_i} \times R^{r_i} \to R^{n_i},$$

$$h_{ij} : R^{n_j} \times R^{m_j} \times R^{r_j} \to R^{r_i}$$

are continuous functions on the sets $X_i \times U_i \times V_i$ and $X_j \times U_j \times V_j$, respectively, while $z \doteq (z_i)_{i=1}^N$ is the control process of the IC system.

The objective functions $G_i(z_i)$ for the particular systems are combined from such quantities as, for example, the averaged yield of the major products and the by-products, the averaged selectivity of the production process, and the averaged energy consumption or its dissipation.

The dynamics of the subsystems is governed by the $\tau_i$-periodic state equations (3), the periods of which can be chosen independently for each of the subsystems according to their dynamic properties. This is guaranteed by the flexible inventory couplings between the subsystems, which enable to stock up on some output products of the subsystems to recycle them in a complex production system. The inequalities (7) restrict the averaged outflows of the inventory couplings by their averaged inflows.

The constraints (4) mirror the averaged availability of the resources used in the process operation. The relationships (5) are responsible for the local asymptotic stability of periodic control processes for particular subsystems. The constraints for the local stability F-levels are important to ensure practical applicability of optimized processes.

To depict the ideal point evolutionary algorithm implementable for the POMC problem we apply the time scaling $t := \tau_i t$ independently to each subsystem. We reduce this way the IC system to the computationally convenient unit time interval $[0,1]$. We convert the continuous-time control of the $i$-th subsystem $u_i(t)$ and its inventory interaction $v_i(t)$ to the discrete-time form $\tilde{u}_i^k$ and $\tilde{v}_i^k$ for $t \in [k/K, (k+1)/K)$, where $u_i^k \in R^{m_i}$ and $v_i^k \in R^{r_i}$ $(k = 0, 1, ..., K-1)$. We set $u_i^K \doteq (u_i^0, u_i^1, ..., u_i^{K-1})$ and $v_i^K \doteq (v_i^0, v_i^1, ..., v_i^{K-1})$.

We assume that the normalized nonlinear state equation of each subsystem

$$\dot{x}_i(t) = \tau_i f_i(x_i(t), u_i(t), v_i(t)), \;\; t \in [0,1] \tag{8}$$

has the uniquely determined solution $x_i(t, \tau_i, x_i(0), u_i, v_i)$ for every optimization argument $(\tau_i, x_i(0), u_i, v_i)$ satisfying the constraints (6). Thus we can treat the states of the subsystems as the resolvable variables $x_i(t, z_i)$ found by high accuracy integration procedures for nonlinear differential equations with the given initial state and the input functions.

We convert this way the POMC problem to the following normalized and disretized form: minimize the vector objective function

$$G(z) \doteq (G_1(z), G_2(z), ..., G_N(z)) \tag{9}$$

composed of the normalized objective functions of the subsystems

$$G_i(z_i) \doteq \int_0^1 g_i(x_i(t, z_i), \tilde{u}_i(t), \tilde{v}_i(t))dt \quad (i = 1, 2, ..., N) \tag{10}$$

and subject for $i = 1, 2, ..., N$ to the normalized process periodicity constraints

$$x_i(0) - x_i(1, z_i) = 0, \tag{11}$$

to the normalized and discretized resource constraints

$$\frac{1}{K}\sum_{k=0}^{K-1} u_i^k = b_i, \tag{12}$$

to the normalized stability constraints

$$|s_i(\Phi_i(1, z_i))|_\infty \leq \alpha_i, \tag{13}$$

to the normalized box constraints

$$\tau_i \in \mathcal{T}_i,\ x_i(0) \in X_i,\ u_i^k \in U_i,\ v_i^k \in V_i \quad (k = 0, 1, ..., K-1),$$
$$x_i(t_l) \in X_i \quad (t_l \doteq l/L,\ \ l = 1, 2, ..., L), \tag{14}$$

and to the normalized and discretized inventory interaction constraints

$$\frac{1}{K}\sum_{k=0}^{K-1} v_i^k \leq \sum_{j=1}^{N}\int_0^1 h_{ij}(x_j(t, z_j), \tilde{u}_j(t), \tilde{v}_j(t))dt, \tag{15}$$

where an additional dense time grid $\{t_l\}_{l=1}^{L}$ is used in (14) to approximate sufficiently exactly the state constraints within the normalized control horizon, and

$$z_i \doteq (\tau_i, x_i(0), u_i^K, v_i^K) \in R^{M_i} \quad (M_i \doteq 1 + n_i + (m_i + r_i)K)$$

is the discrete representation of a controlled cycle of the $i$-th subsystem encompassing its period, its initial state, its discretized control, and its discretized inventory interaction, while

$$z \doteq (z_i)_{i=1}^{N} \in \prod_{i=1}^{N} R^{M_i}$$

is the normalized discretized control process of the IC system.

Let $\widetilde{Z}$ be the set of all the admissible solutions of the POMC problem, i.e. the set of all the multiperiodic cycles $\widetilde{z}$ satisfying the constraints (11)-(15). We determine the ideal point $\widetilde{G}^* \doteq (\widetilde{G}_1^*, \widetilde{G}_2^*, ..., \widetilde{G}_N^*)$ in the objective space of the POMC problem by the computation of $N$ optimal values of the objective functions of the particular subsystems for the multiperiodic operation of the IC system:

$$\widetilde{G}_i^* = \min_{\widetilde{z} \in \widetilde{Z}} G_i(z) \quad (i = 1, 2, ..., N).$$

We define the compromise multiperiodic solution $\widetilde{z}^*$ for the POMC problem as the solution minimizing the distance to the ideal point

$$\widetilde{z}^* = arg \min_{\widetilde{z} \in \widetilde{Z}} |G(z) - \widetilde{G}^*|_\infty,$$

where the distance is defined with the help of the uniform norm $|G(z) - \widetilde{G}^*|_\infty \doteq \max_{i=1,2,...,N} |G_i(z) - \widetilde{G}_i^*|$.

The multiperiodic control process of the IC system may ensure high productivity of particular subsystems and it may be characterized as the high intensity production process with different operation periods for the subsystems adjusted to their particular dynamic properties. On the other hand its implementation is connected with increased requirements for inventory capacities and their maintenance. It may also have low stability margins for the periodic state trajectories, which involve the need of the design of high quality stabilizing loops for the subsystems.

For such reasons we consider the nested control processes "sitting inside" the multiperiodic control process i.e. the periodic control process and the static control process.

The periodic control process may be interpreted as the synchronized operation mode of the IC system. It requires moderate inventory capacities and facilitates the balancing of the inventory interactions.

The POMC problem is converted to the polyoptimal periodic control (POPC) problem by the setting $\tau_i = \tau \ (i = 1, ..., N)$. Such a choice of the operation periods may be convenient for the balancing of the inventory interactions. It reduces, however, the set of admissible solutions to the set $\tilde{Z}$ of all the periodic cycles $\tilde{z}$ satisfying the constraints (11)-(15) with the equal periods $\tau_i = \tau$. We determine the ideal point $\tilde{G}^* \doteq (\tilde{G}_1^*, \tilde{G}_2^*, ..., \tilde{G}_N^*)$ in the objective space of the POPC problem by the computation of $N$ optimal values of the objectives functions connected with the $\tau$-periodic operation of particular subsystems:

$$\tilde{G}_i^* = \min_{\tilde{z} \in \tilde{Z}} G_i(z) \quad (i = 1, 2, ..., N).$$

We define the compromise periodic solution $\tilde{z}^*$ for the POPC problem as the solution minimizing the uniform distance to the ideal point

$$\tilde{z}^* = arg \min_{z \in \tilde{Z}} |G(z) - \tilde{G}^*|_\infty.$$

Fixing in time all the process variables leads to the simplified system with direct interconnections and without inventories. The steady-state control processes may be implemented with the help of simple stabilization loops, for example, of relay type. However, such processes ignore the optimization potential underlying in the process dynamics.

The POPC problem is converted to the polyoptimal steady-state control (POSS) problem by the fixing in time all the process variables, which is equivalent to the minimization of the vector steady-state objective function

$$G(\bar{z}) \doteq (G_1(\bar{z}), G_2(\bar{z}), ..., G_N(\bar{z})) \tag{16}$$

having the components

$$G_i(\bar{z}) \doteq g_i(\bar{x}_i, \bar{u}_i, \bar{v}_i) \tag{17}$$

and subject for $i = 1, ..., N$ to the steady-state constraints

$$f_i(\bar{x}_i, \bar{u}_i, \bar{v}_i) = 0, \tag{18}$$

$$B_i \bar{u}_i = b_i, \tag{19}$$

$$|s_i(e^{f_{i,x_i}(\bar{x}_i, \bar{u}_i, \bar{v}_i)})|_\infty \leq \alpha_i, \tag{20}$$

$$\bar{x}_i \in X_i, \;\; \bar{u}_i \subset U_i, \;\; \bar{v}_i \subset V_i, \tag{21}$$

$$\bar{v}_i \leq \sum_{j=1}^{N} g_j(\bar{x}_j, \bar{u}_j, \bar{v}_j), \tag{22}$$

where

$$\bar{z}_i \doteq (\bar{x}_i, \bar{u}_i, \bar{v}_i)_{i=1}^N \in R^{n_i} \times R^{m_i} \times R^{r_i}$$

is the steady-state control process of the $i$-th subsystem, and

$$\bar{z} \doteq (\bar{z}_i)_{i=1}^N \in \prod_{i=1}^{N} R^{n_i} \times R^{m_i} \times R^{r_i}$$

is the steady-state control process for the IC system.

Let $\bar{Z}$ be the set of all the admissible solutions of the POSS problem, i.e. the set of all the steady-state processes $\bar{z}$ satisfying the constraints (18)-(22). We determine the ideal point $\bar{G}^* \doteq (\bar{G}_1^*, \bar{G}_2^*, ..., \bar{G}_N^*)$ in the objective space of the POSS problem by the computation of $N$ optimal values of the objectives functions connected with the steady-state operation of particular subsystems:

$$\bar{G}_i^* = \min_{\bar{z} \in \bar{Z}} \bar{G}_i(\bar{z}) \;\; (i = 1, 2, ..., N).$$

We define the compromise steady-state solution $\bar{z}^*$ for the POSS problem as the solution minimizing the distance to the ideal point

$$\bar{z}^* = arg \min_{\bar{z} \in \bar{Z}} |G(\bar{z}) - \bar{G}^*|_\infty.$$

**Definition 1:** The triple of compromise nested control processes $z^* \doteq (\tilde{z}^*, \hat{z}^*, \bar{z}^*)$ is said to be

- strongly proper if it satisfies the relationships

$$G(\tilde{z}^*) < G(\hat{z}^*) < G(\bar{z}^*),$$

- partially strongly proper if it satisfies the relationships

$$G(\tilde{z}^*) \leqslant G(\hat{z}^*) < G(\bar{z}^*) \;\text{ or }\; G(\tilde{z}^*) < G(\hat{z}^*) \leqslant G(\bar{z}^*),$$

- proper if it satisfies the relationships

$$G(\tilde{z}^*) \leqslant G(\hat{z}^*) \leqslant G(\bar{z}^*),$$

- weakly proper if it satisfies the relationships

$$G(\tilde{z}^*) \leq G(\hat{z}^*) \leqslant G(\bar{z}^*) \;\text{ or }\; G(\tilde{z}^*) \leqslant G(\hat{z}^*) \leq G(\bar{z}^*),$$

• improper if it satisfies the relationship

$$G(\bar{z}^*) \lesseqgtr G(\tilde{z}^*),$$

where the vector inequality $\leqslant$ means that the inequality $\leq$ holds for all the components with the strict inequality for some of them, and $\lesseqgtr$ means that higher level compromise solutions may improve objective functions for some subsystems at polyoptimal static solution, but deteriorate for other subsystems at this solution.

We are aimed at the comparison of the ideal point compromise solutions of the POMC problem for the steady-state processes, for the periodic processes, and for the multiperiodic processes. The finding of the strongly proper nested triple $z^*$ means the uniform improvement of the ideal point compromise solutions between all the levels of nested optimization problem. It may be the basis for the application of the compromise multiperiodic control process. The other types of the nested triple $z^*$ determine weaker possibilities of the polyoptimal nested optimization of the IC system. The practitioner choosing a definitive process for the implementation takes into account the degree of the improvement of the objective functions for the subsystems between the nested compromise solutions.

## 3. Ideal point evolutionary polyoptimal multiperiodic optimization

The general scheme of the ideal point evolutionary algorithm for the nested polyopimal multiperiodic optimization can be stated as follows :

**Algorithm 1:** Finding of the ideal point compromise solution for the POMC problem.

*Step 1:* Choose randomly an initial steady-state control process population $\bar{z}^0 \doteq (\bar{x}_i^0, \bar{u}_i^0, \bar{v}_i^0)_{i=1}^N$ and apply the evolutionary global optimization (EGO) algorithm of (Skowron & Styczeń, 2009) to solve $N$ single objective static optimization problems

$$\min_{\bar{z}\in\bar{Z}} \bar{G}_i(\bar{z}) \quad (i = 1, 2, ..., N)$$

to obtain the ideal point $\bar{G}^* \doteq (\bar{G}_1^*, \bar{G}_2^*, ..., \bar{G}_N^*)$ in the objective functions space of the POSS problem. Apply the EGO algorithm to find the ideal point compromise solution for the POSS problem

$$\bar{z}^* = arg \min_{\bar{z}\in\bar{Z}} |G(\bar{z}) - \bar{G}^*|_\infty.$$

*Step 2:* Using the $\pi$-test ((Bernstein & Gilbert, 1980; Sterman & Ydstie, 1991)) evaluate the intervals $[\tau_{i-}, \tau_{i+}]$ of period values guaranteeing the local improvement of the subsystems objective functions by the periodic operation, and choose $\tau^0 \in [\tau_{i-}, \tau_{i+}]$ $(i = 1, 2, ..., N)$.

*Step 3:* Choose randomly an initial periodic control process population $\tilde{z}^0 = (\tau^0, \tilde{x}_i^0, \tilde{u}_i^0, \tilde{v}_i^0)_{i=1}^N$ and apply the evolutionary global optimization (EGO) algorithm of (Skowron & Styczeń, 2009) to solve $N$ single objective periodic optimization problems

$$\min_{\tilde{z}\in\tilde{Z}} \tilde{G}_i(\tilde{z}) \quad (i = 1, 2, ..., N)$$

to obtain the ideal point $\tilde{G}^* \doteq (\tilde{G}_1^*, \tilde{G}_2^*, ..., \tilde{G}_N^*)$ in the objective functions space of the POPC problem. Apply the EGO algorithm to find the ideal point compromise solution $\tilde{z}^* = (\tau^*, \tilde{x}_i^*, \tilde{u}_i^*, \tilde{v}_i^*)_{i=1}^N$ for the POPC problem as

$$\tilde{z}^* = arg \min_{\tilde{z}\in\tilde{Z}} |G(\tilde{z}) - \tilde{G}^*|_\infty.$$

*Step 4:* Choose randomly an initial multiperiodic control process population $\tilde{z}^0 = (\tau^0, \tilde{x}_i^0, \tilde{u}_i^0, \tilde{v}_i^0)_{i=1}^N$ and apply the evolutionary global optimization (EGO) algorithm to solve $N$ single objective multiperiodic optimization problems

$$\min_{\tilde{z} \in \tilde{Z}} \tilde{G}_i(\tilde{z}) \ \ (i = 1, 2, ..., N)$$

to obtain the ideal point $\tilde{G}^* \doteq (\tilde{G}_1^*, \tilde{G}_2^*, ..., \tilde{G}_N^*)$ in the objective functions space of the POMC problem. Apply the EGO algorithm to find the ideal point compromise solution for the POMC problem

$$\tilde{z}^* = arg \min_{\tilde{z} \in \tilde{Z}} |G(\tilde{z}) - \tilde{G}^*|_\infty.$$

*Step 5:* Determine the properness of the determined nested triple $\mathbf{z}^*$ on the basis of the Definition 1.

If the nested triple $\mathbf{z}^*$ turns out to be improper the following regularization may improve its properness.

**Algorithm 2:** Combined the ideal point compromise solution and aspiration levels approach for the POMC problem.

*Step 1:* Modify the set of admissible solutions for the POPC problem as follows:

$$\underline{\bar{Z}} \doteq \{\bar{z} \in \bar{Z} : G_i(\bar{z}) \leq G_i(\bar{z}) - \bar{\Delta}_i \ \ (i \in \bar{\mathcal{N}}),$$

where $\bar{\mathcal{N}} \subset \{1, 2, ..., N\}$ is the set of indices of the subsystems, the objective functions of which are deteriorated by the ideal point compromise solution of the POPC problem at the point $\bar{\mathbf{z}}^*$, and the corrections $\bar{\Delta}_i > 0$ determine the aspiration levels for the subsystems with deteriorated objective functions on the periodic optimization level.

*Step 2:* Apply the EGO algorithm to find the combined ideal point compromise and aspiration levels solution $\underline{\bar{z}}^* = (\underline{\tau}^*, \underline{\bar{x}}_i^*, \underline{\bar{u}}_i^*, \underline{\bar{v}}_i^*)_{i=1}^N$ for the POPC problem as

$$\underline{\bar{z}}^* = arg \min_{\underline{\bar{z}} \in \underline{\bar{Z}}} |G(\underline{\bar{z}}) - \bar{G}^*|_\infty.$$

*Step 3:* Modify the set of admissible solutions for the POMC problem as follows:

$$\underline{\tilde{Z}} \doteq \{\tilde{z} \in \tilde{Z} : G_i(\tilde{z}) \leq G_i(\tilde{z}) - \tilde{\Delta}_i \ \ (i \in \tilde{\mathcal{N}}),$$

where $\tilde{\mathcal{N}} \subset \{1, 2, ..., N\}$ is the set of indices of the subsystems, the objective functions of which are deteriorated by the ideal point compromise solution of the POMC problem at the point $\tilde{\mathbf{z}}^*$, and the corrections $\tilde{\Delta}_i > 0$ determine the aspiration levels for the subsystems with deteriorated objective functions on the multiperiodic optimization level.

*Step 4:* Apply the EGO algorithm to find the combined ideal point compromise and aspiration levels solution $\underline{\tilde{z}}^* = (\underline{\tau}_i^*, \underline{\tilde{x}}_i^*, \underline{\tilde{u}}_i^*, \underline{\tilde{v}}_i^*)_{i=1}^N$ for the POMC problem as

$$\underline{\tilde{z}}^* = arg \min_{\underline{\tilde{z}} \in \underline{\tilde{Z}}} |G(\underline{\tilde{z}}) - \tilde{G}^*|_\infty.$$

Of course it may suffice to solve one of the corrected problems POPC or POMC.

From the evolutionary algorithm perspective, the most important thing is the way of coding an individual. We propose to represent an individual of the problem (9)-(15) by the vector (Skowron & Styczeń, 2009) $\check{z} \doteq (\check{z}_\mu)_{\mu=1}^{\check{N}}$, where $\check{N} = \sum_{i=1}^{N} M_i$, $\check{z}_i \doteq \tau_i$ $(i = 1,2,...,N)$ are operation periods of all subsystems, $\check{z}_{j+N+\sum_{l=1}^{i-1} n_l} \doteq x_{ij}(0)$ $(j = 1,2,...,n_i;\ i = 1,2,...,N)$ are coordinates of initial states of all subsystems, $\check{z}_{k+1+(j-1)K+K\sum_{l=1}^{i-1} m_l+N+\sum_{l=1}^{N} n_l} \doteq u_{ij}^k$ $(k = 0,1,...,K-1;\ j = 1,2,...,m_i;\ i = 1,2,...,N)$ are discrete-time control coordinates of all subsystems, $\check{z}_{k+1+(j-1)K+K\sum_{l=1}^{i-1} r_l+K\sum_{l=1}^{N} m_l+N+\sum_{l=1}^{N} n_l} \doteq v_{ij}^k$ $(k = 0,1,...,K-1;\ j = 1,2,...,r_i;\ i = 1,2,...,N)$ are discrete-time inventory interactions of all subsystems. Values of the individual's genes are bounded by the set $Z \doteq [\check{z}^-, \check{z}^+]$ $(\check{z}^\pm \in R^{\check{N}})$, which results from a set of inclusion constraints $\mathcal{T} \times X \times U \times V$.

The form of the individual $\check{z}$ allows to use known crossing and mutation operators (Michalewicz, 1996). However, basing on several experiments we performed, we propose to use a uniform crossing operator and a non-uniform mutation operator. Unfortunately the available operators deliver an individual which violate the constraints (11)-(15). In case of the periodic constraint (11) we propose to use the Newton method as the reconstruction algorithm. The averaged constraints (12) and (15) can be preserved with the help of the reconstruction algorithm described by Skowron and Styczeń (Skowron & Styczeń, 2009).This reconstruction algorithm is not sufficient for the inventory interaction (15) constraint and that's why it shall be used together with the penalty term. The penalty term shall be also applied for stability (13) and box state (14) constraints (Skowron & Styczeń, 2006).

## 4. Illustrative example

Let two continuous stirred tank reactors cooperate with the help of the inventory interactions of the mixed catalytic-resource type. The series reaction $A_1 \leftrightarrows B_1 \rightarrow C_1$ takes place in the first reactor, and the parallel reactions $A_2 \leftrightarrows B_2$ and $A_2 \rightarrow C_2$ take place in the second reactor, where the main reactions are reversible, and $A_i$ is the raw material of the $i$th reactor, $B_i$ is its desired product, and $C_i$ is its by-product. Assume that the $i$-th reactor is $\tau_i$-periodically operated, and denote by $x_{i1}(t), x_{i2}(t), x_{i3}(t)$ its concentrations of $A_i, B_i, C_i$, respectively, by $u_{i1}(t)$ its input concentration of $A_i$, by $u_{i2}(t)$ its flow rate, by $v_i(t)$ its inventory interaction. The interaction of the first reactor uses the by-product of the second reactor as the catalyst of the own reactions, while the interaction of the second reactor uses the by-product of the first reactor as the supplement of the own raw material. Consider the following POMC problem for the discussed system: minimize the vector objective function

$$G(z) \doteq (G_1(z), G_2(z))$$

having the components

$$G_i(z) \doteq -\frac{1}{\tau_i}\int_0^{\tau_i} u_{i2}(t)x_{i2}(t)dt,$$

and subject to the $\tau_1$-periodic state equations of the first subsystem

$$\dot{x}_{11}(t) = u_{12}(t)(u_{11}(t) - x_{11}(t)) - \kappa_{11}v_1(t)^2 x_{11}^{p_{11}}(t) + \kappa_{12}v_1(t)x_{12}^{p_{12}}(t),$$

$$\dot{x}_{12}(t) = -u_{12}(t)x_{12}(t) + \kappa_{11}v_1(t)^2 x_{11}^{p_{11}}(t) - \kappa_{12}v_1(t)x_{12}^{p_{12}}(t) - \kappa_{13}v_1(t)x_{12}^{p_{13}}(t),$$

$$\dot{x}_{13}(t) = -u_{12}(t)x_{13}(t) + \kappa_{12}v_1(t)x_{12}^{p_{13}}(t),$$

to the $\tau_2$-periodic state equations of the second subsystem

$$\dot{x}_{21}(t) = u_{22}(t)(u_{21}(t) + v_2(t) - x_{21}(t)) - \kappa_{21}x_{21}^{p_{21}}(t) + \kappa_{22}x_{22}^{p_{22}}(t) - \kappa_{23}x_{21}/(1 + x_{21}),$$

$$\dot{x}_{22}(t) = -u_{22}(t)x_{22}(t) + \kappa_{21}x_{21}^{p_{21}}(t) - \kappa_{22}x_{22}^{p_{22}}(t),$$
$$\dot{x}_{23}(t) = -u_{22}(t)x_{23}(t) + \kappa_{23}x_{21}/(1 + x_{21}),$$

to the resource constraints

$$\frac{1}{\tau_i}\int_0^{\tau_j} u_{ij}(t)dt = 1 \ \ (i,j = 1,2),$$

to the box constraints

$$\tau_i \in [0.1, 20], \ \ 0 \le x_{ik}(t) \ \ (i = 1,2; k = 1,2,3),$$

$$0 \le u_{ij}(t) \le 2, \ \ 0 \le v_i(t) \le 2 \ \ (j = 1,2), \ \ t \in [0, \tau_i],$$

to the stability constraints

$$|s_i(\Phi_i(\tau_i, x_i, u_i, v_i))|_\infty \le 0.8 \ \ (i = 1,2),$$

and to the inventory interaction constraints

$$\frac{1}{\tau_1}\int_0^{\tau_1} v_1(t)dt \le \frac{1}{\tau_2}\int_0^{\tau_2} x_{23}(t)dt,$$
$$\frac{1}{\tau_2}\int_0^{\tau_2} v_2(t)dt \le \frac{1}{\tau_1}\int_0^{\tau_1} x_{13}(t)dt,$$

where the reactions obey the power law with the exponents $p_{ij}$. The optimization goal is equivalent to the maximization of the averaged yield of the useful product for each of the reactors. We compare the nested polyoptimal steady-state, periodic, and multiperiodic control processes for such cross-recycled reactors.

The evaluation of the initial advantageous duration of the operation periods for the subsystems can be found with the help of the $\pi$-test. Assuming the unit mean value of the inventory interactions and the subsystem parameters $n_i = 3, m_i = 2, p_{i1} = 2, p_{i2} = 1, p_{13} = 1.5, \kappa_{11} = 40, \kappa_{12} = 12, \kappa_{13} = 10, \kappa_{21} = 30, \kappa_{22} = 15, \kappa_{23} = 2$ we obtain the $\pi$-curves for the subsystems (Fig. 1) with the suboptimal operation periods $\tau_1 = 2.5$, $\tau_2 = 3.1$. The form of the $\pi$-curves shows that optimal operation periods for the case considered should be searched within the intermediate frequencies.

| | Ideal point | | | Compromise solution | | |
|---|---|---|---|---|---|---|
| | POSS | POPC | POMC | POSS | POPC | POMC |
| $G_1^*$ | -0.3399 | -0.7352 | -0.7442 | -0.2846 | -0.5143 | -0.6461 |
| $G_2^*$ | -0.4693 | -2.1912 | -2.1914 | -0.4109 | -2.0758 | -2.0854 |

Table 1. The values of the objective functions obtained with the help of evolutionary algorithm

Table 1 shows the results which were achieved with the aid of the evolutionary algorithm described Skowron and Styczeń (Skowron & Styczeń, 2009). It is easy to notice that applying the periodic control for the considered system of two continuous stirred tank reactors cooperated with the help of the inventory interactions significantly improves the productivity

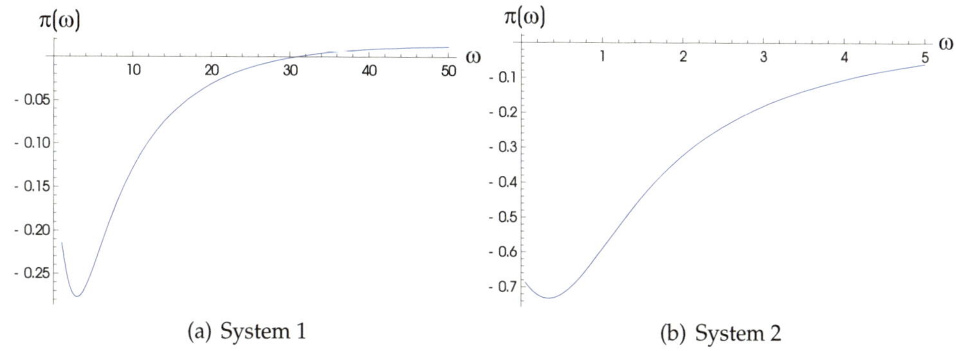


Fig. 1. The results of $\pi$–test

of the system in comparison to the steady-state approach. Comparing ideal points for POSS and POPC problems we can see that productivity of the first system is improved about 116%. Much greater improvement is observed for the second system. Applying the periodic control improves that the productivity of the second about 366%.

The results confirm also that efficiency of the system can be increased by applying multiperiodic control. The ideal point of the first system is improved about 1.2% and for the second system we see improvement equal 0.009%. The improvement after applying the multiperiodic control is not such spectacular like for the case when the steady-state control is replaced by the periodic control. But for some systems improvement of the efficiency of the process about 1-2% can give very huge economical profits.

Encouraged by observed improvement after applying multiperiodic control we calculated also the compromise solution (Table 1, Fig. 2-5). For the first system the improvement is about 25% and for the second system is about 0.46%. We see that received compromise solution for POMC problem is strongly proper according Definition 1. Thus these results confirm that for the considered system of two continuous stirred tank reactors cooperated with the help of the inventory interactions it is wise to apply multiperiodic approach.

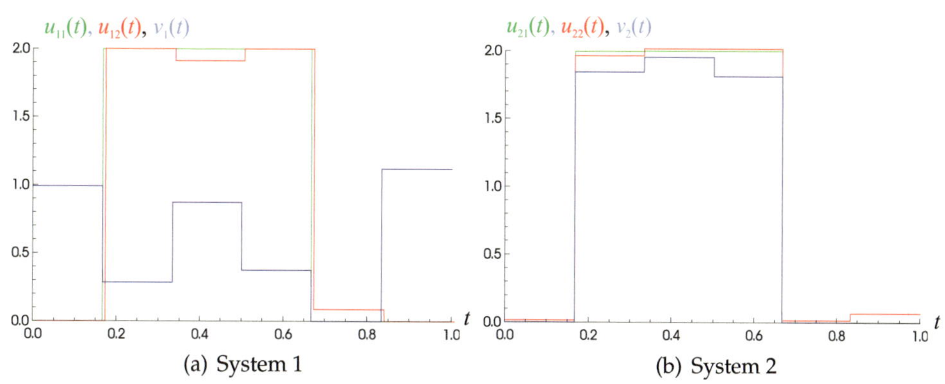


Fig. 2. The optimal control $\tilde{u}_{ij}^*(t)$ and the inventory interaction $\tilde{v}_i^*(t)$ $(i,j = 1,2)$ for POPC problem

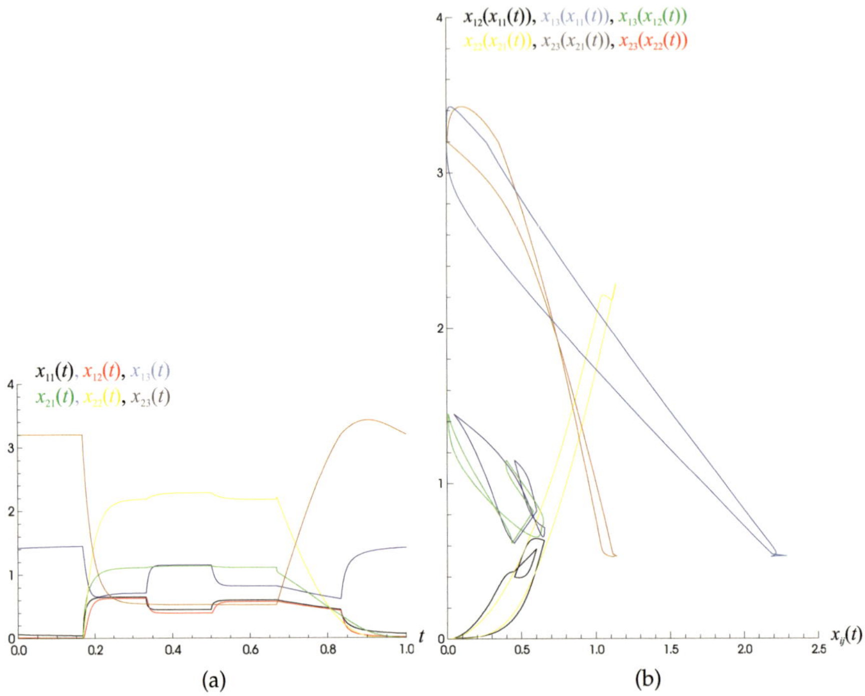


Fig. 3. The optimal state $\tilde{x}_{ij}^{*}(t)$ $(i = 1, 2; j = 1, 2, 3)$ for POPC problem

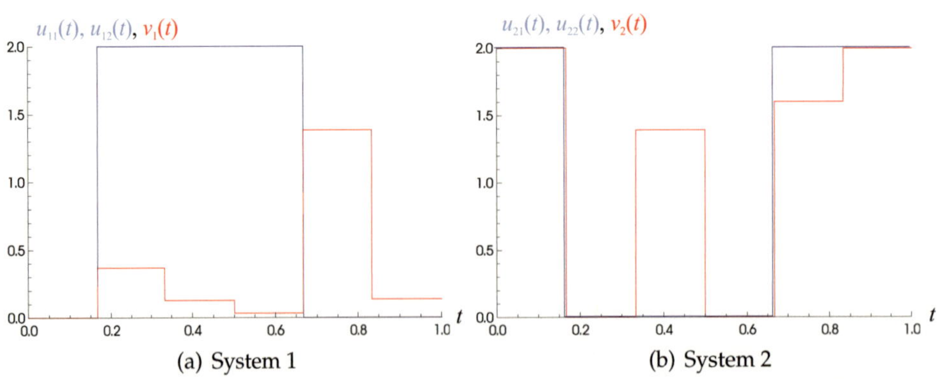


Fig. 4. The optimal control $\widetilde{u}_{ij}^{*}(t)$ and the inventory interaction $\widetilde{v}_{i}^{*}(t)$ $(i, j = 1, 2)$ for POMC problem

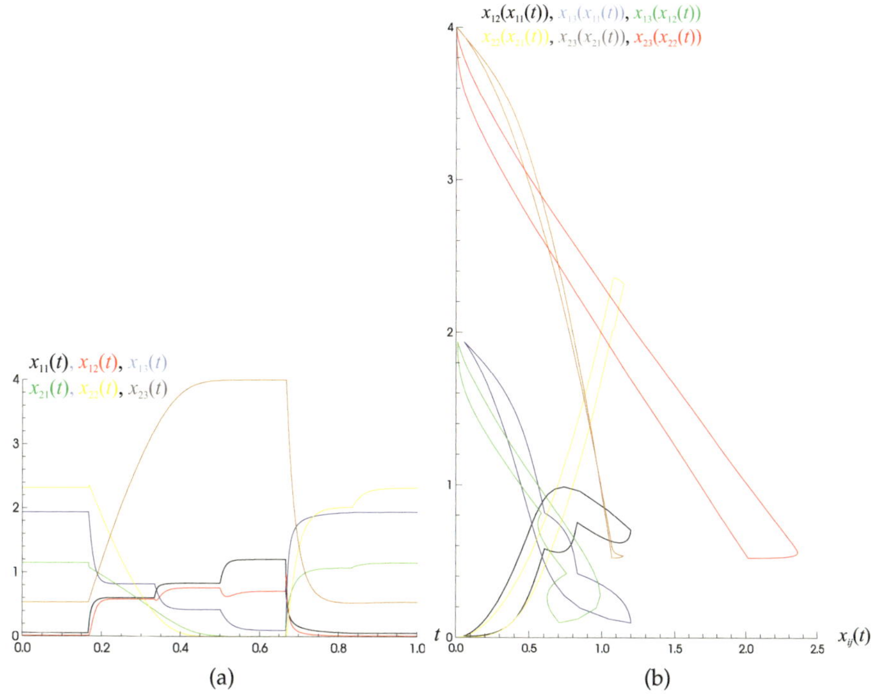

Fig. 5. The optimal state $\tilde{x}^*_{ij}(t)$ $(i = 1,2; j = 1,2,30$ for POMC problem

## 5. Conclusion

The polyoptimal multiperiodic control problem for complex systems with the inventory couplings was analysed. The ideal point evolutionary algorithm was proposed for the solving of this problem. It has been shown that the multiperiodic operation of the complex cross-recycled chemical production systems may ensure the uniform improvement of the vector objective function as compared with the steady-state operation, and with the periodic operation. Such polyoptimal solution may be preferred by practitioners. The method applied shows the advantage of the moderate extent of the computational effort necessary for the finding of a best compromise solution. The solution obtained this way may be further exploited as the starting point for the implementation of some improved nested multiobjective optimization based, for example, on the verification of the attainability of the given aspiration levels for particular objective functions.

## 6. References

Audet, C., Savard, G. & Zghal, W. (2008). Multiobjective optimization through a series of single-objective formulations., *SIAM Journal on Optimization* 19: 188–210.

Bernstein, D. S. & Gilbert, E. G. (1980). Optimal periodic control: The $\pi$ test revisited, *IEEE Transactions on Automatic Control* AC-25: 673–684.

Diaconescu, R., Tudose, R. Z. & Curteanu, S. (2002). A case study for optimal reactor networks synthesis: Styrene polymerization, *Polymer-Plastics Technology and Engineering* 41: 297–326.

Huang, X. X. & Yang, X. Q. (2001). Asymptotic analysis of a class of nonlinear penalty methods for constrained multiobjective optimization., *Nonlinear Analysis* 47: 5573–5584.

Ignatenko, O., Schaik, A. & Reuter, M. (2007). Exergy as a tool for evaluation of the resource efficiency of recycling systems., *Minerals Engineering* 20: 862–874.

Michalewicz, Z. (1996). *Genetic Algorithms + Data Structures = Evolution programs.*, Springer–Verlag, Berlin.

Russo, L., Altimari, P., Mancusi, E., Maffettone, P. L. & Crescitelli, S. (2006). Complex dynamics and spatio-temporal patterns in a network of three distributed chemical reactors with periodical feed switching, *Chaos, Solitons & Fractals* 28: 682–706.

Salmiaton, A. & Garforth, A. (2007). Waste catalysts for waste polymer, *Waste Management* 27: 1891–1896.

Sanchis, J., Martinez, M., Blasco, X. & V.Salcedo, J. (2008). A new perspective on multiobjective optimization by enhanced normalized normal constraint method., *Structural Multidisciplinary Optimization* 36: 537–546.

Sarkar, D. & Modak, J. M. (2005). Pareto-optimal solutions for multi-objective optimization of fed-batch bioreactors using nondominated sorting genetic algorithm., *Chemical Engineering Science* 60: 481–492.

Sawaragi, Y., Nakayama, H. & Tanino, T. (1985). *Theory of Multiobjective Optimization.*, Academic Press, Inc., London.

Skowron, M. & Styczeń, K. (2006). Evolutionary search for globally optimal constrained stable cycles., *Chemical Engineering Science* 61: 7924–7932.

Skowron, M. & Styczeń, K. (2009). Evolutionary search for globally optimal stable multicycles in complex systems with inventory couplings., *International Journal of Chemical Engineering* 2009: 10.

Smith, H. & Waltman, P. (1995). *The Theory of the Chemostat. Dynamics of Microbial Competition*, University Press, Cambridge.

Sterman, L. & Ydstie, B. (1991). Periodic forcing of the cstr: an application of generalized $\pi$-criterion, *A.I.Ch.E. Journal* 37: 986–996.

Tan, R. R., Col-long, K. J., Foo, D. C. Y., Hul, S. & Ng, D. K. S. (2008). A methodology for the design of efficient resource conservation networks using adaptive swarm intelligence, *Journal of Cleaner Production* 16: 822–832.

Tarafder, A., Rangaiah, G. P. & Ray, A. K. (2005). Multiobjective optimization of an industrial styrene monomer manufacturing process., *Chemical Engineering Science* 60: 347–363.

Tarafder, A., Rangaiah, G. P. & Ray, A. K. (2007). A study of finding many desirable solutions in multiobjective optimization of chemical processes., *Computers and Chemical Engineering* 31: 1257–1271.

Tatara, E., Teymour, F. & Cinar, A. (2007). Control of complex distributed systems with distributed intelligent agents, *Journal of Process Control* 17: 415–427.

Yi, C. K. & Luyben, W. L. (1996). Design and control of coupled reactor/column systems - part 3. a reactor/stripper with two columns and recycle., *Computers and Chemical Engineering* 21: 69–86.

Zhang, Q. & Li, H. (2007). MOEA/D: A multiobjective evolutionary algorithm based on decomposition., *IEEE Transactions on Evolutionary Computation* 11: 712–731.

Zitzler, E. & Thiele, L. (1999). Multiobjective evolutionary algorithms: A comparative case study and the strength Pareto approach., *IEEE Transactions on Evolutionary Computation* 3: 257–271.

# 12

# Acidogenic Valorisation of High Strength Waste Products from Food Industry

Luís Arroja[1], Isabel Capela[1], Helena Nadais[1],
Luísa S. Serafim[2] and Flávio Silva[1]
[1]*CESAM, Department of Environment and Planning,*
[2]*CICECO, Department of Chemistry,*
*University of Aveiro,*
*Portugal*

## 1. Introduction

Awareness concerning the environment and the use of clean technologies is growing worldwide. As a consequence, research on biodegradability and use of renewable resources for industrial processes has been intensive in the last years. In addition, product and process innovation in food industries is widely regarded as an essential element of competition between food companies, as it will improve their business performance. To face these new challenges, relevant changes in the food manufacturing processes may occur, involving for example the use of different raw materials as an ingredient in new food products. As a result, food processing wastewater will contain complex organic compounds with refractory properties that may compromise the biological treatment processes already existing in the wastewater treatment plants (WWTP) and consequently the fulfilment of legal requirements for wastewater discharge. To overcome this situation, segregation of specific wastewater process streams with refractory characteristics and its individual pre-treatment will be important for a viable and stable treatment in the already existing WWTP.

In this context, acidogenic fermentation is gaining scientific and commercial interest, since it allows improved wastewater treatability as well as additional recovery of bioproducts. Phase separation in anaerobic systems has been studied to enhance treatment efficiencies and/or generate bioproducts from different industrial wastewaters. Several studies contributed to a first technological shift, since optimum growing conditions for acidogenic microorganisms have been developed to treat complex organic substrates that demand for high hydrolytic efforts.

Phase separation in the anaerobic digestion process implies a process configuration employing separate reactors for acidification and methanogenesis connected in series, allowing optimisation of each process separately. Many types of microorganisms and biochemical pathways are involved in the acidogenesis and consequently a large number of bioproducts are usually formed, including carboxylic acids and alcohols.

Currently, acidogenic processes are widely spread among bioreactor engineered technology, and are usually designed for customised applications and for particular high strength waste products. Although methane is usually considered the final product of anaerobic degradation, there are some other valuable by-products which can compete with methane and has a market for itself, such as volatile fatty acids (VFA) that can constitute a valuable resource for biodegradable polymer production (poly-hydroxyalkanoates - PHA) or biological nutrient removal.

The acidogenic microorganisms have specific physiology, nutritional requirements, optimal pH, growth and substrate uptake kinetics. Although global and two-phase anaerobic processes are well studied, little information exists on adequate design and operation of acidogenic digesters. However, higher kinetic rates for acidogenic metabolism may be the most interesting feature on industrial exploitation of this process.

It has been demonstrated that the composition of VFA produced under acidogenic fermentation can be affected by environmental conditions such as pH, retention time, and temperature. In the two-phase anaerobic processes, the physical separation between acid-formers and methanogenic microorganisms leads to the absence of hydrogen uptake by the methanogens, which will lead to hydrogen accumulation and subsequent alteration of the fermentation pathways during the acidogenic phase. So, there will be a shift towards the utilization of a metabolic pathway that does not use proton reduction as a mean of disposing electrons, hence resulting in the production of high amounts of butyrate as an important electron sink product (Fernandes, 1986). Nowadays, this is one of the starting points for the scientific and technological interest on the acidogenic phase in order to understand the nature of metabolic pathways and the energetic content of key-intermediates, which will result in different specific turn-over rates. So, further research is needed in order to establish relations between process conditions during the acidogenic fermentation and VFA production for different potential feedstock.

This chapter aims to contribute to the use of acid-phase anaerobic digestion of high strength waste products as a practical and economically feasible waste management option together with the recovery of valuable organic acids. Hence, the objective of this chapter is to examine acidogenic fermentation of four organic residues (sugar cane molasses, wastewater containing spent coffee grounds, dairy processing fatty slurry and cheese whey), from the point of view of their potential use as feedstock, because VFA produced in the process have a variety of industrial uses. Therefore this chapter intends to highlight acidification as a powerful tool for organic waste streams treatment and management.

### 1.1 Acidified products as raw materials for PHA production

PHA are polyesters of hydroxyalkanoates stored by more than 100 bacterial genus as sources of carbon/energy or reducing-power. PHA are synthesized and accumulated as intracellular granules usually when there is a limitation in essential component for growth in the presence of excess carbon source. Depending on the substrate provided, microorganisms can include a wide variety of 3-hydroxy fatty acids in the PHA (Braunegg et al., 1998). One of the most widely studied PHA is poly-3-hydroxybutyrate (PHB). PHB is very brittle and crystalline, which makes difficult its utilization in applications that require more flexibility. The incorporation of different monomer units, other than hydroxybutyrate

(HB), in the polymer chain, results in the synthesis of copolymers with improved mechanical properties (Braunegg et al., 1998). PHA are currently produced at the industrial scale using pure microbial cultures and expensive substrates, both contributing to high selling prices. The substrate is one of the main cost factors of PHA production corresponding to 40% of production costs (Reis et al., 2003). The industrial processes developed so far utilize pure cultures or recombinant microorganisms that require very tight sterility conditions and control contributing for high production costs. The use of waste organic carbon and mixed cultures can significantly reduce the price of PHA. The utilization of mixed microbial cultures (MMC) facilitates the use of complex substrates, since microbial population can adapt continuously to changes in substrate. Selection of microorganisms occurs on the basis of its high capacity for PHA storage. Consequently, there is no need for sterile fermentation systems, which contributes to the reduction of the final PHA price (Serafim et al., 2008a).

PHA production in MMC occurs when unbalanced growth conditions are verified, especially those resulting from the periodic absence of an essential nutrient. The most studied processes result from the alternation of oxygen or carbon availability. The former situation, typical in biological nutrient removal systems, usually occurs when the microbial community is submitted to alternating anaerobic (AN) and aerobic (AE) conditions with the carbon substrate supplied at beginning of the AN period. Under AN conditions the microorganisms cannot grow due to the lack of electronic acceptor and some of the bacteria present start to store the external carbon as PHA with simultaneous glycogen degradation. Glycogen, the second internal polymer involved, works as a sink of equivalent reductors. When the AE period begins, stored PHA will be consumed for microbial growth and to replenish the glycogen pool (Serafim et al., 2008a).

Storage of PHA by MMC occurs also when they are subject to transient carbon supply resulting from the alternation of long periods of lack of substrate (famine) with periods with a short time of excess of substrate (feast). During the latter period, substrate uptake is mainly directed to PHA storage and, to a lesser extent, to the biomass growth. After substrate exhaustion, the stored polymer is used as energy and carbon source. This process is known as aerobic dynamic feeding (ADF) or feast and famine. Carbon limitation for a long period causes changes in the macromolecular composition of cells, requiring a physiological adaptation of microorganisms when exposed to high substrate concentration. The dominance of PHA-storing organisms will result from the selective pressure imposed by the operational conditions (Serafim et al., 2008a).

In both situations the carbon substrate should be composed by VFA. In order to obtain a stream rich in these compounds, acidification of the raw material selected for PHA production by MMC is required. The composition of VFA obtained after the acidification step is critical for the development of the process, because not only influences the type of polymer produced but also the microbial composition of the selected population.

By supplying different types of VFA to the culture, PHA with different monomer composition were obtained. Not only in the ADF system using synthetic substrates (Lemos et al., 2006; Serafim et al., 2008b) or a real fermented substrate, like fermented sugar cane molasses (Albuquerque et al., 2007) but also in the AN/AE process with fermented molasses (Pisco et al., 2009). In most of the cases, a feeding stream rich in VFA with even carbon atoms, such as acetate and butyrate resulted most of the times in the formation of PHB homopolymers. The predominance of VFA with odd carbon atoms composition, such as propionate or valerate led to a PHA copolymer with a higher content in 3-hydroxyvalerate (3HV), which is much more interesting from a commercial point of view (Lemos et al., 2006). Other monomers were also observed, like 2-methyl-3-hydroxyvalerate when synthetic media containing propionate or a mixture of acetate and propionate were fed to a MMC selected under ADF conditions (Lemos et al., 2006). Pisco et al. (2009) reported for the first time the presence of 3-hydroxyhexanoate in PHA produced by a MMC, in this case produced from fermented molasses under AN/AE alternating conditions. Moreover, Albuquerque et al. (2007) found a direct correlation between the VFA profile of the feeding and the polymer final monomer composition. All of these monomers contribute to enhance the polymer properties and it was verified by Serafim et al. (2008b) and Albuquerque et al., (2011) that the properties of PHA produced by MMC follow the same trend as verified for PHA obtained by pure cultures.

Lemos et al. (2008) showed that by supplying different VFA composition were selected MMC with difference microbial composition. Despite the differences in the population, PHA produced by these populations showed no significant differences in their properties, except those that related with the monomer compositions (Serafim et al., 2008b). The stability of the polymer composition is an important aspect of PHA production from real complex wastes by mixed cultures. Because of organic waste streams are often sensitive to seasonal and process variations, acidification processes for their subsequent valorisation into PHA must accomplish manipulation of the operational parameters, in order to achieve a coherent and appropriate VFA profile (Albuquerque et al., 2007, 2011; Serafim et al., 2008a).

## 2. Methodology

### 2.1 Waste currents characteristics

In this chapter, acidification processes of four industrial waste products are presented and discussed - sugarcane molasses, spent coffee grounds, dairy processing fatty slurry, and cheese whey. These industrial waste products were selected because they currently constitute environmental problems for their producing industries. Due to their high organic content, these wastes often lead to instability issues inside the conventional bioreactors used for their treatment. Since treatment is mandatory prior to discharge, simultaneous valorisation through acidogenic fermentation was here investigated, from a wastewater treatment point of view with recovery of valuable constituents. The choice of these organic currents covers a large spectrum in the acidification field, since particular composition of each one leads to different biodegradation behaviours.

Organic waste products were collected in different Portuguese food industries. Molasses is a by-product of a sugar refinery industry, and consists mostly of carbohydrates. Spent coffee grounds are a slurry material resulting from instant coffee beverages manufacture, and consist mostly of structural polysaccharides. Dairy processing fatty slurry was collected from a Dissolved Air Flotation system treating dairy industrial effluent and consists mostly of oils and fats. Cheese whey was collected in a medium scale cheese production factory, consisting mostly of proteins. Their organic content is presented in Table 1.

| Waste stream | tCOD (g $L^{-1}$) | sCOD (g $L^{-1}$) | pH | tVFA (mg $L^{-1}$) |
|---|---|---|---|---|
| Sugarcane molasses | 952±130 | 896±102 | 6.8±0.4 | 31±4.9 |
| Spent coffee grounds | 61±15 | 45±16 | 4.1±0.3 | 116±10 |
| Dairy fatty slurry | 324±71 | 38±11 | 5.2±0.1 | - |
| Cheese whey | 103±13 | 98±10 | 6.2±0.5 | 22±5.3 |

Table 1. Main characteristics of the organic waste products from food industry (average ± standard deviation)

### 2.2 Experimental design and hydraulic modes

Acidogenic fermentation of each waste product was evaluated in order to determine optimum operational conditions for the maximisation of VFA yield. The degree of acidification was generally used to assess acidogenic potential for each current, and it was calculated from the total VFA in COD equivalents relative to the influent soluble COD (Bengtsson et al., 2008). Different approaches and results permitted to investigate their acidogenic potential. Two different types of hydraulic modes of operation and reactors were considered: batch and continuous. While batch experiments permitted a preliminary estimation of the acidogenic potential of each waste product, continuous reactors were operated under long term to ensure and validate continuous production of acidified effluents. Upflow Anaerobic Sludge Blanket (UASB) reactor was operated with dairy fatty slurry and Moving Bed Biofilm Reactors (MBBR) type units were used for molasses, spent coffee grounds and cheese whey. Figure 1 illustrates the main aspects of each bioreactor.

Start-up of the 5 L batch reactors consisted of addition of established volumes of anaerobic biomass and of selected effluents, as well as inorganic nutrients (Capela et al., 1999), and distilled water to make up all the volumes to 5 L. The reactors were purged with a stream of nitrogen for about 1 min, to remove any oxygen content, and then sealed. Stirring was performed with magnetic stirrers and thermostatic condition was ensured with hot water baths. Samples were collected with a syringe at the top of the reactors.

For the test of acidification of dairy processing fatty slurry in continuous mode a UASB reactor made of Perspex with a height of 86.4 cm and a working volume of 6 L was used. The reactor had a gas-solid-liquid separator at the top. The temperature was kept at 35±1 °C by means of a water jacket. Biogas production was monitored with wet gas meters. The feed was mixed constantly with a mechanical stirrer and pumped to the bottom of the reactor through a peristaltic pump. Samples of the treated waste were collected from the exit in the top settler.

A 2.54 L anaerobic MBBR reactor permitted combination of suspended growth and fixed film processes, taking the advantages of both without being restrained by their disadvantages. Before assemblage, about 40% of the reactor volume was filled with biofilm carrier elements made from polyethylene (Bioflow 9), with a specific surface area of 800 $m^2$ $m^{-3}$. Stirring was performed with a mechanical stirrer, and feeding with a peristaltic pump. Thermostatic condition was ensured with hot water baths. Mixed liquor was collected by overflow in a settler with a volume of 0.84 L.

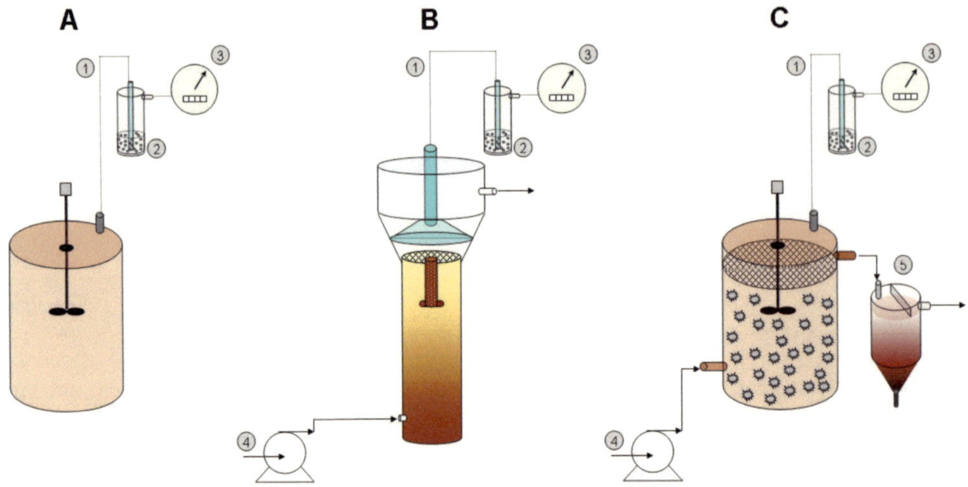


Fig. 1. Anaerobic reactors used in acidification studies: (**A**) Batch anaerobic reactor; (**B**) Upflow anaerobic sludge blanket, UASB; (**C**) Anaerobic moving bed biofilm reactor, MBBR; (1) gas line; (2) gas bubbler; (3) wet gas meter; (4) feedstock; (5) settler

### 2.3 Analytical methods

All the analyses were performed in accordance with standard analytical procedures (APHA, 2005). Alkalinity and pH were measured according to methods 2320B and 4500-$H^+$B. Chemical oxygen demand (COD) was measured by colorimetric method (method 5220D) and the solids analyses were performed with glass microfiber filters (Reeve Angel grade 403), analytical balance, drying oven and muffle furnace (methods 2540B, 2540D and 2540E).

Biogas was measured using a wet gas meter and the gas content analysis was done by collecting a gas sample with a gas tight syringe and injection on-column in a gas chromatograph SRI 8610C with thermal conductivity detector set to 75°C, and a 80/10-2.5 m CRS Hayesep column set to 61°C. Helium was used as carrier gas. Calibration curves were obtained by injecting standards of methane and carbon dioxide.

The VFA were analysed by gas chromatography by injection of 0.5 μL of filtered sample containing 10% (v/v) of formic acid on-column in a gas chromatograph Chrompack CP9001 with flame ionisation detector set to 220°C, and a 25 m x 0.25 mm Chrompack CPSIL-5CB column. Helium and nitrogen were used as the carrier and the make-up gases respectively. Calibration curves were obtained by injecting standards of acetic, n-butyric, iso-valeric, n-caproic (even carboxylic chain VFA), propionic, iso-butyric, and n-valeric acids (odd carboxylic chain VFA).

## 3. Results and discussion

### 3.1 Sugarcane molasses

Molasses is a by-product of sugarcane processing and it contains a high sugar content ranging from 48 to 50%, mainly sucrose, glucose and fructose, a water content of 17 to 25% and polysaccharides (dextrin, pentosans, polyuronic acids) content of 2 to 5%. Its reduced polymeric sugars can further react to form fermentable sugar during enzymatic hydrolysis. So far many researchers have mentioned molasses as a carbon source for several biotechnological processes, such as ethanol and citric acid fermentation (Najafpour & Poi Shan, 2003), alcohol and amino acid production, baker's yeast fermentation and improvement of biological denitrification (Quan et al., 2005). However, the use of high loads molasses as an external carbon source for bio-treatment processes has proven to promote VFA build-up containing heavier carboxylic chains (Silva et al., 2009). Espinosa et al. (1995) observed that a molasses OLR = 17 gCOD $L^{-1}d^{-1}$ applied to an UASB led to the accumulation of VFA, in particular propionic acid, being it attributed to the lack of trace nutrients such as Fe, Ni, Mo and Co. Therefore those accumulated VFA can be faced as a valuable recovery product.

Molasses acidification was performed in two MBBR reactors during 160 days at mesophilic temperature (37±2 °C). In order to study the effect of inoculums type, each reactor was inoculated with distinct anaerobic biomass: AR reactor started operation with acidogenic sludge previously adapted to glucose for 20 days, whereas MR reactor was inoculated with conventional anaerobic biomass from a full-scale municipal digester working on sludge digestion with methane recovery. Hydraulic retention time (HRT) and organic loading rate (OLR) were also investigated to determine optimum operational conditions. Both reactors were submitted to increasing OLR (1 to 70 gCOD $L^{-1}d^{-1}$) and two HRT (6 and 12 h).

In general, output COD varied proportionally to the applied OLR in both reactors. Figure 2A describes soluble COD monitored at the outlet of reactors. Conventional biomass reactor (MR) presented lower COD values than acidogenic biomass reactor (AR), due to the fact that an important fraction of substrates inside reactor followed methanisation pathways. However, when OLR = 33 gCOD $L^{-1}d^{-1}$ was applied, COD values dropped significantly when compared to the previous organic loadings, which is most likely due to HRT change from 12 to 6 h, corresponding to a decrease on COD input. COD removals of 40±9 and 60±7% were achieved under that OLR for AR and MR reactors respectively, and removal in AR reactor is similar to that achieved by Ren et al. (2006) using a bio-hydrogen producing reactor (COD removals in the range 20-40% when varying OLR between 1 and 80 gCOD $L^{-1}d^{-1}$). When OLR changed from 33 gCOD $L^{-1}d^{-1}$ to higher values (52-70 gCOD $L^{-1}d^{-1}$), fermented effluent presenting higher COD content was collected in both reactors, with a significant drop in the COD removal (only 17 and 10% for AR and MR reactors respectively).

As it would be expected, produced methane (Figure 2B) was always higher for MR reactor than for AR during operation at HRT = 12 h, due to the presence of higher amounts of methanogens within the conventional microbial consortium. Moreover, changing of HRT from 12 to 6 h resulted in significant methane production in MR, revealing an effective potential of molasses for methanogenesis and ability of the microbial consortium to deal with moderate OLR (33 gCOD $L^{-1}d^{-1}$), although unsteady-state in methane production was observed at this OLR. Biogas methane content varied between 40 and 55% during longer HRT (12 h) operation, whereas shorter HRT (6 h) at OLR = 33 OLR even resulted in a slight rise on methane content (from 40 to 55%) in AR reactor. An additional increase on OLR from 33 to values above 50 resulted in severe methanogenic inhibition, particularly in MR reactor that exhibited a drop on methane content to values below 5%. Wijekoon et al. (2011) also observed a rising of methane production when the OLR was increased, and it was probably attributed to the ability of the reactor to work in even higher loading rates.

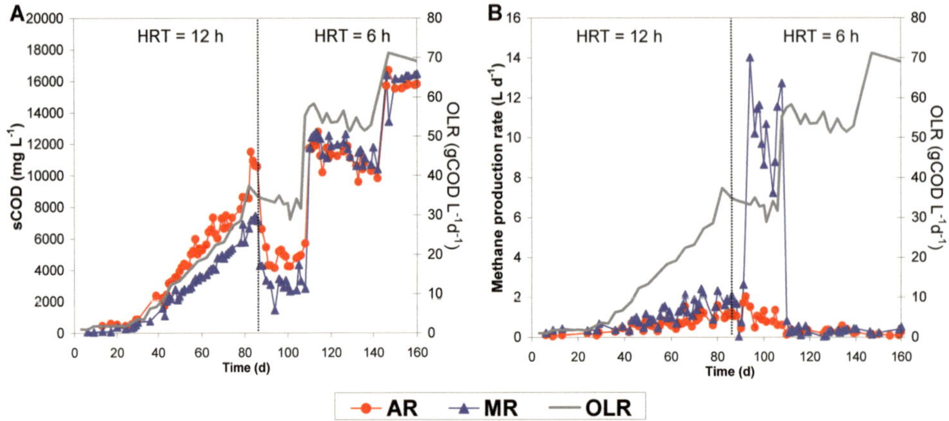


Fig. 2. Performance of MBBR reactors (acidogenic biomass, AR, and conventional biomass, MR) processing sugarcane molasses: A) sCOD in the fermented effluent; B) methane production rate

Figure 3A presents the degree of acidification (DA), which is of fundamental importance when evaluating acidogenic systems. For the tested conditions, it was observed that AR reactor yielded a much higher degree of acidification when compared to the MR reactor inoculated with conventional mixed anaerobic biomass. Values of pH in the acidogenic biomass reactor were also lower than those achieved in the conventional mixed biomass reactor (Figure 3B).

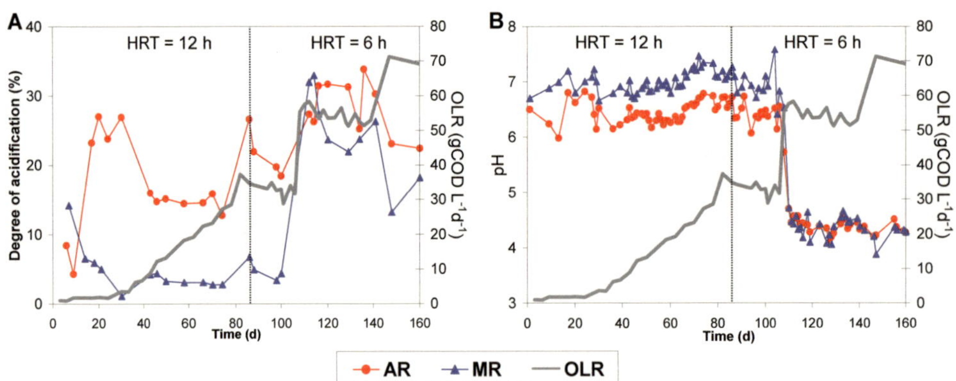


Fig. 3. Performance of MBBR reactors (acidogenic biomass, AR, and conventional biomass, MR) processing sugarcane molasses: A) degree of acidification ($VFA/sCOD_{in}$); B) pH

Higher degrees of acidification achieved in the AR reactor highlight the strategy of using pre-acclimated biomass for VFA production, and it was quite notorious for operation at HRT = 12 h. A DA of 27% was achieved at OLR lower than 8 gCOD $L^{-1}d^{-1}$, while a DA of 16% was achieved at OLR in the range 8-30 gCOD $L^{-1}d^{-1}$. During this period (low OLR), DA for MR reactor was generally 3 to 5 times lower than that for AR reactor. Maximum net production of VFA (30-34% acidification) was achieved under an organic load rate of 52 gCOD $L^{-1}d^{-1}$ for both reactors, with a significant drop in pH (to values around 4 - 4.5), although there had been applied a corresponding increase of alkalinity with increasing OLR

(0.1 g $CaCO_3$ $g^{-1}COD_{in}$). During operation at an OLR in the range 52-55 gCOD $L^{-1}d^{-1}$, some unbalanced conditions in the MR reactor were noticed, as some unsteady condition was observed in the degree of acidification. Conversely, AR reactor kept the acidification degree at 34% and it showed to be well acclimated to those moderate OLR. For higher organic load (70 gCOD $L^{-1}d^{-1}$), it was observed a decrease on the acidification yield in both reactors.

Production of VFA as a function of OLR applied is presented in Figure 4, where the better ability of AR reactor for organic acids production is well emphasised. During operation at HRT = 12 h, VFA production raised with the increasing OLR up to 29 gCOD $L^{-1}d^{-1}$, (from 250 to 2700 mgCOD $L^{-1}$), while the same OLR increase in the MR reactor yielded an almost constant VFA production (100-400 mgCOD $L^{-1}$). During that period, acetic (C2), propionic (C3) and iso-butyric (C3) acids were the main forms of VFA. Below OLR = 8 gCOD $L^{-1}d^{-1}$, acetic acid was predominant, thus contributing to low odd-to-even ratios of VFA carboxylic chains (0.16-0.31). Between OLR of 8 and 29 gCOD $L^{-1}d^{-1}$, odd-to-even ratio presented a remarkable increase (1.10-2.85), mainly due to the build-up of propionic and iso-butyric acids that surpassed acetic acid concentration. Similar shift was reported by Ren et al. (2006) when OLR tested was changed from 4 to 13 gCOD $L^{-1}d^{-1}$, with an increasing from 600 to 2200 mg $L^{-1}$ on total VFA. Conversely, Wijekoon et al. (2011) found an increase of n-butyric and a constant propionic acid concentration when the OLR increased, but those loading rates applied (between 5 and 12 gCOD $L^{-1}d^{-1}$) were much lower than in the present study.

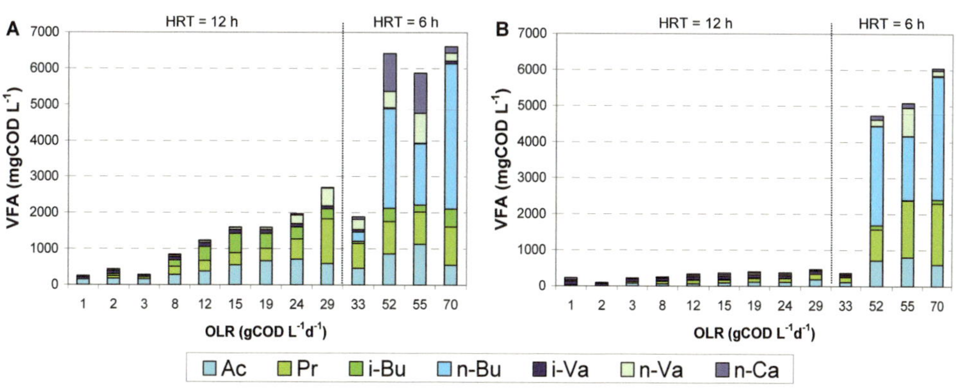


Fig. 4. VFA production as a function of organic loading rate applied: A) Acidogenic biomass reactor, AR; B) conventional mixed biomass reactor, MR

When HRT was set to 6 h, total VFA decreased from 2600 to 1900 mgCOD $L^{-1}$ inside AR reactor, mainly due to decrease on propionic acid concentration (from 1250 to 700 mgCOD $L^{-1}$). In spite of being suitable for feedstock for microbial PHA production, propionic acid build-up has been pointed as key indicator of methanogenic instability due to its worse biodegradation kinetics (Nielsen et al., 2007; Wijekoon et al., 2011). Shift on OLR from 33 to 52 gCOD $L^{-1}d^{-1}$ led to an increase on DA (Figure 4A) with concomitant build-up of longer carboxylic chain VFA, namely n-butyric (C4) acid that achieved an individual concentration of 2800 mgCOD $L^{-1}$, despite the biomass tested (AR or MR). That build-up of longer chain VFA matched the severe pH drop observed in both reactors (Figure 4B), which was also referred in other studies. For acidogenic fermentation of molasses at different pH values, Albuquerque et al. (2007) reported that acetate and propionate concentrations decreased when pH was decreased from 7 to 5, while butyrate and valerate concentrations significantly increased in the same pH shift. For acidogenic fermentation of glucose, Zoetemeyer et al. (1982) referred that lower operating pH favoured the production of longer chain fatty acids, since under these more acidic conditions there are more reducing equivalents available to be incorporated into the fatty acid chains.

Production and profile of VFA from sugarcane molasses is closely correlated with the OLR applied (Ren et al., 2006). It should be mentioned that the massive production of organic acids observed in both reactors during the highest OLR (70 gCOD $L^{-1}d^{-1}$) did not correspond to higher degrees of acidification. Accordingly, high VFA production observed at OLR = 70 gCOD $L^{-1}d^{-1}$ (mainly due to n-butyric acid build-up) did not lead to higher volumetric productivities of VFA, whereas a shift on the HRT from 12 to 6 h combined with a load increase from 33 to 52 gCOD $L^{-1}d^{-1}$ permitted to obtain the maximum amounts of acidified effluent. This was considered the optimum operating condition since it allows for preservation of organic resources (low COD removal) that are demanded for valorisation into VFA.

### 3.2 Spent coffee grounds

Wastewater from instant coffee substitutes production was collected at a Portuguese food processing industry. Because instant coffee substitutes production implicates hot water solubilization of coffee and substitutes roasted beans, this effluent is a dark coloured liquid with acidic properties and is generated at thermophilic temperatures (>45ºC). This effluent presents high solids content with a significant fraction of organic solids (≈98%). The considerable fraction of insoluble solids (24 to 28%) found in this effluent is consistent with the values for instant coffee processing wastewaters found in literature.

The $BOD_5$/COD ratio is less than 0.3 which may indicate the presence of low biodegradable compounds at aerobic conditions. In fact, the presence of organic refractory substances like lignin (Dinsdale et al., 1996) tannins and humic acids (Zayas et al., 2007) in coffee and coffee substitutes wastewaters is pointed out as being responsible for the difficulty in treating these wastewaters by biological treatment processes. Nitrogen is deficient in this wastewater: $BOD_5$:N:P ratio varies between 100:1.5:0.7 - 100:2.8:1.4 which is quite inferior to 100:5:1 considered to be the adequate ratio for aerobic biological treatment.

Acidification of a wastewater containing spent coffee grounds was performed in two MBBR reactors at two temperatures (37 and 55ºC), increasing OLR (2.5-10 gCOD $L^{-1}d^{-1}$) and two values for the HRT (0.5 d and 1 d). The reactors were fed with this effluent previously diluted with tap water. To avoid operational problems in the reactors due to the high solids content of the wastewater in study, TSS level was reduced by settling the effluent for 3 hours and siphoning off the top layer. With this procedure the effluent solids content ranged from 2.2 to 0.8 gTSS $L^{-1}$. Nitrogen and phosphorus were added to give a COD:N:P ratio of 100:1.75:0.25, according to Dinsdale et al. (1997). Alkalinity (0.1g of $NaHCO_3$ for each g of COD present in feed) was added to provide the system with some buffer capacity. This allowed that reactors pH to vary between 4.5-5.

Figure 5 shows the acidification behaviour of each reactor (A-mesophilic (R1) and B-thermophilic (R2)) during all period of experiment, with respect to the acidified effluent pH, organic loading rate and hydraulic retention time. The concentration of the sCOD of the acidified effluent is also presented as an indicative parameter to recognize steady state for each operational condition.

In the beginning of the experiment, at the mesophilic temperature (reactor R1), methane was always present, showing methanogenic activity, with pH reaching values higher than 6. It was necessary to cause adverse conditions in the reactor to avoid this situation by a decrease in pH to values lower than 4. This situation leads to an extent of the experimental period at a load of 2.5 gCOD $L^{-1}d^{-1}$. In general, the sCOD out varied proportionally to the applied OLR in both reactors. Mesophilic reactor presented higher amounts of total VFA when compared with the thermophilic reactor for the same organic loads (Figure 5).

For all the experimental conditions, the reactors presented acidification degrees between 20% and 55%, with pH kept at 4.5-5, except for reactor R1 at the beginning of the experiment. HRT had a strong effect on the behaviour of both reactors, mainly for the sCOD, due to the fact that to maintain the some organic load it was necessary to double the flow rate (decrease of HRT from 1 d to 0.5 d) and decrease the input COD in the same proportion.

At mesophilic temperatures (reactor R1), the process yielded an average acidification degree of 40-50%, with values always higher than 30% (exception for the initial time). The equivalent values obtained in all thermophilic assays (Figure 5) were mostly lower, although also varying between 40% and 50%. For a load of 5 gCOD $L^{-1}d^{-1}$, the acidification degree was 41-43% for both HRT (1 d and 0.5 d). At thermophilic operation, the acidification degree decreased to 28% for the highest OLR (10 gCOD $L^{-1}d^{-1}$). These results are in agreement with the results of Dinsdale et al. (1997) where they obtained acidification degrees between 22% and 38% in CSTR reactor treating coffee wastewater at organic loads ranging from 10 to 16 gCOD $L^{-1}d^{-1}$, HRT between 1 and 0.5 d and pH = 5.

In the literature it is agreed that anaerobic acidification process is affected by several operational parameters such as pH, HRT, organic load and temperature. Hence, the effect of HRT, temperature and organic load in the acidification of a wastewater containing spent coffee grounds used in this work is presents in Figure 6.

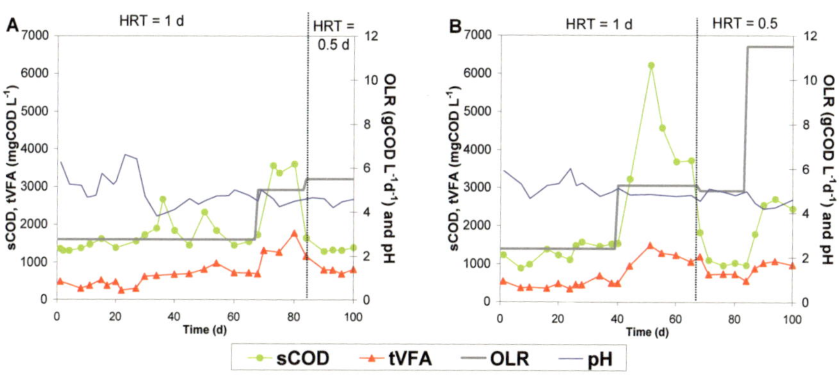


Fig. 5. Performance of MBBR reactors processing a wastewater with spent coffee grounds: A) R1 - mesophilic reactor; B) R2 - thermophilic reactor

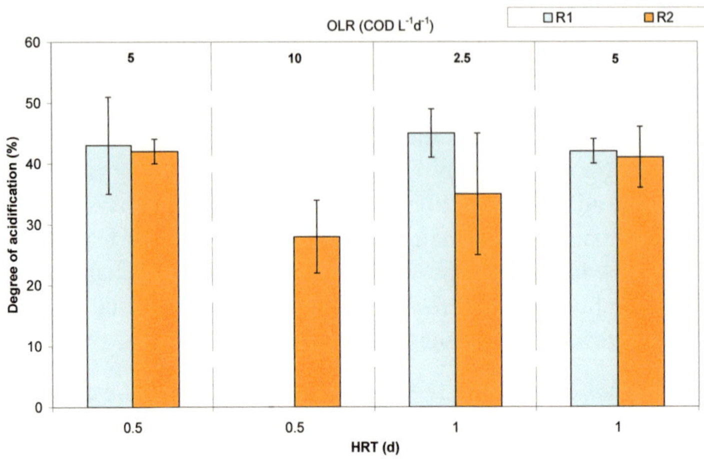


Fig. 6. Acidification degree for the different experimental phases for both reactors (R1 - mesophilic; R2 - thermophilic)

In terms of acidification degree, the operation of reactor R1 at a mesophilic temperature (Figure 6) is not significantly affected by the organic load increase (HRT=1 d and OLR = 2.5→5 gCOD $L^{-1}d^{-1}$), showing a decrease from 44.5% to 42.2%. The decrease of HRT from 1 d to 0.5 d at the same load of 5 gCOD $L^{-1}d^{-1}$, doesn't affect the acidification degree which changed from 42.2% to 42.9%. This behaviour shows that for the operational conditions tested, the HRT of 0.5 d doesn't favour the conversion to VFA.

The thermophilic reactor R2 showed an increase of the acidification degree from 35% to 40% with the increase of the organic load from 2.5 to 5 gCOD $L^{-1}d^{-1}$ (Figure 6, HRT=1 d), which represents an increase of 14%. As already showed for reactor R1, the change in HRT from 1d to 0.5d at the same load of 5 gCOD $L^{-1}d^{-1}$, doesn't affect the acidification process, as the acidification degree changed from 41 to 42%. However, an increase on the OLR from 5 to 10 gCOD $L^{-1}d^{-1}$, affected strongly the acidification degree which changed from 42% to 27%, accounting for a 36% decrease. This value is similar to the one obtained by Dinsdale et al. (1997) for an acidification process of coffee wastewater at an HRT=0.5 d and OLR=16 gCOD $L^{-1}d^{-1}$, without pH control (22% acidification degree).

In conclusion, for the same conditions tested in both reactors R1 and R2, the temperature only affected the first condition (HRT=1 d and OLR=2.5 gCOD $L^{-1}d^{-1}$) with a decrease of 29% when the temperature increased from 36ºC to 55ºC. For the conditions tested the HRT decrease from 1 d to 0.5 d doesn't affect significantly the acidification degree for both temperatures. In the other hand, the increase on the loading rate affected mostly the acidification process at thermophilic temperature, especially for the HRT = 0.5 d.

The effect of HRT and organic load on the composition of VFA is presented in Figure 7. At 37ºC (Figure 7A), the VFA present in the acidified effluent are the acetic acid, propionic, n-butyric and n-valeric, being the acetic acid the one who appears in higher concentrations (52 to 61% m/m), followed by the propionic acid with 25 to 32% (m/m) and after the n-butyric acid with 12 to 13% (m/m). These results are not totally in agreement with the results of Alexiou (1998), most probably because in this work the highest organic load tested was 5 gCOD $L^{-1}d^{-1}$ whereas he studied organic loads higher than 18 gCOD $L^{-1}d^{-1}$. At a mesophilic temperature, the increase of the organic load from 2.5 to 5 gCOD $L^{-1}d^{-1}$, increased all individual volatile acids concentration (total VFA from 615-1335 mg $L^{-1}$ as COD), although in proportion to the total, acetic acid increases in 7% in the acidified effluent in detriment of the propionic acid which is reduced in 13%. The decrease on the HRT decreases the amount of VFA produced (1335-743 mg $L^{-1}$ as COD) and favours the increase on the percentage of acetic acid to the total in 7%, a decrease in the percentage of propionic acid in 10% and the disappearance of the valeric acid which was present before in very low percentages (1.5-2.5%).

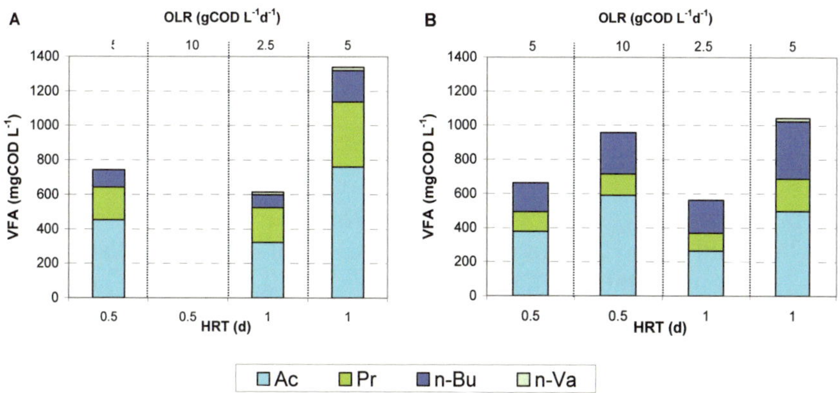


Fig. 7. VFA distribution as a function of HRT and organic load. A) reactor R1 at mesophilic temperature; B) reactor R2 at thermophilic temperature

For the assays at 55°C, the type of volatile acids present in the acidified effluent is the same that the ones found at mesophilic temperature (acetic, propionic, n-butyric and n-valeric) although with a different distribution. Hence, in terms of VFA composition, it was observed that a temperature shift from 37 to 55°C enhanced the presence of n-butyric acid rather than propionic acid, but still with the predominance of acetic acid. At thermophilic temperatures (55°C) Alexiou (1998) found the presence of the same four acids in the acidified coffee effluent at a load of 17.3 gCOD $L^{-1}d^{-1}$, HRT=0.54 d and pH=4.5. The acidified effluent obtained in the experiments of Alexiou (1998) presented 58% acetic, 28% n-butyric, 8% propionic and 2% n-valeric (m/m) which is comparable with the results obtained for reactor R2 from this work during the entire period: 47-62% acetic acid, 34-35% n-butyric, 13-19% propionic acid and 0-1.8% of n-valeric acid (%m/m). A decrease in the HRT stimulates the production of acetic acid (20% increases in the proportion to the total) and inhibits the presence of propionic acid (4% decrease), n-butyric (22% decrease) and the disappearance of n-valeric acid.

In conclusion, the differences found in this work regarding the metabolic pathway for VFA production, i.e. the predominant volatile acids present in the acidified wastewater containing spent coffee grounds are dependent mainly on temperature. It was found that the increase in temperature favours the presence of butyric and acetic acids. Hence, for this type of effluent at thermophilic temperatures doesn't lead to an increase of propionic acid but to an increase of the butyric acid. This behaviour is in accordance with the results by Yu et al. (2002) which studied dairy effluents and concluded that the increase in temperature didn't led to the increase of the ratio of propionic acid to the total. Regarding the effect of

HRT, it was found that the decrease of this parameter favours the increase in the relative percentage of acetic acid and a decrease on the percentage propionic for both temperatures. At thermophilic temperatures the presence of butyric acid is also inhibited. In these conditions, these metabolic pathways are not appropriate for PHA production, but rather for energy production through biogas. So, not only the amount of VFA produced and acidification degree are important but also de distribution of VFA, in order to choose a practical and economically feasible waste management option together with the recovery of valuable organic acids. In this sense, and given the very low biodegradability of this industrial waste, the acidified effluent obtained for each steady state condition was also assessed in terms of biodegradability besides VFA yields, using aerobic BOD tests procedure. For the tested conditions, experimental results indicated that the acidified effluent reached much higher biodegradability degrees with increases as high as 80%.

### 3.3 Dairy processing fatty slurry

The dairy processing fatty slurry used in this work was collected at a Dissolved Air Flotation (DAF) tank in a dairy factory wastewater treatment plant. Batch assays were performed either at mesophilic (37ºC) or thermophilic (55ºC) temperature ranges in order to investigate the influence of temperature, pH control, added alkalinity and organic load on the acidification of dairy processing fatty slurry. At mesophilic temperature (37ºC) the batch tests were conducted with loads of 5, 10, 15 and 30 gCOD $L^{-1}$ both with and without added alkalinity. Added alkalinity varied in the range of 0-10 $gCaCO_3$ $L^{-1}$. For the tests at thermophilic temperature (55ºC) two loads were tested (10 and 30 gCOD $L^{-1}$) with and without added alkalinity and with and without pH control. The tests at 37ºC lasted between 20 and 50 days, the longer tests being those with the higher loads. The tests at 55ºC with no pH control lasted between 110 and 150 days and the tests at 55ºC with pH control lasted 35 days. All the tests were terminated when constancy in the cumulative methane production value was observed for longer than 10 days. The biological anaerobic sludge used in the tests was collected from a wastewater treatment plant in a dairy factory. Due to the characteristics of the fatty slurry the acidification was computed based on feed tCOD.

Cumulative methane production reached at the end of each batch test was a clear function of the applied load (Figure 8A) and mesophilic tests resulted in higher methane production compared to the thermophilic tests. The effect of added alkalinity was not pronounced but pH control (pH4 and pH5) at thermophilic temperature resulted in very low or null methane yield. The factors that most clearly influence the maximum attainable concentration of VFA were the COD load and alkalinity addition (Figure 8B), the VFA concentration peak rising with the rise of any of those two factors. The effect of alkalinity was most pronounced for the higher loads and also the effects of load rise in the rise of the VFA peak was more pronounced in the tests with added alkalinity. For all the tests conducted at loads above 5 gCOD $L^{-1}$ at 37ºC and with no pH control addition of alkalinity caused a rise in the maximum VFA peak of more than 100% compared to the results of similar tests with no added alkalinity.

In all the tests performed at 55ºC the maximum amount of VFA attained was always lower compared to what was observed in the corresponding tests at mesophilic temperature (37ºC). It was observed that for the lowest load tested at 55ºC (10 gCOD $L^{-1}$) the effect of temperature was not significant whilst for the load of 30 gCOD $L^{-1}$ the negative effect of temperature rise was only observable in the tests performed with added alkalinity. For the tests conducted at 55°C it was also noted that the rise in the applied load did not significantly influence the VFA peak and that pH control resulted in slightly lower maximum VFA peaks compared with the equivalent tests conducted with no pH control, even for the test with added alkalinity. Yu and Fang (2001) reported a VFA concentration of approximately 950 mg Ac $L^{-1}$ and around 38% acidification of feed COD for thermophilic acidification of dairy wastewater at pH=5.5 and a load of 4 g COD $L^{-1}$. Yu and Fang (2002) observed a decrease in VFA production for loads above 8 g COD $L^{-1}$ for batch thermophilic acidification of dairy wastewater. In all the batch tests conducted in the present study with dairy processing fatty slurry in both temperature ranges the highest VFA concentration peak (5356 mg Ac $L^{-1}$) was reached at 37°C for a load of 30 g COD $L^{-1}$ with added alkalinity and no pH control. The operating times needed to reach the VFA peaks were 8-15 days at 37°C and 6-23 days at 55°C without pH control. Effects of added alkalinity were not clear. In the tests at 55°C with pH control the times needed to reach the VFA concentration peak were 26-33 days.

The highest acidification (26% of the feed tCOD) was reached in the test conducted with a load of 15 gCOD $L^{-1}$ at 37°C with added alkalinity (Figure 9). At mesophilic temperature the main VFA produced was acetic acid followed by propionic acid that was detected only for loads above 10 g COD $L^{-1}$. For thermophilic temperature the highest acidification degree (15% of the feed tCOD) was attained for the lower COD load (10 gCOD $L^{-1}$) with added alkalinity and no pH control. Acetic and propionic acids were the major VFA produced although butyric and valeric acids were also detected.

COD balances at the end of each batch test (Figure 10) show that in mesophilic temperature the main fraction of the feed tCOD (56%-96%) was converted to methane and that the non-solubilised COD was always below 39% of the feed tCOD. At the end of the tests conducted in thermophilic temperature the main fraction of the feed tCOD remains in non-solubilised form (above 68%) and methanised fraction was always below 23%. Apparently at 55°C the anaerobic biomass was not sufficiently adapted to the fatty substrate and hydrolysis was the rate limiting step of the acidification process.

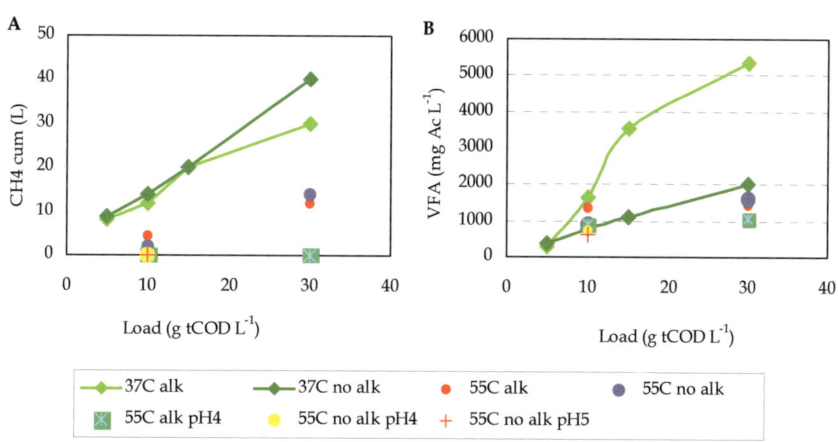


Fig. 8. A) Cumulative methane production in each batch test; B) maximum VFA concentration peak reached in each batch test.

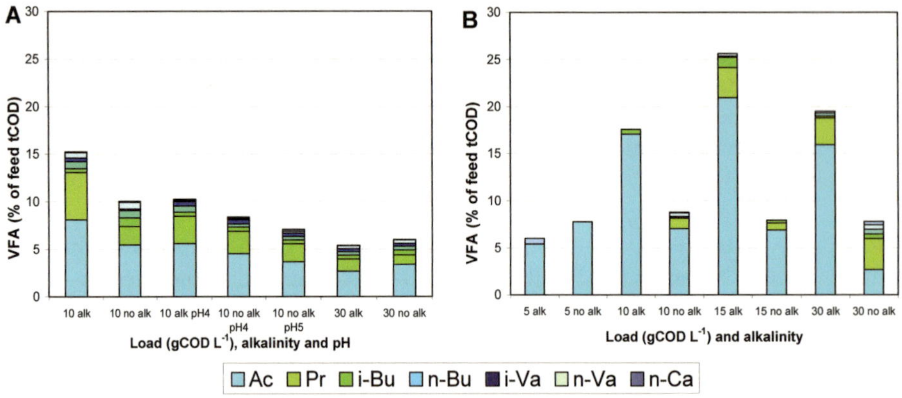


Fig. 9. Maximum degree of acidification reached in the batch tests and corresponding VFA composition. A) mesophilic temperature; B) thermophilic temperature.

A test was also conducted in a mesophilic (35±1)°C UASB reactor to determine the acidification potential in continuous operation. The reactor was inoculated with anaerobic flocculent sludge from an industrial reactor treating dairy wastewater. The feed was prepared every 4 days diluting the dairy processing fatty slurry and nutrients with tap water; from day 81 onwards alkalinity was supplemented to the feed (1 g $NaHCO_3$ L$^{-1}$ and 1 g $KHCO_3$ L$^{-1}$). The reactor was operated for a period of 200 days with a constant load of 10 g COD L$^{-1}$d$^{-1}$ and a hydraulic retention time (HRT) of 12 hours.

The operation of the reactor was divided in four distinct phases (Figure 11) although the COD removal efficiency was similarly low in all phases (average 33%). In the first phase (80 days) no alkalinity was supplemented to the reactor feed and methane production was very low, around 0.4 L d$^{-1}$, as was the degree of acidification (average 5% of the feed tCOD). In the second phase (days 81 to 125) alkalinity was added to the feed and $CH_4$ production and acidification improved slightly (3.3 L d$^{-1}$ and 7.6% of the feed tCOD, respectively). From day 126 to 167 the feed to the reactor was suspended whilst the temperature was kept constant (35±1)°C and the methane production was continuously monitored. This feed interruption was established in order to promote an adaptation of the microbial population present in the biological sludge towards a more complete degradation of the fatty substrates. This adaptation strategy has been previously used in UASB reactors fed with fatty wastewater (Nadais et al., 2011) since it was found that a period without feed improves the biomass capacity to degrade complex fatty substrates. During days 126 to 150 of this feedless period average $CH_4$ production reached 6.3 L d$^{-1}$ and for the remaining of the feedless period (days 151 to 167) $CH_4$ production was negligible (Figure 11A). In the fourth phase of operation (days 168 to 200) the average $CH_4$ production was around 5 L d$^{-1}$ and acidification reached the highest value (33% of the feed tCOD). In all the feeding phases (I, II and IV) the main VFA produced was acetic acid, the concentration of other acids being negligible. The 33% acidification coupled with 22% methanisation of the feed tCOD observed in this work (Table 2) are in accordance with the value of 54% reported by Fang and Yu (2000) for the mesophilic acidification of dairy wastewater at 12 HRT. From Table 2 and Figure 11B it can be seen that hydrolysis improved significantly along the operation period as well as the methanisation of the removed COD. Apparently in the first and second phases acidogenesis was the rate limiting step whilst in the fourth phase acidification improved considerably. This means that the feedless period (phase III) served the adaptation of acidifying microorganisms.

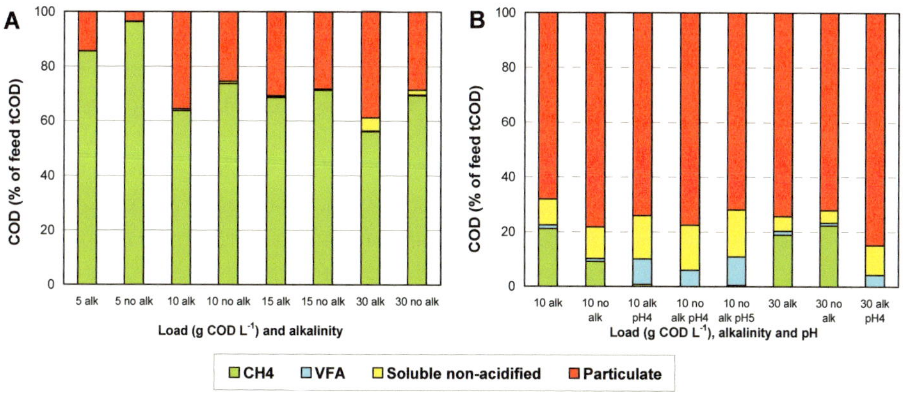


Fig. 10. COD balances at the end of the batch tests. A) mesophilic temperature; B) thermophilic temperature

For continuous operation hydrolysis of the fatty substrate improved compared to the batch experiments and the limiting step was acidification but biomass adaptation towards higher acidification yields was observed. The rate limiting step in batch experiments was hydrolysis, particularly at thermophilic temperature. In batch tests a wider variety of VFA was detected compared to the continuous test. The highest acidification yield (33% of the feed tCOD) was attained in continuous mode although VFA produced consisted only of acetic acid.

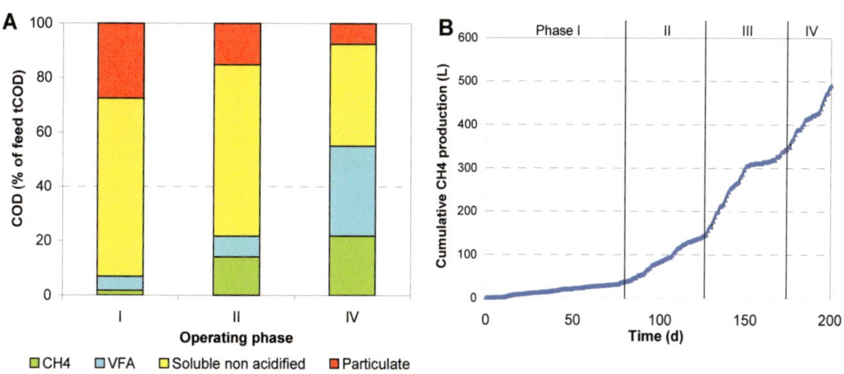


Fig. 11. A) cumulative methane production; B) COD balances for the operation of the UASB reactor.

| Operating phase | Hydrolysis of particulate COD (%) | Acidification of feed tCOD (%) | Methanization of feed tCOD (%) | Methanization of removed tCOD (%) |
|---|---|---|---|---|
| I | 50 | 5 | 2 | 6 |
| II | 72 | 8 | 14 | 48 |
| IV | 86 | 33 | 22 | 72 |

Table 2. Data from COD balances in UASB reactor acidifying dairy processing fatty slurry.

### 3.4 Cheese whey

Cheese whey is a lactose-rich by-product of cheese manufacturing. It represents about 85–95% of the milk volume processed and retains 55% of milk nutrients. To use these nutrients, whey is usually processed to separate it into protein, lactose, lipids, and mineral salts. Alternative utilizations for whey include either disposal as waste or use in fertilisers or animal feed (de Wit, 2001). By using whey nutrients as substrates for value-added by-production, both the pollution problem and disposal cost can be simultaneously abated. Even though there have been a number of technological developments in the transformation of whey into other useful products, utilization or disposal of whey is still one of the most significant problems faced by the dairy industry. Whey's high organic content, combined with low buffer capacity, makes anaerobic treatment in high rate reactors difficult, due to tendency to acidify the environment very quickly (Malaspina et al., 1996). Therefore, anaerobic processing of cheese whey is here exploited in order to recover organic acids rather than conventional treatment with methanisation.

Anaerobic batch assays were performed on a factorial basis at 37±2°C, to optimise alkalinity input and OLR (F/M ratio) on cheese whey fermentation. Ranges of the studied factors varied (at four levels) between 1 and 8 g$CaCO_3$ $L^{-1}$ for alkalinity, and 2 and 10 gCOD $g^{-1}$VSS for food-to-microorganism (F/M) ratio. Sixteen reactors were seeded with an initial concentration of 2 gVSS $L^{-1}$ of anaerobic mixed biomass collected from a full-scale municipal digester working on conventional digestion of waste activated sludge. In order to first investigate the effect of reaction time over VFA production, batch reactors ran for 20 days and VFA profiles for each assay are presented in Figure 12. Response Surface Methodology was applied to predict scenarios for maximisation of VFA production.

Regardless of the alkalinity added or the F/M tested, initial pH values were always recorded in the range of 6.9-7.2 for all experiments and VFA production was clearly achieved in a period of 5 days of incubation. A conservative trend in VFA concentration can be observed, mainly for higher F/M tests (4, 7 and 10 gCOD $g^{-1}$VSS), whereas the lowest F/M tested (2 gCOD $g^{-1}$VSS) showed an unexpected disappearance of VFA in all assays after 5 days of reaction. Moreover, maximum VFA plateau increased with F/M ratio. Because of the maximum VFA productions were achieved at 5 days of reaction, response surface methodology (RSM) was applied to results attained in that period (tVFA, degree of acidification and odd-to-even ratio in carboxylic chains). Computation was done using software package StatSoft Statistica. Equation 1 presents the mathematical form of the quadratic model applied (Myers et al., 2009).

$$E(z) = \beta_0 + \beta_1 x_1 + \beta_2 x_2 + \beta_{1,2} x_1 x_2 + \beta_{1,1} {x_1}^2 + \beta_{2,2} {x_2}^2 \quad (1)$$

where E(z) is the response variable, $x_1$ is the F/M ratio in gCOD $g^{-1}$VSS, $x_2$ is the alkalinity added in g $L^{-1}$ as $CaCO_3$, $\beta_0$ is the model constant, $\beta_1$ and $\beta_2$ are linear coefficients (main effects), $\beta_{1,2}$ is a cross-product coefficient (interaction) and $\beta_{1,1}$ and $\beta_{2,2}$ are quadratic coefficients. Table 3 presents the parameters obtained in the quadratic modelling, as well as regression coefficients ($r^2$) and P-value for each regression. Figure 13 presents the quadratic surface predicted to total VFA production, degree of acidification and odd-to-even ratio of carboxylic chains. Dots represent experimental data.

| *Parameter* | Response variables | | |
|---|---|---|---|
| | tVFA | DA | Odd-to-even ratio |
| $\beta_0$ | 342.8 | 4.672x10$^{-1}$ | 2.277x10$^{-1}$ |
| $\beta_1$ | 1399 | 2.020x10$^{-2}$ | -4.937x10$^{-2}$ |
| $\beta_2$ | -628.8 | 6.489x10$^{-3}$ | 9.418x10$^{-3}$ |
| $\beta_{1,2}$ | 88.59 | 2.045x10$^{-3}$ | -8.118x10$^{-4}$ |
| $\beta_{1,1}$ | -96.18 | -4.588x10$^{-3}$ | 3.089x10$^{-3}$ |
| $\beta_{2,2}$ | 55.76 | 1.250x10$^{-3}$ | 4.167x10$^{-4}$ |
| $r^2$ | 0.865 | 0.838 | 0.861 |
| P-value | 8.11 x10$^{-5}$ | 1.35 x10$^{-3}$ | 9.10 x10$^{-5}$ |

Table 3. Summary of the parameters adjusted for response variables

All the P-values were below 0.005, so experimental behaviours are well described by the computed models. Maximum experimental value for tVFA production (10.9 gCOD $L^{-1}$) was achieved in the assay combining both the highest F/M ratio (10 gCOD $g^{-1}$VSS) and alkalinity (8 $gCaCO_3$ $L^{-1}$). Gross production of organic acids increased with the organic load, and the increasing was more marked when alkalinity amount was the highest (increasing from 2450 to 10900 mgCOD $L^{-1}$). The positive value of the cross-product coefficient ($\beta_{1,2}$ = 88.59) verifies that combined increasing on F/M ratio and alkalinity maximises VFA production. On the other hand, when F/M ratio was the lowest, added alkalinity yielded a much lower raise on the VFA production (from 1300 to 2400 mgCOD $L^{-1}$).

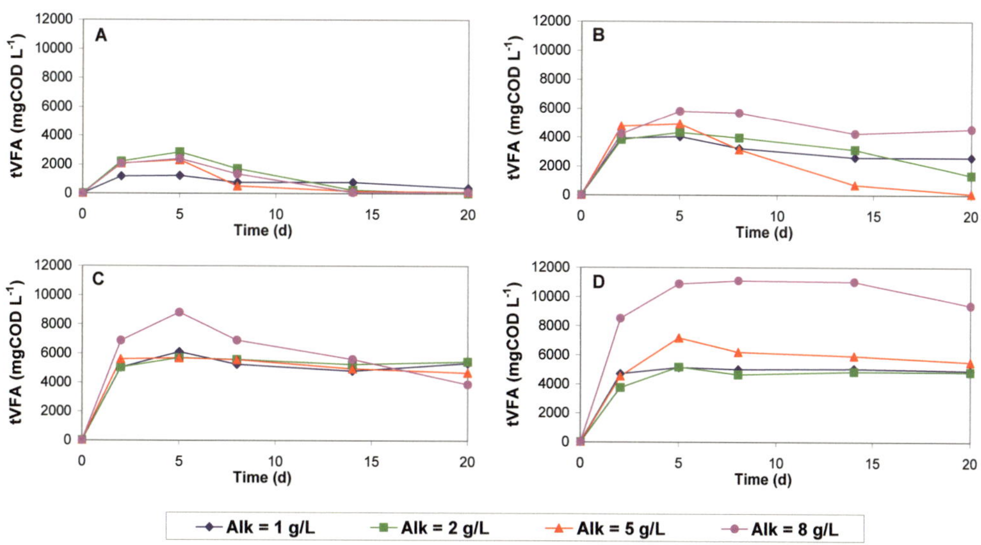


Fig. 12. VFA production from cheese whey as a function of reaction time: A) F/M = 2 gCOD $g^{-1}$VSS; B) F/M = 4 gCOD $g^{-1}$VSS; C) F/M = 7 gCOD $g^{-1}$VSS; D) F/M = 10 gCOD $g^{-1}$VSS

Calculation of degree of acidification (DA) reflects the net yield of VFA on COD fed (Figure 13B). Experimental DA presented a reverse trend from that obtained for gross VFA production. It is reasonable to argue that higher organic loads yield higher amounts of organic acids. However maximum net yields were observed in small F/M ratios (4 gCOD $g^{-1}$VSS). For DA, alkalinity presented lower influence than for gross production as well. Prediction for maximum DA was 0.68±0.25 $gCOD_{VFA}\ g^{-1}COD_{fed}$ at an F/M = 4.3 and Alk = 8 $gCaCO_3\ L^{-1}$. F/M ratio revealed an antagonistic effect, mainly when combined with low alkalinity. Alkalinity present in the system is quickly consumed due to rapid conversion of lactose into VFA, making it necessary to constantly impose pH control to the media (Backus et al., 1988).

Final values for pH were in the range of 6.8-7.9 for lower F/M series (2 and 4 gCOD $g^{-1}$VSS), while for higher F/M ratios (7 and 10 gCOD $g^{-1}$VSS) pH was recorded in the range of 4.2-6.0. The lowest pH values in both ranges were observed in the lowest alkalinity-containing assays, in which higher degrees of acidification were achieved. Similar tendency was also observed by Davila-Vazquez et al. (2008), which obtained final values in the range 3.7-3.9 when initial pH was set to the lowest value (4.5) in batch experiments, also corresponding to low gross VFA productions.

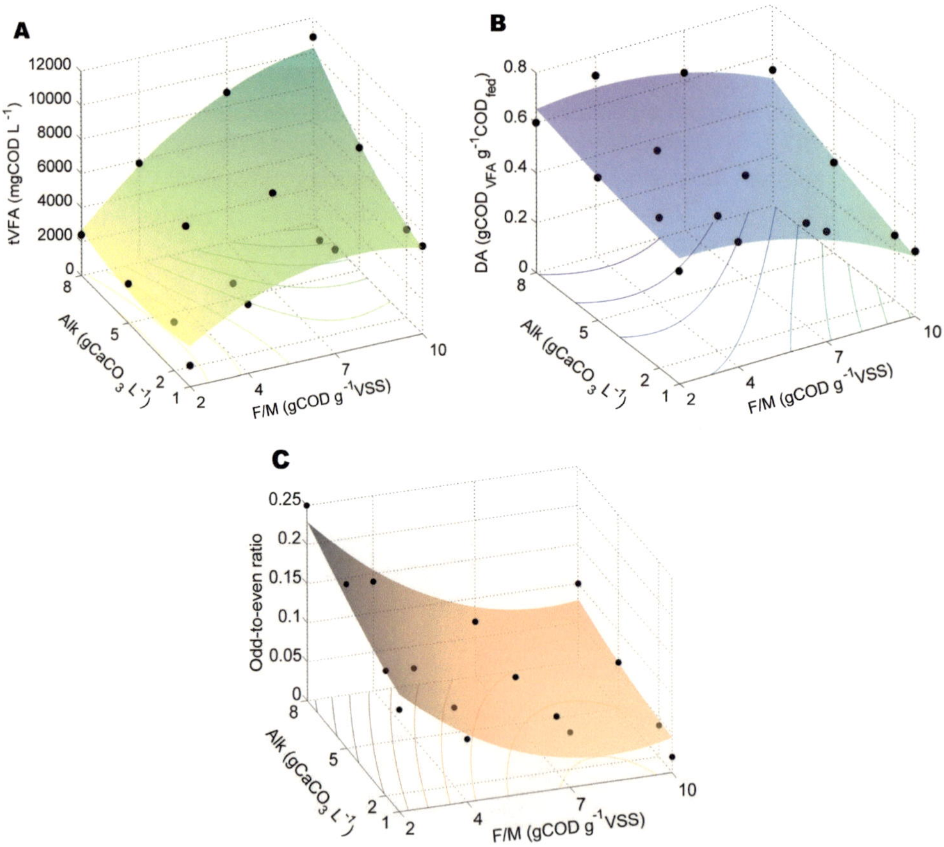


Fig. 13. Response surfaces for whey batch acidification with varying F/M and alkalinity: A) Gross VFA production; B) degree of acidification; C) odd-to-even ratio of carboxylic chains

VFA profiles varied significantly during the batch experiments and it can be illustrated by the odd-to-even ratio of carboxylic chains (Figure 13C). Although n-butyric and acetic acids were always predominant, lower F/M series assays (2 gCOD $g^{-1}$VSS) yielded few amounts of odd VFA (propionic and n-valeric acids in the range of 150-500 mgCOD $L^{-1}$) that contributed to an increase in the odd-to-even ratio up to 25%. Regarding the shape of the response surface, it is quite visible the antagonistic effect provided by the increase of F/M ratio, which is also verified by the negative value of $\beta_1$. In a less extent, a positive value for $\beta_2$ highlights a synergistic effect of alkalinity. The lowest experimental odd-to-even ratios (6-7%) were achieved for the F/M = 7 gCOD $g^{-1}$VSS tests, where n-butyric acid was predominant (54±5% of tVFA).

Butyric acid build-up with an increasing on F/M ratio was also noticed by Yang et al. (2007) and Bengtsson et al. (2008) that reported the predominance of n-butyric and acetic acids in batch experiments with cheese processing wastewater and cheese whey respectively. n-Butyric acid build-up has also been observed in other fermentation studies for cheese whey (Castelló et al., 2009; Davila-Vazquez et al., 2009; Yang et al., 2003). n-Butyric acid occurrence is such magnitude results from an increase on $H_2$ production, which slows down the overall acid production because of stoichiometric shifts (one mole of butyrate produced instead of two moles of acetate) (Fernandes, 1986). This may be the reason why lower degrees of acidification observed at higher F/M ratios are concomitant with higher n- butyric acid productions achieved at those organic loads, contributing to lower odd-to-even ratios.

Cheese whey was then fed to a MBBR reactor operated for 149 days, for continuous acidification of this organic waste. HRT was maintained at 12 h with OLR varying in the range of 30-100 gCOD $L^{-1}d^{-1}$. Alkalinity added was set to 6 g $L^{-1}$ as $CaCO_3$ and biomass concentration inside the reactor (carrier-attached plus suspended solids) varied between 2 and 24 gVSS $L^{-1}$. Due to the difficulty for controlling biomass concentration inside high-rate hybrid type reactors (fixed film plus suspended growth) such as MBBR, and consequently the F/M ratio, organic loads and HRT were chosen taking into account not only the previous results for batch acidification tests but also previous studies on cheese whey acidification (Azbar et al., 2009; Castelló et al., 2009; Davila-Vazquez et al., 2009).

Operation at the lowest OLR (30 gCOD $L^{-1}d^{-1}$) produced effluent with high acidification degree (65.6%) (Table 4), in which acetic and n-butyric were the main VFA forms. Moreover during OLR = 30 gCOD $L^{-1}d^{-1}$, almost all of the soluble COD (90.0±7.1%) was present as VFA after fermentation (Figure 14A), which confirms the very high readily fermentable organic content of whey, as also observed by Bengtsson et al. (2008). Similarly to what happened with molasses acidification, the load increase only permitted to get higher amounts of acidified effluent, as acidification degree dropped to values of 36% when OLR was set between 50 and 100 gCOD $L^{-1}d^{-1}$. VFA content in the soluble COD outlet also decreased to values in the range 70-76% at those higher OLR, in spite of gross VFA production being the highest (14.8 gCOD $L^{-1}d^{-1}$) at the highest OLR.

| OLR (gCOD $L^{-1}d^{-1}$) | COD removal (%) | DA (%) | pH | Alkalinity (mg $L^{-1}$ as $CaCO_3$) | tVFA (gCOD $L^{-1}$) | Odd-to-even ratio |
|---|---|---|---|---|---|---|
| 30 | 26±7 | 65.6±6.8 | 5.07±0.13 | 1366±203 | 10.3±0.9 | 0.74±0.20 |
| 50 | 50±6 | 36.7±12.8 | 4.86±0.13 | 837±328 | 8.5±2.4 | 0.62±1.46 |
| 100 | 55±13 | 36.1±13.8 | 4.55±0.15 | 492±397 | 14.8±2.0 | 0.37±1.02 |

Table 4. Monitored variables at different organic loading rates (average ± standard deviation)

VFA productions observed in this study were similar to those reported by Davila-Vazquez et al. (2009), which achieved a total VFA production of 9.7 g $L^{-1}$ with an OLR of 55 g lactose $L^{-1}d^{-1}$ and an HRT of 10 h. On the other hand when those authors reduced HRT to 6 h and increased OLR to 92, only 10.6 g $L^{-1}$ of VFA were collected, which are lower than the value achieved in the present study with an OLR of 100 gCOD $L^{-1}d^{-1}$ (14.8 g $L^{-1}$). The main reason for this difference may be the higher retention time (12 h) used in this study that permitted to better exploit acidification skills of the reactor.

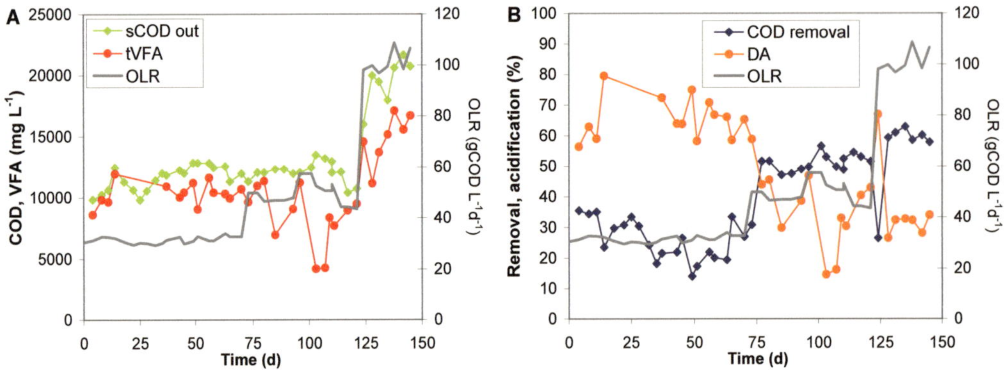


Fig. 14. Performance of MBBR reactors processing cheese whey: A) $sCOD_{out}$ and tVFA; B) COD removal and degree of acidification

Values of pH achieved in all the OLR tested were always higher than 4.5. Bengtsson et al. (2008) reported significant drops in VFA yields for pH values below 4.5, in spite of observing a more stable VFA composition at low values. Those authors also concluded that an increase in pH from 5.25 to 6 produced a shift in VFA profile (odd-to-even ratio ranged from of 25% to 100%), with the predominance of propionic acid. In the present study, a similar trend was detected, since a decrease on the odd-to-even ratio (from 74 to 37%) was observed with the load increase, with the major shifts attributed to a rising of heavier carboxylic chain acids (n-butyric, i-valeric and n-caproic acids) at the higher OLR tested (100 gCOD $L^{-1}d^{-1}$). At last, both degree of acidification and odd-to-even ratio followed trends observed in the batch experiments.

## 4. Conclusions

The main conclusions are:

High strength waste products from food industries can simultaneously be converted to VFA in acidogenic anaerobic reactor with considerable acidification degree (30-65%) depending on type of waste and reach COD reductions of 40-60%.

Complex and/or slowly biodegradable waste products can be transformed in compounds with higher biodegradability which will improve treatment efficiencies and bioreactor stability.

The optimum operational conditions depend on the waste selected. OLR and pH are two parameters that affect the acidification degree. According to the type of wastes used, the OLR shall be 2.5-20 gCOD $L^{-1}d^{-1}$ for slowly biodegradable wastes (spent coffee grounds and dairy fatty slurry) and 30-50 gCOD $L^{-1}d^{-1}$ for high strength wastes such as molasses and cheese whey.

Temperature is the main parameter which affects the composition of volatile acids in the acidified wastewater containing spent coffee grounds. Acetic acid was always predominant in all experiments, with propionic acid being the second at mesophilic temperatures and butyric acid at thermophilic temperatures.

Adaptation of biomass is a feasible strategy for treatment efficiency improvement and/or recovery of bioproducts.

## 5. Acknowledgements

The authors gratefully acknowledge financial support from FCT -Fundação para a Ciência e a Tecnologia, under the project PTDC/AMB-AAC/101050/2008. Flávio Silva also acknowledges his Ph.D. grant (SFRH/BD/46845/2008) from FCT - Fundação para a Ciência e a Tecnologia.

## 6. References

Albuquerque, M. G. E.; Eiroa, M.; Torres, C.; Nunes, B. R. & Reis, M. A. M. (2007). Strategies for the development of a side stream process for polyhydroxyalkanoate (PHA) production from sugar cane molasses. *Journal of Biotechnology*, Vol. 130, No. 4, pp. 411-421, 0168-1656

Albuquerque, M. G. E.; Martino, V.; Pollet, E.; Avérous, L. & Reis, M. A. M. (2011). Mixed culture polyhydroxyalkanoate (PHA) production from volatile fatty acid (VFA)-rich streams: Effect of substrate composition and feeding regime on PHA productivity, composition and properties. *Journal of Biotechnology*, Vol. 151, No. 1, pp. 66-76, 0168-1656

Alexiou, L. (1998). A study of pre-acidification reactor design for anaerobic treatment of high strength industrial wastewaters. PhD Thesis, University of Newcastle upon - tyne, London, England

APHA (2005). *Standard Methods for the Examination of Water and Wastewater* (21st edition), APHA, AWWA, WEF, 0875530478, Washington, D.C.

Azbar, N.; Çetinkaya Dokgöz, F. T.; Keskin, T.; Korkmaz, K. S. & Syed, H. M. (2009). Continuous fermentative hydrogen production from cheese whey wastewater under thermophilic anaerobic conditions. *International Journal of Hydrogen Energy*, Vol. 34, No. 17, pp. 7441-7447, 0360-3199

Backus, B. D.; Clanton, C. J.; Goodrich, P. R. & Morris, H. A. (1988). Carbon-to-nitrogen ratio and hydraulic retention time effect on the anaerobic digestion of cheese whey. *Transactions of the ASAE*, Vol. 31, pp. 1274-1282

Bengtsson, S.; Hallquist, J.; Werker, A. & Welander, T. (2008). Acidogenic fermentation of industrial wastewaters: Effects of chemostat retention time and pH on volatile fatty acids production. *Biochemical Engineering Journal*, Vol. 40, No. 3, pp. 492-499, 1369-703X

Braunegg, G.; Lefebvre, G. & Genser, K. F. (1998). Polyhydroxyalkanoates, biopolyesters from renewable resources: Physiological and engineering aspects. *Journal of Biotechnology*, Vol. 65, No. 2-3, pp. 127-161, 0168-1656

Capela, I. F.; Azeiteiro, C.; Arroja, L. & Duarte, A. C. (1999). Effects of pre-treatment (composting) on anaerobic digestion of primary sludges from a bleached kraft pulp mill, *Proceedings of* Second international symposium on anaerobic digestion of solid wastes, Barcelona, Spain.

Castelló, E.; García y Santos, C.; Iglesias, T.; Paolino, G.; Wenzel, J.; Borzacconi, L. & Etchebehere, C. (2009). Feasibility of biohydrogen production from cheese whey using a UASB reactor: Links between microbial community and reactor performance. *International Journal of Hydrogen Energy*, Vol. 34, No. 14, pp. 5674-5682, 0360-3199

Davila-Vazquez, G.; Alatriste-Mondragón, F.; de León-Rodríguez, A. & Razo-Flores, E. (2008). Fermentative hydrogen production in batch experiments using lactose, cheese whey and glucose: Influence of initial substrate concentration and pH. *International Journal of Hydrogen Energy*, Vol. 33, No. 19, pp. 4989-4997, 0360-3199

Davila-Vazquez, G.; Cota-Navarro, C. B.; Rosales-Colunga, L. M.; de León-Rodríguez, A. & Razo-Flores, E. (2009). Continuous biohydrogen production using cheese whey: Improving the hydrogen production rate. *International Journal of Hydrogen Energy*, Vol. 34, No. 10, pp. 4296-4304, 0360-3199

de Wit, J. N. (2001). *Lecturer's Handbook on whey and whey products*, European Whey Products Association. Retrieved from: http://www.euromilk.org/

Dinsdale, R. M.; Hawkes, F. R. & Hawkes, D. L. (1996). The mesophilic and thermophilic anaerobic digestion of coffee waste containing cofree grounds. *Water Research*, Vol. 30, No. 2, pp. 371-377, 0043-1354

Dinsdale, R. M.; Hawkes, F. R. & Hawkes, D. L. (1997). Mesophilic and thermophilic anaerobic digestion with thermophilic pre-acidification of instant coffee production wastewater. *Water Research*, Vol. 31, No. 8, pp. 1931-1938, 0043-1354

Espinosa, A.; Rosas, L.; Ilangovan, K. & Noyola, A. (1995). Effect of trace metals on the anaerobic degradation of volatile fatty acids in molasses stillage. *Water Science and Technology*, Vol. 32, No. 12, pp. 121-129, 0273-1223

Fang, Herbert H. P. & Yu, H. Q. (2000). Effect of HRT on mesophilic acidogenesis of dairy wastewater. *Journal of Environmental Engineering*, Vol. 123, No. 12, pp. 1145-1148, 0733-9372

Fernandes, M. I. A. P. (1986), *Application of porous membranes for biomass retention in a two-phase anaerobic reactor*, PhD thesis, Department of Public Health Engineering, University of Newcastle upon Tyne, Newcastle, 201 p.

Lemos, P. C.; Levantesi, C.; Serafim, L. S.; Rossetti, S.; Reis, M. A. M. & Tandoi, V. (2008). Microbial characterisation of polyhydroxyalkanoates storing populations selected under different operating conditions using a cell-sorting RT-PCR approach. *Applied Microbiology and Biotechnology*, Vol. 78, No. 2, pp. 351-360, 0175-7598

Lemos, P. C.; Serafim, L. S. & Reis, M. A. M. (2006). Synthesis of polyhydroxyalkanoates from different short-chain fatty acids by mixed cultures submitted to aerobic dynamic feeding. *Journal of Biotechnology*, Vol. 122, No. 2, pp. 226-238, 0168-1656

Malaspina, F.; Cellamare, C. M.; Stante, L. & Tilche, A. (1996). Anaerobic treatment of cheese whey with a downflow-upflow hybrid reactor. *Bioresource Technology*, Vol. 55, No. 2, pp. 131-139, 0960-8524

Myers, R. H.; Montgomery, D. C. & Anderson-Cook, C. (2009). *Response Surface Methodology: Process and Product Optimization Using Designed Experiments* (3rd edition), Wiley, 0470174463, New York.

Najafpour, G. D. & Poi Shan, C. (2003). Enzymatic hydrolysis of molasses. *Bioresource Technology*, Vol. 86, No. 1, pp. 91-94, 0960-8524

Nadais, H.; Barbosa, M. L.; Ramos, C. G.; Grilo, A.; Sousa, S. A.; Capela, I.; Arroja, L. & Leitão, J. H. (2011). Enhancing wastewater degradation and biogas production by intermittent operation of UASB reactors. *Energy*, Vol. 36, No. 4, pp. 2164-2168

Nielsen, H. B.; Uellendahl, H. & Ahring, B. K. (2007). Regulation and optimization of the biogas process: Propionate as a key parameter. *Biomass & Bioenergy*, Vol. 31, No. 11-12, pp. 820-830, 0961-9534

Pisco, A. R.; Bengtsson, S.; Werker, A.; Reis, M. A. M. & Lemos, P. C. (2009). Community Structure Evolution and Enrichment of Glycogen-Accumulating Organisms Producing Polyhydroxyalkanoates from Fermented Molasses. *Applied and Environmental Microbiology*, Vol. 75, No. 14, pp. 4676-4686, 0099-2240

Quan, Z.-X.; Jin, Y.-S.; Yin, C.-R.; Lee, J. J. & Lee, S.-T. (2005). Hydrolyzed molasses as an external carbon source in biological nitrogen removal. *Bioresource Technology*, Vol. 96, No. 15, pp. 1690-1695, 0960-8524

Reis, M. A. M.; Serafim, L. S.; Lemos, P. C.; Ramos, A. M.; Aguiar, F. R. & Van Loosdrecht, M. C. M. (2003). Production of polyhydroxyalkanoates by mixed microbial cultures. *Bioprocess and Biosystems Engineering*, Vol. 25, No. 6, pp. 377-385, 1615-7591

Ren, N.; Li, J.; Li, B.; Wang, Y. & Liu, S. (2006). Biohydrogen production from molasses by anaerobic fermentation with a pilot-scale bioreactor system. *International Journal of Hydrogen Energy*, Vol. 31, No. 15, pp. 2147-2157, 0360-3199

Serafim, L. S.; Lemos, P. C.; Albuquerque, M. G. E. & Reis, M. A. M. (2008a). Strategies for PHA production by mixed cultures and renewable waste materials. *Applied Microbiology and Biotechnology*, Vol. 81, No. 4, pp. 615-628, 0175-7598

Serafim, L. S.; Lemos, P. C.; Torres, C.; Reis, M. A. M. & Ramos, A. M. (2008b). The influence of process parameters on the characteristics of polyhydroxyalkanoates produced by mixed cultures. *Macromolecular Bioscience*, Vol. 8, No. 4, pp. 355-366, 1616-5187

Silva, F.; Nadais, H.; Prates, A.; Arroja, L. & Capela, I. (2009). Molasses as an external carbon source for anaerobic treatment of sulphite evaporator condensate. *Bioresource Technology*, Vol. 100, No. 6, pp. 1943-1950, 0960-8524

Wijekoon, K. C.; Visvanathan, C. & Abeynayaka, A. (2011). Effect of organic loading rate on VFA production, organic matter removal and microbial activity of a two-stage thermophilic anaerobic membrane bioreactor. *Bioresource Technology*, Vol. 102, No. 9, pp. 5353-5360, 0960-8524

Yang, K.; Yu, Y. & Hwang, S. (2003). Selective optimization in thermophilic acidogenesis of cheese-whey wastewater to acetic and butyric acids: partial acidification and methanation. *Water Research*, Vol. 37, No. 10, pp. 2467-2477, 0043-1354

Yang, P.; Zhang, R.; McGarvey, J. A. & Benemann, J. R. (2007). Biohydrogen production from cheese processing wastewater by anaerobic fermentation using mixed microbial communities. *International Journal of Hydrogen Energy*, Vol. 32, No. 18, pp. 4761-4771, 0360-3199

Yu, H. Q. & Fang, H. H. P. (2001). Acidification of mid- and high-strength dairy wastewaters. *Water Research*, Vol. 35, No. 15, pp. 3697-3705, 0043-1354

Yu, H. Q. & Fang, H. H. P. (2002). Anaerobic acidification of a synthetic wastewater in batch reactors at 55 °C. *Water Science and Technology*, Vol. 46, No. 11-12, pp. 153-157

Yu, H. Q.; Fang, H. H. P. & Gu, G. (2002). Comparative performance of mesophilic and thermophilic acidogenic uplflow reactors. *Process Biochemistry*, Vol. 38, No. 3, pp. 447-454, 0032-9592

Zayas, T.; Geissler, G. & Hernandez, F. (2007). Chemical oxygen demand reduction in coffee wastewater through chemical flocculation and advanced oxidation processes. *Journal of Environmental Sciences*, Vol. 19, No. 3, pp. 300-305, 1001-0742

Zoetemeyer, R. J.; van den Heuvel, J. C. & Cohen, A. (1982). pH influence on acidogenic dissimilation of glucose in an anaerobic digestor. *Water Research*, Vol. 16, No. 3, pp. 303-311, 0043-1354

# 13

# Status and Prospects of Concentrated Organic Wastewater Degradation

Wei Liu[1,2], Dushu Huang[1,2], Ping Yi[1,2], Ying Li[1]
[1]*College of Science, Honghe University,*
[2]*Key Laboratory of Natural Pharmaceutical and Chemical Biology of Yunnan Province, China*

## 1. Introduction

Clean fresh water is essential to our life. Unfortunately, in our consumer-based economy, a large amounts of wastewater having been generated by industrial plants. Wastewaters produced in many industrial plants often contain organic compounds which are toxic and not amenable to direct biological treatment. These wastewaters must be treated in order to meet the specifications for discharge or for recycling in the process. Processes that enable the recycling of industrial wastewaters are becoming increasingly important as the amount and quality of freshwater available in certain regions of the world continues to decrease due to long periods of drought and growing water demands. In addition, increasingly stricter wastewater discharge standards continue to be introduced worldwide in an effort to reduce the environmental impacts of industrial processes. This has led to an increasing amount of research being conducted on processes for removing organic and inorganic pollutants from wastewaters.

Currently, biological and chemical methods are almost used to degrade the industrial organic wastewater. The biological methods can not achieve the desired results in most situations because most organic waste is toxic and difficult to degrade. For example, the toxic aqueous phenol solutions with concentrations exceeding 0.5 g/L should not be treated in biological plants, even though laboratory tests have performed degradation of solutions with up to 2 g/L of phenol (Foster, 1985). The incapability of conventional methods to effectively remove organic pollutants has made it evident that new, compact and more efficient methods with low energetic and operation costs are needed.

The chemical methods include wet oxidation processes, photocatalytic oxidation, sonochemical degradation technique and electrochemical oxidation technique. These alternative methods are the so-called advanced oxidation processes which have been reported to be effective for degradation of organic contaminants from wastewater. They are based on the initial formation of radicals such as hydroxyl radicals that later act as non-selective oxidation agents (Glaze, 1994; Ledakowicz, 1998; Scott & Ollis, 1995). These methods are limited to treat waters which contain low concentrations of organic substances, and they are almost used in acidic, neutral or weak basic industrial organic wastewater treatment but not basic especially strong basic organic wastewater.

## 2. Wet oxidation and catalytic wet oxidation

Wet oxidation (WO) which includes wet oxidation and catalytic wet oxidation (CWO), wet air oxidation (WAO) and catalytic wet air oxidation (CWAO) and catalytic wet peroxide (CWPO) can be distinguished in the oxidation processes which are carried out at high temperature and pressure conditions. Among the chemical methods of organic wastewater degradation, CWO is considered most prospective to achieve large-scale industrialization in a short time. WO and CWO processes that are used for removing vast variety of organic compounds from solution have been the subjects of considerable studies over the last six decades as numerous researchers continue to investigate the ability of these processes to remove different types of organic compounds from a wide variety of simple and complex wastewaters.

### 2.1 Fundamentals of WO and CWO

The WO oxidation process, which was first patented by Zimmerman in the late 1950s (Zimmerman, 1950), removes organic compounds in the liquid phase by oxidizing them completely using an oxidant such as oxygen or air. The process is extremely clean because it does not involve any harmful chemical reagents and the final products are carbon dioxide and water. However, one of the significant drawbacks of the WO process is that high temperatures and pressures are usually required to achieve a high degree of oxidation of many organic compounds within a reasonable amount of time.

Solution pH can influence WO process if a carboxylic acid and its corresponding salt have significantly different chemical stabilities. The difference in WO between acetic acid and the sodium salt of acetic acid is an example. In the research of Mishra (Mishra et al. 1995), the resonance stabilization energy makes the sodium salt more difficult to oxidize. The difference in WO process between oxalic acid and the sodium salt of oxalic acid can also be explained. Another possible effect that solution pH can have on the WO of some carboxylic acids involves the type of intermediates formed. In some cases, the solution pH can alter the reaction pathway of a carboxylic acid, leading to greater or lesser formation of certain intermediates.

The high temperatures, pressures, and reaction times usually required to achieve complete oxidation of organic compounds that are present in various wastewaters has led to a considerable amount of research on WO catalysts to overcome the costly, high-pressure, energy-intensive conditions. The number of different industrial waste streams requiring organics removal and the diversity of organic compounds has resulted in the investigation of a wide range of homogeneous and heterogeneous catalysts over the last three decades.

Catalytic wet oxidation (CWO) is an upgraded wet oxidation with the incorporation of suitable catalysts. It is a reaction involving an organic compound in water and oxygen over a catalyst and it can be used at much milder temperature and pressure conditions so that it can reduce the capital costs and the corrosion. Heterogeneous oxidation involves intensive contacting of an organic compound in solution with oxygen over a solid catalyst. Heterogeneous systems have the advantages over homogenous systems because the catalysts can be separated much more easily after the process (Luck, 1999; Matatov-Meytal & Sheintuch, 1998; Mishra, 1995). The process of CWO and WO are considered to be very similar because acetic acid can be identified at the end of the reaction (Debellefontaine et al., 1996).

The wet-air oxidation (WAO) process which generates active oxygen species such as hydroxyl radicals is also known to have a great potential for the treatment of wastewater containing a high content of organic compounds, for which traditional biological process is unfeasible (Mishra, 1995). The WAO process often takes place at high temperatures and pressures. In WAO processes, the organic contaminants dissolved in water are degraded by means of an oxidizing agent into biodegradable intermediates or inorganic compounds such as $CO_2$, $H_2O$ and inorganic salts, which remain in the aqueous phase. One of the main drawbacks of the WAO process is its inability to achieve complete mineralization of organics. According to the WAO process, the organic compounds are oxidized in the liquid phase by some gaseous source of pure oxygen or air. Sulphur is converted to sulphate, halogens to halides and phosphorus to phosphates. Organic nitrogen may produce ammonia, nitrate and nitrogen. WAO produces no $NO_x$, $SO_2$, HCl, dioxins, furans, fly ash, etc. Typical operating conditions are in the range of 100~300℃ and 0.5~20MPa. The main differences between the processes consist in the reactor type and the catalyst used.

Compared to conventional wet-air oxidation, catalytic wet-air oxidation (CWAO) has lower energy requirements (Alejandre et al., 2001; Lin et al., 2002; Silva et al., 2003; Bhargava et al., 2006). One can use less severe reaction conditions to reduce chemical oxygen and achieve much higher oxidation rates due to the presence of a catalyst. In the CWAO process, organics are oxidized to innocuous inorganic compounds such as $CO_2$, $H_2O$ and heteroatom dissolved ions at much lower temperatures and pressures than in uncatalyzed processes. Interest in innovative methods of wastewater treatment based on catalytic oxidation has been growing rapidly, as this technique has been confirmed as a powerful method of purifying wastewaters in numerous studies (Silva et al., 2003; Abecassis-Wolfovich et al., 2004; Garcia et al., 2006; Gomes et al., 2005; Minh et al., 2006). The catalysts that have been used in CWAO include metal oxides, supported noble metals and active carbon. The metal oxides such as CuO, ZnO, CoO, $Al_2O_3$, MnO and CeO etc. are all be used for aqueous effluents oxidation (Pintar & Levec, 1994; Hamoudi et al., 1998; Chen et al., 2001; Pintar & Levec, 1992). The noble metals of Pd, Pt, Ru and Ir have been very effective in the treatment of various compound, and the deposition of noble metals on hydrophobic supports such as active carbons and styrene divinyl benzene co-polymer, is also effective for oxidation of pollutants such as ammonia (Kim & Ihm, 2002; Barbier et al., 2002; Pintar et al., 2001; Huang et al., 2001).

Wet peroxide oxidation (WPO) was proposed to use hydrogen peroxide as the oxidizer instead of molecular oxygen. As opposed to WAO-WO, which employ a gaseous source of oxidizing agent and which is a two-step process including mass transfer and oxidation, WPO uses a liquid oxidizing agent of hydrogen peroxide which eliminates the mass transfer problems. The reports about the WPO are scare, but its good efficiency lately proved has attracted attention. It has been proved to be effective under mild conditions for the treatment of phenolic compounds. The promising results about the degradation of phenolic compounds over several catalysts, such as iron aluminium combinations and Fe-Zeolites respectively have been obtained (Debellefontaine et al., 1996; García-Molina et al., 2006; Okawa et al., 2005; Najjar et al., 2005; Nikolopoulus et al., 2006).

### 2.2 Application of WO and CWO

There are many kinds of wastewater generated in chemical industries, such as pulp mill wastewater, dyeing and printing industrial wastewater, distillery wastewater, coking process wastewater and wastewater of paper mill and leather industries.

The pulp wastewater contains the phenols and other organic compounds such as chloro-, thio-, and alkali-lignin. These compounds are resistant to biological degradation. Prasad (Prasad & Joshi, 1987) studied the CWO of black liquor from the Kraft pulp industry, which is traditionally treated by incineration with cupric and zinc oxides, as well as manganese and selenium dioxides as catalysts. Sonnen et al. (Sonnen et al., 1997) found that the addition of transition metals to the reaction mixture can alleviate the formation of refractory compounds without compromising the rate of COD removal. An et al. found that alumina-supported palladium catalysts were very effective in removing pollutants and color from a Kraft pulp mill (An et al., 2001).

Wastewater from the dyeing and textile industries causes environmental problems because of the high COD loading and the color of these wastes. The color of these wastes is often strong enough to adversely affect the transmission of light, and the high COD loading depletes the oxygen content of the waterway, thus destroying marine life and the fragile ecosystems in waterways. WO has been used to these wastes. Lei et al. studied five different types of wet oxidative processes: WAO, CWAO, promoted CWAO, WPO and catalytic wet peroxide oxidation (CWPO) (Lei et al., 1998). Oxygen in excess is used for the first three of these processes, whereas no oxygen is required in the latter two. Chen et al. found that WAO and CWAO process which using $Cu^{2+}$ ions could not be applied successfully to high-concentration reactive dyes at 200℃. $H_2O_2$ was used as a promoter at a lower temperature of 150℃ in order to improve COD reductions (Chen et al., 1999). Raffainer and van Rohr used promoted WO to the catalytic destruction of Orange II which is a representative contaminant from dyeing industries (Raffainer & van Rohr, 2001). This process using $Fe^{2+}$ ions at mild temperatures gave promising results. WO at a partial oxygen pressure of 1 MPa, a temperature of 190℃, and a pH of 2 showed complete conversion of the dye in just an hour.

The process of converting biomass to ethanol produces wastewater, known as stillage, which is highly loaded with oxygen-demanding materials. Belkacemi et al. evaluated three different methods such as thermolysis, noncatalytic WO and CWO for the degradation of organics (Belkacemi et al., 1999). Dhale et al. selected wet oxidative treatment after the waste was treated by a thermal membrane pretreatment process (Dhale & Mahajani, 2000). The pretreatment, which reduced the COD by 40% and the color by 30%, was then followed by WO at 180-225℃ and $PO_2$ 0.69-1.38 MPa. A homogeneous $FeSO_4$ catalyst at 210℃ reduced the COD by 60% within 2 h and removed 95% of the color. This is only a slight improvement over noncatalytic wet oxidation, which achieved the same result at 220 ℃.

Leather-processing plants are required to evaluate their wastewater and reduce the concentration of toxic compounds. Wastewater from leather production contains high concentrations of inorganic and organic pollutants, which are usually removed by a combination of physical separation and WAO methods. Comadran et al. studied the deodorization and decolorization of wastewater from the leather and hide industries. A heterogeneous catalytic system was used in a single-stage process to effectively destroy odors caused by volatile organic compounds (Comadran & Comella, 2002 ).

### 2.3 Mechanism of degradation reactions

In most systems, both WO and CWO occur simultaneously and reaction pathways are usually very similar. In this section, the chemistry of reaction kinetics and the types of chemical reactions that can lead to oxidation of various organic compounds are discussed.

### 2.3.1 Free-radical and non-free-radical reactions of WO process

In WO process, in order to improve the rates of the chemical reactions, a better understanding of the types of reactions is required. The general consensus among researchers in the field is that the chemical reaction occurs mostly via free-radical chemical reactions. Once the radicals are generated, they virtually attack all organic compounds. According to the nature of the organic compounds, two types of initial attack are possible. One is that the radical abstracts a hydrogen atom to form water as with alkanes or alcohols. The other one possibility consists of an electrophylic addition of the radical to double bonds. After the addition of the radical, free organic radicals are generated that react with oxygen molecules generating a peroxiradical and allowing the initiation of a chain reaction system that ends in the complete mineralization of the compounds (Buxton et al., 1988; Glaze & Kang, 1989; Haag & Yao, 1992). Some examples of the different types of free-radical chemical reactions that can occur are showed by Eqs. 1-26 (Mantzavinos et al., 1997; Ingale et al., 1996; Robert et al., 2002; Rivas et al., 1998; Emanuel et al., 1980; Wakabayashi & Okuwaki, 1988; Eyer, 2001; Li et al., 1991; Farhataziz & Ross, 1977; Patterson et al., 2001; Glaze et al., 1995; Ichinose, S.; Okuwaki, 1990 ):

$$RH + O_2 \rightarrow R\bullet + HO_2\bullet \tag{1}$$

$$H_2O + O_2 \rightarrow H_2O_2 + O\bullet \tag{2}$$

$$H_2O + O_2 \rightarrow HO_2\bullet + \bullet OH \tag{3}$$

$$RH \rightarrow R\bullet + H\bullet \tag{4}$$

$$O_2 \rightarrow O\bullet + O\bullet \tag{5}$$

$$H_2O \rightarrow \bullet OH + H\bullet \tag{6}$$

$$RH + O_2 + RH \rightarrow 2R\bullet + H_2O_2 \tag{7}$$

$$XCH_2^- + O_2 \rightarrow XCH_2\bullet + O_2\bullet^- \tag{8}$$

$$C_6H_5O^- + O_2 \rightarrow C_6H_5O\bullet + \bullet O_2^- \tag{9}$$

$$O_2 + e^- \rightarrow \bullet O_2^- \tag{10}$$

$$R\bullet + O_2 \rightarrow ROO\bullet \tag{11a}$$

$$ROO\bullet + RH \rightarrow ROOH + R\bullet \tag{11b}$$

$$O\bullet + H_2O \rightarrow HO\bullet + HO\bullet \tag{12}$$

$$RH + HO_2\bullet \rightarrow R\bullet + H_2O_2 \tag{13}$$

$$H\bullet + H_2O \rightarrow HO\bullet + H_2 \tag{14}$$

$$ROOH \rightarrow RO\bullet + HO\bullet \tag{15}$$

$$RH + HO\bullet \rightarrow R\bullet + H_2O \tag{16}$$

$$XCH_2OO\bullet + XCH_2^- \rightarrow XCH_2OO^- + XCH_2\bullet \tag{17}$$

$$2ROO\bullet \rightarrow ROOR + O_2 \tag{18}$$

$$2HO_2\bullet \rightarrow H_2O_2 + O_2 \tag{19}$$

$$HO_2\bullet + OH\bullet \rightarrow H_2O + O_2 \tag{20}$$

$$XCH_2OO\bullet + O_2\bullet^- \rightarrow XCH_2^- + 2O_2 \tag{21}$$

$$2O_2\bullet^- + H_2O \rightarrow HOO^- + HO^- + O_2 \tag{22}$$

$$HO\bullet + CO_3^{2-} \rightarrow {}^-OH + \bullet CO_3^- \tag{23}$$

$$O_2\bullet^- + \bullet CO_3^- \rightarrow O_2 + CO_3^{2-} \tag{24}$$

$$HO\bullet + HCO_3^- \rightarrow H_2O + \bullet CO_3^- \tag{25}$$

$$ROO\bullet + O_2\bullet^- \rightarrow ROO^- + O_2 \tag{26}$$

The radical pathway is very complex and it involves three main type chemical reactions such as initiation, propagation, and termination. Eqs. 1-10 above-mentioned belong to the type of initiation, and Eqs. 11-17 the type of propagation, the residual equations belong to the type of termination.

Although the majority of chemical reactions that lead to oxidation of organic compounds during WO are free-radical reactions, there are various other reactions that can lead to oxidation of an organic compound under typical WO conditions. Examples include removal of an α-hydrogen from a carboxylic acid/salt by hydroxide (Eq. 27), removal of an alcoholic hydrogen from phenol (Eq. 28), and base-induced retro-aldol reaction of α-β-hydroxyacids (Eq. 29).

$$XCH_3 + {}^-OH \rightarrow XCH_2^- + H_2O \tag{27}$$

$$C_6H_5OH + {}^-OH \rightarrow C_6H_5O^- + H_2O \tag{28}$$

$$C_6H_8O_7 + {}^-OH \rightarrow CH_3CO_2H + HO_2CCOCH_2CO_2H + {}^-OH \tag{29}$$

#### 2.3.2 Degradation process of CWAO

CWAO is divided into homogeneous and heterogeneous catalytic wet oxidation in industry (Bhargava et al., 2006; Kolaczkowski et al., 1999; Patria et al., 2004; Luck, 1996; Ishii et al.,

1997). Bayer Loprox (Kolaczkowski et al., 1999) and ATHOS (Patria et al., 2004) methods have been used in industry are belong to homogeneous catalytic wet oxidation, which have some shortcomings including high reaction temperature, high reaction pressure and difficult to recover the homogeneous catalysts. On the other hand, Nippon Shokubai (Kolaczkowski et al., 1999; Luck, 1996) and Osaka methods (Luck, 1996; Ishii et al., 1997) are belong to heterogeneous catalytic wet oxidation, which can overcome the shortcomings of heterogeneous catalytic wet oxidation, but the service time and the stability of the catalyst is still not perfect. Although the catalytic wet oxidation has practical applications in industry, it is difficult to be directly applied to the basic organic wastewater because the HO generated in the process can be captured by $OH^-$, $CO_3^{2-}$, $HCO_3^-$ and the loss of oxidative capacity of HO can be achieved in the basic conditions.

$$HO\bullet + CO_3^{2-} \rightarrow {}^-OH + \bullet CO_3^- \tag{30}$$

$$HO\bullet + HCO_3^- \rightarrow H_2O + \bullet CO_3^- \tag{31}$$

$$HO\bullet + OH^- \rightarrow H_2O + O\bullet^- \tag{32}$$

$$2HO\bullet + CO_2^- \rightarrow H_2O + \bullet CO_3^- \tag{33}$$

Eqs. 30-33 shows the probable process of hydroxyl radical captured. It can be seen from the figure that in alkaline conditions, $OH^-$,$CO_3^{2-}$ and $HCO_3^-$ can capture the $HO^\bullet$. It make the $HO^\bullet$ lost the ability to further oxidation of organic compounds, which is now the difficulty of CWO directly used for organic alkaline wastewater.

Wakabashi and Okuwaki found that the metal copper powder can be an effective catalyst for the degradation of sodium acetate at 250 °C, and a plausible but not confirmed catalytic mechanism (see Eqs. 34-38) was proposed (Wakabayashi & Okuwaki, 1988). They considered that the acetate first generated to $XCH_2^-$, then through $XCH_2^\bullet$ and $XCH_2OO^\bullet$, $XCH_2OO^-$ was generated in the reaction process under alkaline conditions, and $XCH_2OO^-$ is prone to further oxidation of oxalate, and carbonate and bicarbonate were finally achieved.

Wakabayashi and Okuwaki also found that increasing the alkalinity resulted in an increase in sodium acetate oxidation, using an iron powder catalyst in a nickel reactor. The nickel reactor was determined to corrode under the experimental conditions here, and, hence, nickel oxide acted as a co-catalyst for the oxidation reaction. They proposed a base-catalyzed oxidation mechanism for the oxidation of acetate in the presence of iron powder and nickel oxide, where the rate of formation of nickel oxide seems to be one of the rate-determining steps. The effect of alkalinity on the CWO reaction mechanism proposed by Wakabayashi and Okuwaki is an example of an indirect effect on the CWO mechanism. In this mechanism, hydroxide is required to increase the rate of formation of an intermediate species with which the catalyst reacts.

$$XCH_3 + OH^- \xrightarrow{K} XCH_2^- + H_2O \tag{34}$$

$$XCH_2^- + O_2 \xrightarrow{k(MeO)} XCH_2\bullet + O_2^- \tag{35}$$

$$XCH_2 \bullet + O_2 \rightarrow XCH_2OO \bullet \tag{36}$$

$$XCH_2OO \bullet + XCH_2^- \rightarrow XCH_2OO^- + XCH_2 \bullet \tag{37}$$

$$XCH_2OO \bullet + O_2^- \rightarrow XCH_2OO^- + O_2 \bullet \tag{38}$$

CWAO has got its success in laboratory applications but not yet the industrial recognition met with non-catalytic WAO. The main reasons are that the homogeneous catalysts have to be removed in a subsequent step, while the heterogeneous catalysts have to maintain their activity for sufficiently long periods. Many types of CWAO processes including NS-LC process, Osaka gas process and Kurita process are used currently (Kolaczkowski et al, 1999; Luck, 1984; Debellefontaine, 2000). The NS-LC process uses a vertical monolith reactor with a Pt–Pd/$TiO_2$ –$ZrO_2$ catalyst. The operating conditions are 220℃ and 4 MPa. The Osaka gas process uses a mixture of precious and base metals on titania or zirconia-titania supports. Typical operating conditions are 250℃ and 6.86 MPa. The Kurita process uses nitrite instead of oxygen, and a similar catalyst, becoming more effective at lower temperatures, around 170℃. Surprisingly, the industrial applications of CWAO operate at temperatures and pressures that are not significantly lower than those encountered in WAO. In addition, they use expensive noble metal catalysts.

### 2.3.3 Mass transfer mechanisms of the WO reaction

The mass transfer mechanisms involved in WO process are schematically illustrated in Fig.1. The following steps are necessary:

1. Oxygen diffusion from gaseous to liquid phase. In this step the oxygen contained in the gaseous phase transport from the bulk of the gaseous phase to the interface with the liquid phase, and then transport crosses the Gaseous-Liquid boundary layers and diffusion into the bulk of the liquid phase. However, this interface offers certain resistance to be crossed and it is necessary to reduce the thickness of the boundary layer as much as possible. By keeping turbulence in the liquid phase the boundary layer becomes thinner and the oxygen mass transfer improves. Despite the efforts to increase the efficiency of this mechanism, it is usually the responsible for the physical resistance in the whole process.
2. Diffusion of organic compounds from solid to liquid phase. The organic compounds of the solid phase transport cross the Solid-Liquid interface and dissolve into the bulk of the liquid phase. Normally this step does not present an important resistance to the whole process in view of the fact that the high temperature provokes a fast diffusion and dissolution of the solids.
3. Organics degradation reaction. As be shown by step 3 in the figure, the organics from the solid degraded by the oxygen from the gaseous phase. This WO reaction takes place in the liquid phase. The rate of the reaction depends on many factors such as temperature, pressure and catalyst.
4. Desorption of gaseous products. The $CO_2$ formed in the course of the reaction transport from the liquid phase to the gaseous phase. This step does not suppose an important resistance to the whole process, not withstanding the fact that high pressure conditions complicate the diffusion of the gas from one phase to the other.

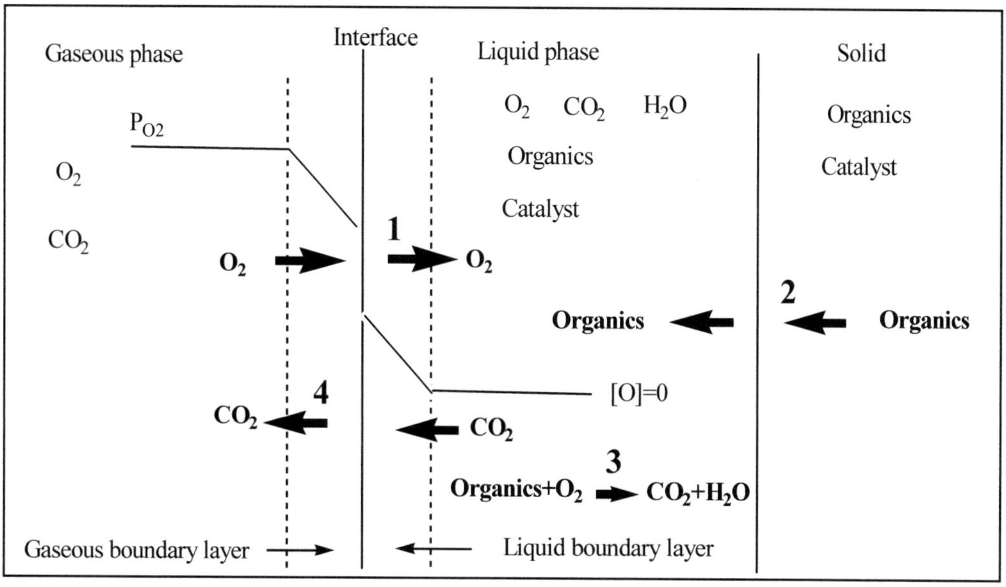


Fig. 1. Schematic representation of the mechanisms involved in wet oxidation processes

From the figure it can be observed that oxygen is transferred from the gaseous phase to the liquid phase (step 1), meanwhile organics diffuse from the solid phase to the liquid phase (step 2). The third step in the reaction mechanism is the oxidation of the organics by means of the oxygen (step 3). As a result of the reaction $CO_2$ is formed which diffuses from the liquid phase to the gaseous phase (step 4).

Attending to the previous description of the mechanisms, it can be deduced that the controlling stages of the WO process are the diffusion of the oxygen from the gaseous phase and the chemical reaction. Three extreme cases are conceivable according to the rates of these mechanisms (Beyrich et al., 1979):

**Case 1.** The reaction rate is far higher than the diffusion rate of oxygen. The oxygen diffusing through the film is completely consumed in the film by a very rapid reaction. The rapid disappearance of the dissolved oxygen gives rise to a high concentration gradient at the interface of the gaseous phase and the liquid phase, and the oxygen is not present in the bulk of the liquid phase.

**Case 2.** The reaction rate is higher than the diffusion rate of oxygen. The reaction takes place essentially in the bulk of the liquid but the liquid dissolved oxygen concentration is low due to the reaction. So, the transfer of oxygen from the gaseous phase to the liquid phase controls the whole process.

**Case 3.** The reaction rate is far lower than the diffusion rate of oxygen. The reaction is so slow that the concentration of the dissolved gas in the bulk of the liquid attains approximately the interface concentration or even the saturation concentration. In this case, the reaction controls the whole process.

According to Beyrich (Beyrich et al., 1979), only the two last cases are of major importance for Wet Air Oxidation.

## 3. Photocatalysis oxidation

Photocatalysis is generally thought of as the catalysis of a photochemical reaction at a solid surface, usually a semiconductor. The early photocehmical studies on aqueous solutions of organic wastewater in the presence of inorganic semiconductors such as silver halides, ZnO, and the like were driven by applications in electro- and photography (Gerischer & Willig, 1976; Memming, 1984). The first example for observing dye instability in the presence of an inorganic semiconductor ($TiO_2$) and illumination appears to be in 1969 when the photocatalytic reduction of Methylene blue to the leuco form was reported (Pamfilov et al., 1969; Yoneyama et al., 1972). This was rapidly followed by other studies.The photocatalytic oxidation of dyes such as Methylene Blue, Rhodamine B, Fluorescein and Methyl Orange was reported (Matthews, 1987, 1991). The early studies were accompanied soon by finding in many laboratories which showed that many organic compounds could be decomposed in aqueous media with a combination of $TiO_2$ and near-UV light.

### 3.1 Mechamisms of photocatalysis

It is generally considered that photocatalysis is based in "back-to-back" or short-circuited photoelectrochemical and electrochemical reactions, involving electrogenerated electrons and holes. At the most global level, these can be written:

$$hv \rightarrow e^-_{CB} + h^+{}_{VB} \tag{39}$$

$$2H_2O + 4h^+{}_{VB} \rightarrow O_2 + 4H^+ \tag{40}$$

$$O_2 + 4H^+ + 4e^-{}_{CB} \rightarrow 2H_2O \tag{41}$$

Reactions (40) and (41) can be designated as the oxygen photoevolution reaction (OPER) and the oxygen reduction reaction (ORR), respectively. Of course, the later can occur either, as shown. On the titania surface itself or on a separate electrode. These processes can be examined sparately with standard electrochemical methods and can provide an overall rationale for the energetics. The point at which the anodic and cathodic currents are equal is a good indication of the local current that could be expected under a given set of conditions. The basic idea has been uesd for many years to quantitatively analyze the energetics of local electrochemical cells on corroding metals (Wagner et al, 1938, 2006).

Most of the organic pollutants in water can be completely decomposed and mineralized at the surface of UV-excited $TiO_2$ photocatalysts; these include alkanes, haloalkanes, aliphatic alcohols, carboxylic acids, alkenes, aromatics, haloaromatics, polymers, surfactants, herbicides, pesticides, and dyes. While only UV light and $O_2$ are necessary for the reactions, many factors such as light intensity, PH, ions, photocatalysts, kinds and concentrations of substrateds, etc., have a great influence on the efficiency of the mineralization process. As in the gas-solod system, photocatalytic reactions in water work best at room temperature, and thus no heating is needed.

Serpone and co-workers studied the quantum yields of liquid-solid photocatalytic reations on $TiO_2$ slurry photocatalysts (Salinaeo et al., 1999; Serpone & Salinaro, 1999; Serpone 1997). Similarly to the study of Ohko et al., they found that maximum quantum yields could be obtained under light-limited conditions, that is, low light intensity and relative high substrate concentration (Emeline et al., 2005), The maximum QY (365 nm) for phenol degradation was measured as 14% using Degussa P25 as a photocatalyst at pH3, and this value varied over a range of 3.5%-30% for six types of $TiO_2$ photocatalysts tested (Salinaeo et al., 1999; Serpone & Salinaro, 1999). Aside from the light-limited condition, over a wide range of experimental conditions, liquid-solid photocatalytic reactions on $TiO_2$ can also be described by the Eq. $r=k\Gamma I^{\alpha}$, ( Here, r is the reaction rate, k the firt-order rate constant, $\Gamma$ the concentration per unit real surface area, and I the light intensity. Under light-rich conditions ,the value of $\alpha$ is between 0 and 1), the same as for gas-solid photocatalytic reactions. The QYs measured in these conditions varied over the wide range of 0.1%-10%, which was several times lower than those of gas-solod photocatalytic reactions. The low QY should restrict the applications of $TiO_2$ photocatalysis in water purification. The intrinsic reasons for the lower QY in water vs. air are yet to be explained in detail.

Pichat et al. Studied the influence of the sintering process on the photocatalytic removal rate of various organic pollutants in water (Enriquea et al., 2007; Agrios & Pichat, 2006; Enriquez et al., 2006). While the increase of sintering temperature is expected to decrease the recombination rate of charge carriers; it does result in a decrease in surface area. Their purpose was to examine the net effect of the sintering process on photocatalytic reactions. In one study, they compared the removal rate of three chlorophenolic compounds and one chloroaliphatic acid compound on four $TiO_2$, samples, which were all obtained identically by $TiOSO_4$ thermohydrolysis with subsequent calcinations at various temperatures (Enriquez et al., 2006). They found that the removal rate increased with the sintering temperatures for the three chlorophenolic compounds, whereas it was the opposite for the aliphatic acid compound. They suggested that the hole attack mechanism for carboxylic acids is much more sensitive to surface area variation than would be the •OH radical mechanism for cholorophenolic compounds ,which can react in the near-surface solution phase. Their studies demonstrate suggest the difficulty in finding a high-efficiency photocatalyst versatile for all types of pollutants in water(Enriquea et al., 2007).

Generally, loading of noble metal co-catalysts such as Pt, Au, and Pd, etc., enables accelerating liquid-solid photocatalytic reactions (Sakthivel et al, 2004). These co-catalysts could enhance the charge separation and catalyze the oxygen reduction reaction. Recent studies suggest that addition of electron receptors, such as hydrogen peroxide, ozone, persulfate, etc., also works well for improving reaction rates (Kositai et al., 2004; Agustina et al, 2005). These electron receptors are more reducible than molecular oxygen. Moreover, when accepting an electron, these chemicals will dissociate into highly reactive radicals in subsequent reactions, which can also participate in the photocatalytic reactions (Akira, 2008).

$$H_2O_2 + e^- \rightarrow OH^- + \bullet OH \tag{42}$$

$$O_3 + e^- + H^+ \rightarrow O_2 + \bullet OH \tag{43}$$

$$S_2O_8^{2-} + e^- \rightarrow SO_4^{2-} + SO_4^- \bullet \tag{44}$$

## 4. Sonochemical degradation

Sonochemical processes have been widely used in chemistry and chemistry and chemical engineering field. Recently, these processes have found new applications in the environmental field, because of advantages in terms of operational simplicity, secondary pollutant formation and safety. Sonochemical treatment had been found to be one of the successful technologies for degradation of organic pollutants such as chlorinated aromatic and aliphatic compounds, phenolic compounds and dyes (Teo et al., 2001; Maleki et al., 2005; Vajnhandl & Marechal, 2007).

### 4.1 Basic principles of sonochemistry

#### 4.1.1 Cavitation

Sonochemistry is principally based on acoustic cavitation which includes the formation, growth, and implosive collapse of bubbles in a liquid. The diffuse energy of sound is enhanced through cavitation (Suslick, 1990). Positive and negative pressures are exerted on a liquid by compression and expansion cycles respectively of ultrasound waves (Suslick, 1989). When a sufficiently large negative pressure is applied to the liquid, the average distance between the molecules would exceed the critical molecular distance necessary to hold the liquid intact, and the liquid will break down and voids or cavities will be created; cavitation bubbles will then be formed (Mason & Lorimer, 1988). The negative pressure works against the tensile strength of the liquid and thus it depends on the type and purity of the liquid. In pure water more than 1000 atm of negative pressure would be required for cavitation whereas in tap water, negative pressure of only a few atmospheres will form bubbles (Suslick, 1990). Once produced, these cavities, voids, or bubbles may grow in size until the maximum of the negative pressure has been reached (Mason & Lorimer, 1988). The growth of cavitation bubble is mainly sound intensity dependent. Sometimes the cavity expansion occurs so quickly by high-intensity ultrasound during the negative-pressure cycle that the positive pressure in the next cycle cannot reduce the size of the cavity. For lowintensity ultrasound the size of the cavity oscillates in phase with the expansion and compression cycles and in this case the bubble surface area is comparatively larger during expansion cycles than during compression cycles. There is a critical size of cavitation bubble which depends on the ultrasound frequency and in this situation the cavity can absorb energy more efficiently from ultrasound and grow more rapidly. But when a cavity experiences a very rapid growth it can no longer absorb energy as efficiently from the sound waves and thus the liquid rushes in and the cavity implodes (Suslick, 1990). Gas and vapors are compressed inside the cavity and this generates heat, which finally produces a short-lived localized hot spot. In brief the cavitational collapse creates an unusual environment for a chemical reaction in terms of enormous local temperatures and pressures (Suslick, 1989). The most sophisticated models predict temperatures of thousands of degrees Celsius, pressures of from hundreds to thousands of atmospheres and heating times of less than a microsecond (Suslick, 1990).

#### 4.1.2 Sonochemical reaction schemes

The sonochemical reactions differ from chemical reactions in many ways. The ultrasonic energy influences the chemical reactions by providing huge heat (pyrolysis) or producing reactive free radicals (Suslick, 1990). Again ultrasonic waves increase the mass transfer rate in an aqueous solution via turbulence (Gondrexon et al, 1997; Weavers & Hoffmann, 1998). There are mainly three reaction sites (Fig.2): a) cavity interior; b) gas-liquid interface; and c) bulk liquid. Inside the cavitation bubble water molecules are pyrolyzed forming •OH and •H radicals in the gas phase. The substrate either reacts with the hydroxyl radical or undergoes pyrolysis. In the interfacial region, a similar reaction occurs but in an aqueous phase. The additional reaction is the recombination of •OH radicals to form $H_2O_2$. In bulk phase the reactions are basically between the substrate and the •OH radical or $H_2O_2$. All these reactions are considered in homogeneous sonochemistry (Adewuyi, 2001). Most of the hydrophobic compounds react inside the cavitation bubble whereas hydrophilic substances react at bulk phase (Adewuyi, 2001; Liang et al., 2007). The heterogeneous systems also follow the same physical mechanism but differ in terms of cavitational threshold, high speed liquid jet etc. (Liang et al., 2007).

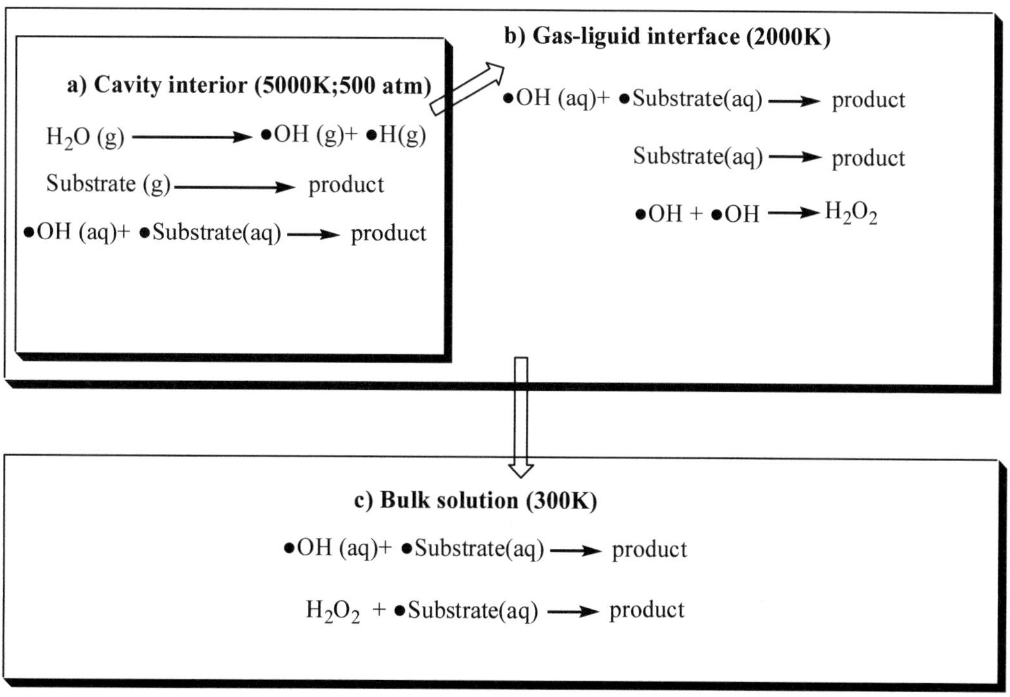


Fig. 2. Sonochemical reaction zones

The degradation of organic compounds mainly follows either thermal decomposition or free radical reaction mechanism. Inside a cavitation bubble only thermal decomposition (pyrolysis) occurs in gas phase with the rupture of C-C, C=C, C-N, C-O bonds at a very high temperature (~5000 K). At the gas-liquid interface both thermal decomposition (~2000 K) and free radical (OH) reaction may occur side by side. The reaction mechanism is basically controlled by the physical properties of the organic compounds. Volatile and hydrophobic organic compounds are degraded mainly by direct thermal decomposition leading to formation of combustion byproducts whereas hydrophobic and less volatile or nonvolatile organic compounds degrade to form oxidation or reduction byproducts by reacting with •HO radicals or hydrogen atoms diffusing out of the cavitation bubbles (Pankaj & Viraraghavan, 2009).

Sonochemical degradation seems to be a promising technology for degradation of several organic compounds. The degradation process is governed by pyrolysis and/or free radical reactions. The extent of degradation depends on both ultrasonic (frequency, power) and experimental (temperature, pH, dissolved gas) conditions. The substrate characteristics (vapor pressure, density, surface tension etc.) also influence sonochemical reactions. In terms of operation simplicity it is an acceptable treatment process but it is efficient with compounds possessing a high vapor pressure (e.g. ClBz, $CCl_4$). Although the initial rate of sonolchemical degradation is fast, complete mineralization is not achieved in most cases. Sometimes it needs much higher energy for complete mineralization which is economically not favorable. Ultrasound can be used with other advanced oxidation processes in combination to provide a higher efficiency. Most of studies are laboratory oriented which cannot be implemented directly in large scale applications. More studies are needed on kinetics, reactor design and scale-up for practical application (Pankaj & Viraraghavan, 2009).

## 5. Electrochemical oxidation

Electrochemistry, as a branch of physical chemistry plays an important role in most areas of science and technology (Grimm et al, 1998). Electrochemistry offers promising approaches for the prevention of pollution problems in the process industry. The inherent advantage is its environmental compatibility, due to the fact that it uses a clean reagent, the electron. The strategies include both the treatment of effluents and waste and the development of new processes or products with less harmful effects, often denoted as process-integrated environmental protection (Bockris, 1972).

Electrochemical technologies have gained importance in the world during the past two decades. There are different companies supplying facilities for metal recoveries, the treatment of drinking water as well as process waters resulting from tannery, electroplating, dairy, textile processing, oil and oil-in-water emulsion, etc. (Chen, 2004). At present, electrochemical technologies have reached such a state that they are not only comparable with other technologies in terms of cost, but sometimes they are more efficient and compact. The development, design and application of electrochemical technologies in water and wastewater treatment has been focused on particularly in some technologies such as electrodeposition, electrocoagulation, electrofloculation and electrooxidation (Rajeshwar, 1994).

Studies on electrochemical oxidation for wastewater treatment go back to the 20th century (Rajeshwar, 1994), when the electrochemical decomposition of cyanide was investigated (Kuhn, 1971). Extensive investigation of this technology commenced in the 70s, when Nilsson et al. In 1973 investigated the anodic oxidation of phenolic compounds (Nilsson, 1973). Mieluch et al. studied for the first time the electrochemical oxidation of phenol compounds in aqueous solutions (Mieluch, 1975). In 1975, Dabrowski et al. studied the electrochemical purification of phenol-containing wastes in a pilot plant (Dabrowski, 1975), while Papouchado et al. investigated the pathways of phenolic compounds anodic oxidation

(Papouchado, 1975). Later, in 1979, Koile and Jonhson examined the electrochemical removal of phenolic films from platinum anodes (Koile & Johnson, 1979); in the same period, Smith de Sucre obtained relevant results in phenol electro-oxidation during wastewater treatment (Smith, 1979), and in the 80s these studies were continued in collaboration with Chettiar(Chettia, 1981; Chettiar & Watkinson, 1983).

During the last two decades, research work has focused on the efficiency in oxidizing various pollutants at different electrodes, on the improvement of the electrocatalytic activity and electrochemical stability of the electrode materials, the investigation of factors affecting the process performance and the exploration of mechanisms and kinetics of pollutant degradation (Chen, 2004). Experimental investigations, focusing on the behaviour of different anodic materials, have been carried out by different research groups, the results of which warrant a detailed description. Attempts for an electrochemical oxidation/destruction treatment for waste or wastewater can be subdivided into two important categories: direct oxidation at the anode, and indirect oxidation using appropriate anodically-formed oxidants (Juttner et al, 2000).

### 5.1 Electrochemical oxidation mechanism

Electrochemical oxidation of pollutants can occur directly at anodes through the generation of physically adsorbed "active oxygen" (adsorbed hydroxyl radicals, •OH) or chemisorbed "active oxygen" (oxygen in the oxide lattice, $MO_{x+1}$) (Comninellis, 1994). This process is usually called "anodic oxidation" or "direct oxidation" and the course for the anodic oxidation was described by Comninellis (Comninellis, 1994); the complete destruction of the organic substrate or its selective conversion into oxidation products is schematically represented in eq. 45– eq. 50.

When a toxic, non-biocompatible pollutant is treated, the electrochemical conversion transforms the organic substrate into a variety of metabolites; often, biocompatible organics are generated, and biological treatment is still required after the electrochemical oxidation. In contrast, electrochemical degradation yields water and $CO_2$, no further purification being necessary. Nevertheless, the feasibility of this process depends on three parameters: (1) the generation of chemically or physically adsorbed hydroxyl radicals, (2) the nature of the anodic material and (3) the process competition with the oxygen evolution reaction.

A mechanism for the electrochemical oxidation of organics, based on intermediates of oxygen evolution reaction in aqueous media, was formerly proposed by Johnson (Feng & Johnson, 1990; Chang & Johnson, 1990, 1990; Vitt & Johnson, 1992; Feng et al, 1994; Treimer et al, 2001). The process involves anodic oxygen transfer from $H_2O$ to organics via hydroxyl radicals formed by water electrolysis.

The electrochemical oxidation of some organics in aqueous media may take place without any loss in electrode activity, except at high potentials, and with concomitant evolution of oxygen (Comninellis & Nerini, 1995; Comninellis & Plattner, 1998; Tahar & Savall, 1998; Comninellis & Battisti, 1996). Furthermore, it has been described that the nature of the electrode material strongly influences both the selectivity and the efficiency of the process (Comninellis, 1994; Comninellis & Battisti, 1996; Foti et al, 1997; Simond, 1997). To interpret these observations, a comprehensive model for the anodic oxidation of organics in acidic medium, including the competition with the oxygen evolution reaction, has been proposed (Comninellis, 1994; Comninellis & Battisti, 1996; Foti et al, 1997; Simond, 1997). More recent results, obtained at conductive diamond electrodes7 (which are characterized by a very high oxygen overpotential), fit the model predictions quite well. Based on these results, Comninellis explained the differences considering two limiting cases, i.e. the so-called "active" and "non-active" anodes (Marselli et al, 2003).

In both cases, the first reaction (eq. 45) is the oxidation of water molecules leading to the formation of adsorbed hydroxyl radicals:

$$M + H_2O \rightarrow M(HO\bullet) + H^+ + e^- \quad (45)$$

Both the electrochemical and chemical reactivities of adsorbed hydroxyl radicals depend strongly on the nature of the used electrode material.

With active electrodes there is a strong interaction between the electrode (M) and the hydroxyl radical (•OH). Adsorbed hydroxyl radicals may interact with the anode, forming a socalled higher oxide MO (eq. 46). This may be the case when higher oxidation states are available, for the electrode material, above the thermo-dynamic potential for the oxygen evolution (1.23 V vs. SHE)[104].

$$M(HO\bullet) \rightarrow MO + H^+ + e^- \quad (46)$$

With active electrodes, the redox couple MO/M acts as a mediator in the oxidation of organics (eq. 47). This reaction is in competition with the side reaction of oxygen evolution, which is due to the chemical decomposition of the higher oxide (eq. 48):

$$MO + R \rightarrow M + RO \quad (47)$$

$$MO \rightarrow M + \frac{1}{2}O_2 \quad (48)$$

The oxidative reaction via the surface redox couple MO/M (eq. 47) may be much more selective than the reaction involving hydroxyl radicals (eq. 49). A typical example of an active electrode is the case of $IrO_2$ (Comninellis, 1994).

With a non-active electrode, weak interactions exist between the hydroxyl radical and the electrode surface. In this case, the oxidation of organics is mediated by hydroxyl radicals (eq. 49) and may result in fully oxidized reaction products such as $CO_2$.

$$M(HO\bullet) + R \rightarrow M + mCO_2 + nH_2O + H^+ + e^- \quad (49)$$

In the above schematic equation, R is a fraction of an organic compound containing no heteroatoms, which needs one oxygen atom to be fully transformed into $CO_2$ (Marselli et al, 2003). This reaction competes with the side reaction of hydroxyl radicals (direct or indirect consumption, through the formation of hydrogen peroxide as intermediate) to oxygen (eq. 50) without any participation of the anode surface:

$$M(HO\bullet) \rightarrow M + \frac{1}{2}O_2 + H^+ + e^- \qquad (50)$$

A non-active electrode does not participate in the anodic reaction and does not provide any catalytic active site for the adsorption of reactants and/or products from the aqueous medium. In this case, the anode serves only as an inert substrate, which can act as a sink for the removal of electrons. In principle, only outer-sphere reactions and water oxidation are possible with this kind of anode. Intermediates produced by the water oxidation are subsequently involved in the oxidation of organics in aqueous medium (Marselli et al, 2003).

The electrochemical activity (which may be related to the overpotential for oxygen evolution) and chemical reactivity (rate of the organics oxidation with electrogenerated hydroxyl radicals) of adsorbed •OH are strongly linked to the strength of the M-OH• interaction. As a general rule, the weaker the interaction, the higher the anode reactivity for organics oxidation (fast chemical reaction); boron-doped diamond electrodes (BDD) are typical non-active electrodes, characterized by high stability and acceptable conductivity. This model assumes that the electrochemical oxidation is mediated by hydroxyl radicals, either adsorbed at the surface (in the case of active electrodes) or free, in the case of the non-active ones (Marselli et al, 2003; Carlos et al., 2006).

## 6. Conclusion and outlook

Three important aspects of WO and CWO process have been thoroughly reviewed: (i) the fundamentals of WO; (ii) the degradation of organic pollutants using WO process; and (iii) the mechanism of degradation reactions. WO and CWO studies on the wide range of industrial process waters and wastewaters illustrate the potential of these technologies. The potential benefits of the CWO process over other conventional water treatment processes, such as low reaction temperatures and residence times and the formation of harmless products, will be a key driver for more research in the field. The main challenge faced in the development of successful industrial-scale CWO processes for treating specific wastewaters seems to be the development of suitable catalysts which is highly active, economical, and environmentally friendly. It is anticipated that further research in the area of catalytic activity, combined with research on catalyst stability, reusability, and environmental friendliness, will lead to the discovery of catalysts that are effective for many industrial-scale CWO processes.

Some of the fascinating new applications of photocatalysis have been indicated. These are not to be considered so very separate from the fundamental work. Both tend to reinforce each other. We believe that people working in both fundamental and applied aspects should try very hard to understand what the others are doing. This will undoubtedly lead to advances in both areas.

The relative high estimated cost for ultrasound treatment can be reduced by developing better transducer technology and increasing the efficiency of electrical to acoustic energy conversion. Future studies can be directed towards the development of efficient transducers, reduction of energy loss in the ultrasonic system and the design of economically feasible full scale ultrasonic reactors. Ultrasound can be sued with other advanced oxidation processes in combination to provide a higher efficiency. Most of studies are laboratory oriented which cannot be implemented directly in large scale applications. More studies are needed on kinetics, reactor design and scale-up for practical application.

The analysis of available literature points out the validity of the electrochemical approach for the elimination of different organic pollutants; in consideration of the specific reactivity of each organic substrate, dedicated tests should be carried out in order to identify the most suitable electrode materials and experimental conditions. Generally speaking, the mediated electrochemical approach can be considered more effective than the direct one, because of the minor problems of electrode fouling and/or corrosion. In contrast with other technologies, the form of the waste (liquid, sludge) and its homogeneity are relatively unimportant: the CerOx approach is surely more appropriate in the case of a sludge, while Bakhir's modular plants are dedicated to clear water solutions when a direct oxidation is chosen, but the limitation disappears in case of a mediated approach.

## 7. Acknowledgment

This work was supported by the Research of National Quality Inspection standards of nonprofit industry (200810998)

## 8. References

Abecassis-Wolfovich, M., Landau, M.V., Brenner, A., Herskowitz, M. Ind.Eng. Chem. Res. 43 (2004) 5089.

Adewuyi Y. G., Ind Eng Chem Res 2001;40:4681–4715.

Agrios A.G., Pichat P., J. Photochem. Photobiol. A: Chem. 180 (2006) 130-135.

Agustina T. E., Ang H.M., Vareek V. K., J. Photochem. Photobiol. C: photochem. Rev. 6 (2005) 264-273.

Akira F., Xintong Z., T. Donald A., Surf. Sci. Rep. 63 (2008) 515-582.

Alejandre, A., Medina, F., Rodriguez, X., Salagre, P., Cesteros, Y. J.E.Sueiras, Appl. Catal. B 30 (2001) 195.

An, W.; Qinglin, Z.; Ma, Y.; Chuang, K. T. Catal. Today 64 (2001) 289.

Barbier Jr., J., Oliviero, L., Renard, B., Duprez, D. Catalysis Today 75 (2002) 29.

Belkacemi, K.; Larachi, F.; Hamoudi, S.; Turcotte, G.; Sayari, Ind. Eng. Chem. Res. 38(1999) 2268.

Beyrich, J., Gautschi, W., Regenass, W., Wiedmann, W. Comp. Chem. Eng. 3 (1979) 161.

Bhargava, S.K., Tardio, J., Prasad, J., Foger, K., Akolekar, D.B., Grocott, S.C. Ind. Eng. Chem. Res. 45 (2006) 1221.

Bockris J. O'M. Electrochemistry of cleaner environments, Plenum, New York, 1972.

Buxton, G.V., Greenstock, C.L., Helman, W.P., Ross, A.B. J. Phys. Chem. 17 (1988) 513.

Carlos A., Martı´nez H., Sergio F.. Chem. Soc. Rev. 2006 (35), 1324-1340

Chang H., Johnson D. C., J. Electrochem. Soc., 1990, 137, 2452–2457.

Chang H., Johnson D. C., J. Electrochem. Soc., 1990, 137, 3108–3113.

Chen, G.; Lei, L.; Yue, P.-L. Ind. Eng. Chem. Res. 38 (1999) 1837.

Chen G., Sep. Purif. Technol., 2004, 38, 11–41.
Chen, H., Sayari, A., Adnot, A., Larachi, F. Appl. Cat. B: Environ. 32 (2001) 195.
Chettiar M., M.A. Sc. Thesis, University of British Columbia, (1981).
Chettiar M.,Watkinson A. P., Can. J. Chem. Eng., 1983, 61, 568–574.
Comadran, G.; Comella, J. Ing. Quim. 34 (2002) 135.
Comninellis Ch., Battisti A. De, J. Chim. Phys., 1996, 93, 673–679.
Comninellis Ch., Electrochem. Acta, 1994, 39, 1857-1862.
Comninellis Ch., Nerini A., J. Appl. Electrochem., 1995, 25, 23–28.
Comninellis Ch., Plattner E., Chimia, 1988, 42, 250–252.
Dabrowski A., Mieluch J., Sadkaoski A., Wild J., Zoltowski P., Prezm. Chem., 1975, 54, 653–655.
Debellefontaine, H., Chakchouk, M., Foussard, J.N., Tissot, D., Striolo, P. Environmental Pollution 92 (2) (1996) 155.
Debellefontaine, H., Foussard, J.N. Waste Management 20(2000)2.
Dhale, A. D., Mahajani, V. V. Indian J. Chem. Technol. 7 (2000) 11.
Emanuel, N. M.; Zaikov, G. E.; Maitus, Z. K. Oxidation of Organic Compounds. Medium Effects in Radical Reactions; Pergamon Press: Oxford, U.K., 1980.
Emeline A.V., Ryabchuk V.K., Serpone N. J. Phys. Chem. B 109 (2005) 18515-18521.
Enriquea R. Agrios A. G., Pichat P., Catal. Today 120 (2007) 196-202.
Enriquez R., Pichat P., Environ J. Science Health A: 41 (2006) 955-966
Eyer, S. L. Investigation of Catalytic Wet Oxidation of Bayer Liquor, Ph.D. Dissertation, Department of Applied Chemistry, RMIT University, Melbourne, Australia, 2001.
Farhataziz, P. C.; Ross, A. B. Selected specific rates of radicals of transients from water in aqueous solutions. National Bureau of Standards, Washington, DC, NSDRS-NBS59, 1977.
Feng J., Johnson D. C.. J. Electrochem. Soc., 1990, 137, 507–510.
Feng J., Johnson D. C., Lowery S. N., Carey J.. J. Electrochem. Soc., 1994, 141, 2708–2711.
Foster, C.F. Biotechnology and wastewater treatment. Cambridge University Press, Cambrage, 1985
Foti G., Gandini D., Comninellis Ch., Curr. Top. Electrochem., 1997, 5, 71–91.
Garcia, J., Gomes, H.T., Serp, Ph., Kalck, Ph., Figueiredo, J.L., Faria, J.L. Carbon 44 (2006) 2384.
García-Molina, V., Esplugas, S., Wintgens, Th., Melin, Th. Desalination 189 (2006) 110.
Gerischer H., Willig F., Top. Vurr. Chem. 61(1976) 31.
Glaze, W. H. Chem. Oxid. 2 (1994) 44.
Glaze, W.H., Kang, J.W. Advanced oxidation processes. Ind. Eng. Chem. Res. 28 (1989) 1573.
Glaze, W. H.; Lay, Y.; Kang, Joon-Wun. Ind. Eng. Chem. Res. 34 (1995) 2314.
Gomes, H.T., Selvam, P., Dapurkar, S.E., Figueiredo, J.L., Faria, J.L. Micropor. Mesopor. Mater. 86 (2005) 287.
Gondrexon N., Renaudin V., Boldo P., Gonthier Y., Bernis A., Petrier C., Chem Eng J 1997;66:21–26.
Grimm J., Bessarabov D. , Sanderson R. D., Desalination, 1998, 115, 285–294.
Haag, W.R., Yao, C.C.D. Environ. Sci. Technol. 26 (1992) 1005.

Hamoudi, S., Larachi, F., Cerrella, G., Cassanello, M. Ind. Eng. Chem. Res. 37 (1998) 3561.
Huang, T.-L., Maccines, J. M., Cliffe, K. R. Wat. Res. 35 (2001) 2113.
Ichinose, S.; Okuwaki, A. Bull. Chem. Soc. Jpn. 63(1990) 159.
Ingale, M. N.; Joshi, J. B.; Mahajani, V. V.; Gada. M. K. Process Safety EnViron. Protect. 74 (1996) 265.
Ishii, T., Mitsui, K., Sano, K., Shishida, K., Shiota,Y. Patent, U.S.5399541, 1997
Juttner K., Galla U., Schmieder H., Electrochim. Acta, 2000, 45, 2575–2594.
Kim, S-K., Ihm, S-K. Ind. Eng. Chem. Res. 41 (2002) 1967.
Koile R. C., Johnson D. C., Anal. Chem., 1979, 51, 741–744.
Kolaczkowski, S.T., Plucinski, P., Beltran, F.J., Rivas, F.J., Mclurgh, D.B. Chem.Eng.J. 73(1999) 143.
Kositai M., Antoniadis A. Poulios I., Kiridis I., Malato S., Solar Energy 77 (2004) 591-600.
Kuhn A., J. Appl. Chem. Biotechnol., 1971, 21, 29–64.
Ledakowicz, S. Environ. Protect. Eng. 24 (1-2) (1998) 35.
Lei, L.; Hu, X.; Yue, P. Water Res. 32 (1998) 2753.
Liang J., Komarov S., Hayashi N., Kasai E.. Ultrason Sonochem 2007a;14:201–207.
Li, L.; Chen, P.; Glkoyna, E. F. AIChE J. 37 (1991) 1687.
Lin, S.S., Chen, C.L., Chang, D.J., Chen, C.C. Water Res. 36 (2002) 3009.
Luck, F. Catal.Today 27(1996) 195
Luck, F., Catal. Today 53 (1999) 81.
Maleki A., Mahvi A.H., Vaezi F., Nabizadeh R. Iran J Environ Health Sci Eng 2005; 2(3):201–206.
Mantzavinos, D.; Hellenbrand, R.; Livingston, A. G.; Metcalfe, I.S. Water Sci. Technol. 36 (1997) 109.
Marselli B., Garcia-Gomez J., Michaud P. A., Rodrigo M. A., Comninellis Ch., J. Electrochem. Soc., 2003, 150, D79–D83.
Mason T. J., Lorimer.J. P. Sonochemistry: Theory, applications and uses of ultrasound in chemistry. New York: Ellis Horwood Ltd.; 1988.
Matatov-Meytal, Yu. I., Sheintuch, M., Ind. Eng. Chem. Re. 37 (1998) 309.
Matthews R. W., J. Phys. Chem. 91 (1987) 3328.
Matthews R. W., Water Res. 25 (1991) 1169.
Memming R., Prog. Surf. Sci. 17 (1984) 7.
Mieluch J., Sadkowski A., Wild J., Zoltowski P., Przem. Chem., 1975, 54(9), 513–516.
Minh, D.P., Gallezot, P., Besson, M. Appl. Catal. B 63 (2006) 68.
Mishra, V. S., Mahajani, V.V., Joshi, J.B., Ind. Eng. Chem. Re. 34 (1995) 2.
Najjar, W., Ghorbel, A., Perathoner, S., Centi, G. Studies in Surface Science and Catalysis 158B (2005) 2009.
Nikolopoulus, A.N., Igglessi-Markopoulu, O., Papayannakos, N. Ultrasound Sonochemistry 13 (2006) 92.
Nilsson A., Ronlan A., Parker V. D.. J. Chem. Soc., Perkin Trans., 1973, 1, 2337–2345.
Okawa, K., Suzuki, K., Takeshita, T., Nakano, K. Journal of Hazardous Materials B127 (2005) 68.
Pamfilov A.V., Mazurkevich YaS. Pakhomova E.P., Kinet. Catal. (USSR) 10 (1969) 915.
Pankaj C., Viraraghavan T., Sci. Total Environ. 407 ( 2009 ) 2474–2492

Papouchado L., Sandford R. W., Petrie G., Adams R. N.. J. Electroanal. Chem., 1975, 65, 275–284.

Patria, L., Maugans, C., Ellis, C., Belkhodja, M., Cretenot, D., Luck, F., Copa, B. Advanced oxidation processes for water and wastewater treatment, Parsons, S., Ed., IWA Publishing, London, 2004, p247

Patterson, D. A.; Metcalfe, I. S.; Xiong, F.; Livingston, A. G. Ind. Eng. Chem. Res. 40(2001) 5517.

Pintar, A., Besson, M., Gallezot, P. Appl. Cat. B: Environ. 30 (2001) 123.

Pintar, A., Levec, J. Chem. Eng. Sci. 47 (1992) 2395.

Pintar, A., Levec, J. Ind. Eng. Chem. Res. 33 (1994) 3070.

Prasad, C. V. S., Joshi, J. B. Indian Chem. Eng., 29 (1987) 46.

Raffainer, I. I.; von Rohr, P. R. Ind. Eng. Chem. Res. 40 (2001) 1083.

Rajeshwar K., Iba´n˜ ez J. G. , Swain G. M., J. Appl. Electrochem., 1994, 24, 1077–1091.

Rivas, F. J.; Kolaczkowski, S. T.; Beltran, F. J.; McLurgh, D. B. Chem. Eng. Sci. 53 (1998) 2575.

Robert, R.; Barbati, S.; Ricq, N.; Ambrosio, M. Water Res. 36 (2002) 4821.

Sakthivel S., Shankar M.V., Palanichamy M., Arabindoo B., Bahnemann D. Murugesan W., V., Water Res. 38 (2004) 3001-3008.

Salinaeo A., Emeline A.V., Zhao J., H. Hidaka, Ryabchuk V.K., Serpone N.. Pure Appl. Chem. 71 (1999) 321-335.

Scott, J. P., Ollis, D. F. Environ. Prog. 14 (2) (1995) 88.

Serpone N., J. Photochem. Photobiol. A: Chem. 104 (1997) 1-12.

Serpone N., Salinaro A., Pure Appl. Chem. 71 (1999) 303-320.

Silva, A.M.T., Castelo-Branco, I.M., Quinta-Ferreira, R.M., Levec, J. Chem. Eng. Sci. 58 (2003) 963.

Silva, A.M.T. Quinta-Ferreira, R.M. J. Levec, Ind. Eng. Chem. Res. 42(2003) 5099.

Simond O., Schaller V., Comninellis Ch., Electrochim. Acta, 1997, 42, 2009–2012.

Smith V., de Sucre, M.A. Sc. Thesis, University of British Columbia, (1979).

Sonnen, D. M.; Reiner, R. S.; Atalla, R. H.; Weinstock, I. A. Ind. Eng. Chem. Prod. Res. DeV. 36(1997) 4134.

Suslick K.S., Sci. Am. 1989;260:80–86.

Suslick K.S., Sonochemistry. Science 1990;247(3):1439–1445.

Tahar N. B., Savall A., J. Electrochem. Soc., 1998, 145, 3427–3434.

Teo K., Xu C., Yang C. Ultrason Sonochem 2001;8:241–246.

Treimer S. E., Feng J., Scholten M. D., Johnson D. C., Davenport A. J., J. Electrochem. Soc., 2001, 148, E459–E463.

Vajnhandl S., Marechal A.M. J Hazard Mater 2007;141:329–335

Vitt J. E., Johnson D. C., J. Electrochem. Soc., 1992, 139, 774–778.

Wagner C., Traud W, Z. Elektrochem. 44 (1938) 391-454.

Wagner C., Traud W, Z. Mansfeld F., Corrosion 62 (2006) 843-855.

Wakabayashi, T.; Okuwaki, A. Bull. Chem. Soc. Jpn. 61(1988) 4329.

Weavers L. K., Hoffmann M. R.. Environ Sci Technol 1998;32:3941–3947.

Yoneyama H., Toyoguchi Y., H. Tamura, J. Phys. Chem. 76 (1972) 3460.

Zimmerman, F. J. Wet air oxidation of hazardous organics in wastewater. U.S. Patent No. 2,665,249, 1950.